A Text Book of
Modern Economic
Botany

A Text Book of
Modern Economic
Botany

By
A.V.S.S. SAMMBAMURTY
M.Sc., Ph.D., F.G.S., F.B.S., Reader in Botany,
Sri Venkateswara College, Dhaula Kuan, New Delhi-110021
&
N.S. SUBRAHMANYAM
M.Sc., Ph.D.
Reader in Botany, Sri Venkateswara College, Dhaula Kuan, New Delhi-110021

CBS

CBS PUBLISHERS & DISTRIBUTORS
4596/1A, 11 Darya Ganj, New Delhi - 110 032 (India)

ISBN : 81-239-0629-3

First Edition : 1998
Reprint: 2008
Copyright © Authors & Publisher

Published by :
S.K. Jain for CBS Publishers & Distributors,
4596/1A, 11 Darya Ganj, New Delhi - 110 002 (India)

Printed at :
Asia Printograph, Shahdara, Delhi - 110 032

Preface

Economic Botany or Ethnobotany, literally "people botany" stresses the past, present and future utility of plants — the impact of plants on humans, their cultures, their industries, and their civilizations and reciprocally, the impact of humans on plants. It is an interdisciplinary study encompassing a core of botany, anthropology, archeology, linguistics, history, sociology, comparative religion and other disciplines.

In an attempt to present the vast subject of *Economic botany* briefly in a text book form useful for B.Sc. degree students, the present book is the result. The authors took pains in presenting the material as modern as possible, touching modern trends in utilizing under-exploited plants for human needs. Majority of the classical crops like cereals, millets, woods, oils, essential oils, medical plants, fibres, flowers, spices and condiments, beverages, gums, and resins, tannins, latex, rubbers etc. were presented with numerous examples. Extensive details are, however, limited and the readers are advised to consult the suggested readings given at the end of the book. Cultivation practices, botany of the plant, origin, cytogenetics, uses, economical aspect of the product are all dealt with in majority of the crops.

The authors greatly acknowledge Dr. A.S. Reddy, Principal, Sri Venkateswara College, for his interest and encouragement. Dr. Sammbamurty acknowledges with thanks the help rendered by his daughter, Jaishrie Sammbamurty, in the preparation of the exhaustive index. The authors also acknowledge CBS Publishers & Distributors for the nice and efficient production of this book in a short time.

A.V.S.S. Sammbamurthy
N.S. Subrahmanyam

About the Book

A Text Book of Modern Economic Botany is a concise text book on Economic Botany, useful for B.Sc. students of Indian Universities. In this book the subject matter is dealt with in 25 chapters, covering cereals, millets, pulses, fibre crops, oil crops, essential oil, wood, fruits, spices and condiments, vegetables, fruits, flowers, rubber, gums and resins, fodder crops, fumitories and masticatories, beverages, giving a modern touch keeping 7 in view of the human needs. The book is amply illustrated practical examples. Numerous genera and species names are included for a wider perspective as can be obtained for a glance at the index. Many appendices are given for example, Ready Reckoner for all the species, classified by their use, microchemical test, adulterants, and antibiotics for extra information.

About the Authors

Dr. A.V.S.S. Sammbamurthy (M.Sc. Ph.D., F.B.S., F.G.S.)

Dr. Sammbamurty is a Senior Reader in Botany in Sri Venkateswara College, teaching various branches of Botany for the past two and half decades to B.Sc. and B.Sc.(H) students. He is a member of several scientific bodies and a Fellow of the Genetics Society of India and a Fellow of The Botanical Society of India. He published 25 research papers in Rice Cytogenetics in different Indian and International Journals. He attended and participated in several Indian and International symposia, seminars and congresses. He published jointly with Dr. N.S. Subrahmanyam, 'A Text Book of Economic Botany' published by Wiley Eastern (1989). He is an author of 'A Text Book of Botany', Vol. I (1990) consisting of *Algae, Fungi* and *Plant Pathology, Bacteria* and *Viruses* and Vol. II (1998) comprising *Bryophytes, Pteridophytes, Gymnosperms* and *Paleobotany,* published by Konark Publishers. He is also the author of 'Genetics', Narosa Publishers, (in Press, 1998) and 'Glossary of Botany' published by CBS Publishers (in Press, 1998).

Dr. N.S. Subrahmanyam (M.Sc. Ph.D.)

Dr. N.S. Subramanyam is a Senior Reader in Sri Venkateswara College, teaching various disciplines in Botany for the past two decades to B.Sc. and B.Sc.(H) students. He published jointly with Dr. A.V.S.S. Sammbamurty, 'A Text Book of Economic Botany' (Wiley Eastern, 1989). He also published 'A Text Book of Modern Taxonomy' (Vikas, 1997), 'Manual of Practical Taxonomy' (Vikas, 1997) and 'Question Bank in Botany' for Medical Entrance Tests (Vikas, 1997).

Suggested Readings

Cobley, L.S. (1976). *An Introduction to the Botany of Tropical Crops*. The English Language Book Society. Longman.

Hill, A.F. (1952). *Economic Botany*. McGraw Hill Book Co., N.Y.

Kocchar, S.L. (1981). *Economic Botany in Tropics*. Macmillan, New Delhi.

Purseglove, J.W. (1974). Tropical Crops, Vol. I (Dicotyledons) : Vol. II (Monocotyledons). The English Lasnguage Book Society, Lond.

Sammbamurty, A.V.S.S. and N.S. Subrahmanyam. (1989). *A Test Book of Economic Botany*. Wiley Eastern, New Delhi.

Schery, R.W. (1972). *Plants for Man*. Prentice Hall. N.J.

Simpson, B.B. and Ogorzaly, M.C. (1986). *Economic Botany*, McGraw Hill Book Co., N.Y.

Contents

Mung bean : Acreage, production, world mungbean production, countries of cultivation, cultivation practices, uses

Rice bean : Distribution, cultivation practices, uses

Winged bean : Distribution, botany, uses, cultivation, tubers

Lablab bean : Acreage, production, cultivation

Bambara ground nut;

Faba bean : Distribution, cultivars, varieties, yield, ecology, uses

Marma bean : Distribution, botany, seed, seed oil

Toxic substances in raw pulses

agarwood oil, camphor, vetiver oil, sandalwood oil, lemon grass oil; citronella oil, geranium oil, rosemary oil, patchouly oil, peppermint oil, thymol, ylang-ylang, champak, jasmine, oil of neroli, attar of roses, lavander oil, palmarosa grass oil
Jasmine flower oil : Importance, concrete, absolute from pomade, uses of concrete, cultivation of jasmine, processing unit, plant and machinery, purification of the solvent, extraction, distillation

1

Scope of Economic Botany

Man's dependence on plants for the essentials of his life like, food, fibre and shelter has been of paramount importance in his life since the human race began. Civilization, however has brought with it an ever-increasing complexity, and has increased man's requirements to an amazing degree. To satisfy his requirements man is still searching, directly or indirectly, for plants useful to him. Thus the science dealing with the study of plants useful to man is known as *Economic Botany*.

It has been estimated that about 3000 species of plants have been used as food by human beings throughout history and about 200 domesticated as food crops. Today, the world virtually depends on about 15 species as staple food, most of which have been highly modified by humans.

Depending on the nature of the useful part available to man, plants are classified into cereals (paddy, wheat, oats, corn etc.), millets (sorghum, pearl millet, ragi millet, Italian millet etc.), pulses (bengal gram, red gram, black gram, green gram etc.), oil seeds (groundnut, coconut, sesame, safflower, sunflower etc.), beverages (tea, coffee, kola, mate, alcoholic drinks etc.), medicinal (belladonna, cinchona, ipecac, quinine, periwinkle etc.), fibre-yielding (cotton, jute, ramie, hemp etc.), wood yielding (teak, mahogany, pine, deodar, shisham, rose wood), rubber-yielding (*Hevea, Castilla, Ficus*, guayule etc.), fruit crops (banana, mango, apple, grapes etc.), and sugar (sugar cane beet) etc.

Cereals form the most important source of food for man. The name cereals was derived from the Roman Goddess named *ceres* whom Romans believed as giver of grain. The true cereals are barley, maize, oats, rice, rye and wheat. All cereal crops belong to the grass family Gramineae or Poaceae. Cereals have high per cent of carbohydrates. The morphology of the useful product in cereals is *caryopsis*. Caryopsis is a single-seeded dry indehiscent fruit commonly referred to as *grain*.

CENTRES OF ORIGIN OF CULTIVATED PLANTS

Alphonse de Candolle (1886) a French botanist and systematist produced a voluminous work *Prodromus systematis naturalis regni Vegetabilis*, (*Origin of Cultivated Plants*), concerning the geography of plants, distributions of wild relatives, history, names, linguistic derivatives, archaeology, variation patterns etc., on the cultivated plants.

N.I. Vavilov, a Russian geneticist and agronomist at the National Institute of Plant Industry, Leningrad, published his work *Studies on the Origin of Cultivated Plants in 1926* and *The origin variation, immunity and breeding of cultivated plants* in 1951. Vavilov launched an ambitious plant breeding programme that was ever attempted to collect and assemble all of the useful germplasm of all crops. A vigorous, worldwide plant exploration programme was launched and a systematic survey for genetic resources of crop plants was started by him. He was interested in the genetic diversity of crop plants and their Centres of Origin. The centre of origin could be determined by an analysis of patterns of variation, and according to Vavilov, the geographical region in which one found the greatest genetic diversity was the centre of origin for that crop (Table 1.1). Vavilov thought that areas of maximum genetic diversity represented centres of origin and that the origin of a crop could be identified by the simple procedure of analysing variation patterns and plotting regions where diversity was concentrated. It turned out that centres of diversity are not the same as centres of origin, yet many crops do exhibit centres of diversity.

It is thought that India is the centre of origin of many crops like *Oryza sativa* (rice), *Saccharum officinarum* (sugarcane) and *Cicer arietinum* (gram). Other important species of Indian origin are *Cocos nucifera* (coconut), *Cajanus cajan* (red gram), *Emblica officinalis* (Indian gooseberry), *Areca catechu* (arecanut), etc. In the Malaysian region *Musa paradisiaca* may have originated. Wheat, barley, onion, pomegranate and pea appear to have their centre of origin in the Middle East and Mediterranean regions. *Litchi chinensis* (litchi), *Camellia sinensis* (tea), *Solanum melongena* (brinjal) and *Papaver somniferum* have originated in China. *Zea mays* (maize), *Gossypium hirsutum* (cotton), *Capsicum annum* (chilly) are thought to belong to Central America and South Mexico while *Arachis hypogea* (groundnut) and *Hevea braziliensis* (rubber tree) originated in Brazil. *Cinchona*, tomato and tobacco are also thought to have originated in South America.

Harlan (1971) developed the idea of centres and non-centres and demonstrated that plant domestication occurred almost everywhere south of Sahara and north of equator from the Atlantic onto the Indian ocean. Such a vast region was called by Harlan as *non-centre*. He further defined 3 centres (places of agricultural origins) each more truly definable or less connected to a very large and diffuse *noncentre* where agriculture has been introduced.

Table 1.1. Vavilov's 8 Centres of Origin of some crops

Centre	Crops (Vavilov N, 1951)
1. **Chinese**	Buckwheat, hemp, mulberry, orange, peach, poppy, soybean, tea, tung (*Aleurites* spp.)
2. **Indian**, including Indo-Malayan	Banana, bread fruit, chickpea, citron, coconut, mango, black pepper, rice, safflower, sesame, sugarcane, yam
3. **Central Asian** (includes MW India)	Apple, carrot, grape, onion, pea, pear, radish, spinach
4. **Near Eastern**	Alfalfa, fig, flax, hazelnut, lentil, melon, oats, quinne, rye, wheat Einkorn (*Triticum monococcum*, emmer (*Triticum turgidum*)
5. **Mediterranean**	Asparagus, beet, cabbage, carob (*Ceratonia siliqua*), lavender, leek (*Allium ampeloprosum*), lettuce, olive
6. **Abyssinian**	Barley, castor bean, coffee, African millet (*Eleusine coracana*), pearl millet, okra (*Abelmoschus esculentus*), sorghum (*Sorghum bicolor*)
7. **Mexican-Central American**	Avocado, common bean (*Phaseolus vulgaris*), cacao, corn, cotton, sweet potato, red pepper, winter squash (*Cucurbita moschata*)
8. **Central Andean** (including most of South America)	Manioc (*Manihot esculentum*), peanut, pineapple, white potato (*Solanum tuberosum*), pumpkin, rubber, tobacco, tomato

Centre	Non-centre
A_1 Near east	A_2 African
B_1 North Chinese	B_2 South Eastern Asian or South Pacific
C_1 Central American or Meso American	C_2 South American

According to Harlan (1971) there are three independent systems with a centre and a noncentre. According to him no single model will explain agricultural origins and he recognises a humanistic *no-model* model to explain the origins of domesticated plants. 'Man took the initiative in modifying his environment, and plants responded genetically to his activities' (Harlan, 1975). Plant domestication is an evolutionary process operating under the influence of human activities. The reasons for the origin of secondary centres of crop plants might be due to:

(i) a long history of continuous cultivation;

(ii) ecological diversity;

(iii) human diversity, different tribes are attracted to different races of a crop, a science known as Ethnobotany;

(iv) introgression with wild or weedy relatives or between different races of the crop, which leads to hybridization, segregation and selection; and

(v) the deliberate introduction of certain exotic plants by man from one continent to another during history.

Conservation of plant wealth

The variability in enormous variety of crop plants appears to have been concentrated in relatively few ecogeographic regions of the world during the whole history of evolutionary changes. These regions were recognised as natural sites of evolution, commonly termed as *Centres of Origin* or *centres of crop diversity*.

Besides, during the last few decades efforts were made to collect the available variation in various economic crop plants alongwith their wild relatives to exploit them for crop improvement. Thus *germ plasm collection* created by human efforts, constituted the reservoir of variability. About 10 per cent of the world's flowering plants numbering 20,000 to 30,000 species is getting to be dangerously rare or under threat. It is therefore, urgent and imperative on the world community that such genetic heritage either be maintained as *living gene pools* or conserved under long-term storage conditions of maximum physical and genetic security and stability.

There are broadly 3 ways by which the genetic variability in crop plants could possibly be conserved so that desirable genes would become available whenever required.

1. Living gene pools

Old varieties and races of crop plants would be maintained as living gene pools at various national and international genetic resources centres (see Table 1.2).

Table 1.2. Living gene pools

Crop	Gene pool centre
1. Crop plants	Vavilov Institute of Plant Industry, Leningrad (Russia); 3,00,000 collections
2. Crop plants	Royal Botanic Gardens, Kew (England), 50,000 collections
3. Crop plants	Gartersleben, Germany 30,000 collections
4. Maize	CIMMYT, Mexico, Columbia, Brazil centres
5. Wheat	Throughout the world 4,00,000 collections
6. Small grains	USA
7. Sorghum	IARI Regional Centre, Hyderabad (India)
8. Pearl millet	IARI Regional Centre, Hyderabad (India)
9. Legumes	NBPGR, New Delhi
10. Sugarcane	Camal Point, Florida, USA; and Coimbatore, India
11. Rice	IRRI, Philippines; CRRI, Cuttack, India
12. Groundnut	Bombay, Senegal (Africa)
13. Potato	Cambridge, UK; Wisconsin, USA and Shimla (India)
14. Cotton	Tashkent (Russia)
15. Grapes	Tashkent (Russia)
16. Soybean	USA
17. Coffee	Ethiopia
18. Sweet potato	New Zealand
19. Colocasia	Fuji
20. Bread fruit	Western Samoa

International Centre for Tropical Agriculture (CIAT) Palmira, Colombia is a germplasm centre for cassava, common beans, maize and rice. International Centre for the Improvement of Maize and Wheat (CIMMYT), El Batan, Mexico is a research and germplasm centre for wheat, triticale, barley and maize. International **Potato Centre (CIP) Lima, Peru, is a germ plasm centre** for potatoes. International Board for Plant Genetic Resources (IBPGR), FAO, Rome, Italy is a centre for the conservation of plant genetic material, especially cereals. International Centre for Agricultural Research in Dry Areas (ICARDA), Lebanon is a centre for the study of mixed farming systems, especially barley, wheat and lentils. International Crops Research Institute for the Semi-Arid Tropics (ICRISAT), Hyderabad, India, is a centre for sorghum, pearl millet, pigeon peas, chick peas, and peanuts. International Institute for Tropical Agriculture (IITA), Ibadan, Nigeria is a centre for forming systems; relay station to IRRI and CIMMYT for rice and maize, cowpeas, soybeans, limabeans, pigeon peas, cassava, sweet potatoes and yams. International Rice Research Institute (IRRI), Los Banos, Philippines, is a centre for irrigated rice, upland rice, multiple cropping systems. West African Development Association (WARDA) Monravia, Liberia, is a centre for regional cooperative effort in rice research.

2. Long-term cold storage laboratory (Gene bank)

The most practicable genetic conservation measure is the establishment of cold-storage laboratory for long term holding of variable seeds, tissues and pollen. With the considerable advance in technology of seed storage, it is now possible to maintain seed of many species for periods of 20-25 years, compared to 3-5 years in the past.

For each 1 per cent reduction in moisture content, and for each 5°C lowering the storage temperature the life of seed is doubled. Seed moisture content ranging from 4 to 6 per cent is reported to be ideal for maximum seed life. The lower the storage temperature, the longer the life of seed. For many species, very low temperatures of −20°C or less are beneficial to the maintenance of maximum viability. It is considered desirable for seed storage laboratories to aim at very long storage periods under sub zero temperatures of about −20°C maintaining low moisture content in seed stored in hermitic containers.

Long-term seed storage laboratories

Fort Collins (United States); The National Seed Storage Laboratory for Genetic Resources at Hiratsuka (Japan); Bari (Italy); Braunschweig (Germany); Izmir (Turkey); CIMMYT (Mexico), IRRI, Manila (Philippines); Kuban (Russia); NBPGR, New Delhi (India). International Board of Plant Genetic Resources (FAO), Rome outlined the procedure for conservation of germplasm collections.

3. Gene sanctuaries : Conservation of natural areas (Biosphere Reserves)

The chief characteristics of a biosphere reserve are that:

1. A biosphere reserve should be a protected area of land and coastal environment.
2. Each biosphere reserve will include representative examples of natural homes, unique communities, a harmonius land scape and ecosystem to be restored.
3. A biosphere reserve should be large enough as an effective conservation unit.
4. A biosphere reserve should provide opportunities for ecological research, education and training.
5. A biosphere reserve must have legal protection.
6. In some cases, biosphere reserves should incorporate natural parks and sanctuaries. India has a vast potential for creation of such biosphere reserves.

Namdapha (Arunachal Pradesh); Uttarakhand (Valley of flowers); Nanda Devi (Uttar Pradesh); Gulf of Mannar (Tamil Nadu); Sunderbans (West Bengal); Thar Desert (Rajasthan); Manas, Kaziranga (Assam), Little Rann of Kutch (Gujarat); North Islands of Andamans (Andaman and Nicobar); Kanha (Madhya Pradesh); Nokrek (Tura Range), Meghalayas, Nilgiris (Karnataka, Kerala and Tamil Nadu) are the biosphere reserves of India.

A species may have several recognisable races of subspecies whose abundance may change at unequal rates. The two basic causes of extinction of a species are (1) Failing to adopt to changing environments; (2) Over specialization, resulting again in a failure to readopt quickly enough to the deterioration of resources due to over-exploitation.

Green revolution

The increasing demand of food for growing populations is posing serious food problem. Agricultural scientists have been successful in increasing the quantity as well as quality of the crop plants. Proper applications of fertilizers to soil, crop rotation practices, weed control etc., have resulted into the so-called *industrialisation of agriculture.* This resulted into the *green revolution* beginning in the late 1960's with an attempt to increase food production in developing countries. New varieties of wheat, rice, and corn were developed that responded better than the traditional varieties to fertilizers, irrigation and chemicals for pest and weed control.

However, alongwith the increase in food production, there came an increased dependence on energy-based expensive technology for producing hybrid seeds, fertilizers, pesticides, and tractors, as well as changes in social and political set ups. Industrialised (fuel powered) agriculture in Japan produces four times the yield as does man and domestic animal powered agriculture in India, but is 100 times as demanding of resources and energy. Agro-industry is one of the chief causes of air and water pollution. Green revolution brought a social revolution in developing countries by widening the gap between rich and poor. In the US, farmers are shifting from routine use of herbicides, pesticides, NPK fertilizers etc., to totally *organic farming* in which no chemicals and weeds or insecticides are used. The only fertilizers used are organic matter, such as manure, and crushed limestone or other rocks.

The International Maize and Wheat Improvement Centre (Centro International De Mejoramiento De Maizy Trigo (CIMMYT) based at Mexico is responsible for collecting promising wheat germplasms from all over the world, breeding high yielding varieties and making them available to different countries. It also provides facility for collaborative research in wheat improvement and training of scientists. Such facilities are availed of by India too. It may be recalled that the HYVs of wheat were evolved in this institute under the guidance of Dr. Normal E. Borlaug who was honoured with the 1970 Nobel Peace Prize for his work in this field.

High yielding variety (HYV) wheat seeds imported from CIMMYT in 1963 was the starting point of the subsequent 'green revolution' or more correctly the 'wheat revolution' in India. This 'green revolution' in India was mainly due to the rigorous efforts of Dr. M.S. Swaminathan a staunch geneticist and agriculturalist. Besides CIMMYT, The International Centre for Agriculture Research in Dryland Areas (IACRDA) is working for improvement of durum wheat.

A great deal of publicity has been given to the so-called Green Revolution, a modernization of agricultural practices that some enthusiasts hoped would enable the less developed countries to keep agricultural production well ahead of population growth for several decades. There are two general components to the Green Revolution : increased use of recently developed high-yield varieties of grain (Primarily wheat and rice), and increased use of "inputs" (especially fertilizers and irrigation water, but also often pesticides), which are required to realize the potentially high yields of those crops.

Since the mid-1960s there has been a rapid expansion of acreage planted to high-yield crops. The area planted to new varieties of wheat and rice in Asia, Latin America, and Africa increased from about 59,000 hectares to over 32 million hectares between 1965 and 1975. By 1975 it had reached 43 million hectares.

Miracle varieties

The new high-yield varieties of wheat and rice are capable of producing yields considerably greater than those of traditional strains. For instance, the first rice variety developed by the International Rice Research Institute in the Philippines, IR-8, can produce two or more times the harvest of traditional rice plants from a given area, if handled correctly. Some newer strains perform even better than IR-8 and have other improved qualities.

All the new grain varieties are extremely responsive to fertilizers. If planted in good soil, given the large amounts of water that are a necessary accompaniment of heavy fertilizer inputs, and given protection from pests, the strong, short stalks of the new dwarf varieties can carry a truly miraculous load of grain.

These new varieties also mature faster and are less sensitive to seasonal variations in day length than are traditional strains. Both characteristics increase the possibility of multiple cropping - that is, growing and harvesting more than one crop per year - when adequate water is available. In Mysore State in India, for example, farmers are growing three corn crops every fourteen months. Where there is a dry season with inadequate water available for growing rice, some farmers are growing new high-yielding grain sorghums (which require less water) in alternation with rice. In some areas of China, northern India, and Pakistan, farmers are planting rice in the summer and wheat in the winter.

It must be emphasized, however, that the full potential yields of the miracle grains can be realized only if an entire complex of conditions is met, especially the proper input of fertilizers, water and pesticides. Without these, yields may be little higher than those of traditional varieties, and in some cases they may be less.

Bacteria, filamentous fungi, yeasts and an alga are capable of obtaining energy and cell carbon source from crude petroleum or its refined products. This relationship has been exploited in the use of microbes as sources of single-cell protein as well as in studies on both the cleanup of oil spills and on the disposal of oily wastes by land spreading.

Single-cell protein

Protein-rich material can be produced by culturing single-celled organisms on petroleum byproducts, sewage sludge, or other substrates. Such single-cell protein (SCP) is made sufficiently pure for human consumption in 1980.

SCP make an indirect contribution as animal feed, especially in Europe and the USSR. It can, for instance, replace feeds such as corn, soybeans, and oilseed cakes.

A related project is the development of a protein supplement from algae grown especially for this purpose. A West German Company is experimenting with a powder form that can be added to milk or other foods to raise protein content. It appears to have been beneficial in treating children with *kwashiorkor* and as a supplement for mothers on inadequate diets. It is expected that the product eventually can be made as cheaply as soy protein.

A high-quality protein and vitamin-rich concentrated food familiar to American health-food buffers is brewer's yeast, which also can be produced quite cheaply. However, it has not been seriously promoted as a food supplement, presumably because it is thought that the strong flavour might not be widely acceptable.

Producing SCP from agricultural wastes by new, simple techniques may be a more practical approach for less developed countries. The product is a concentrated feed-supplement for pigs and poultry, and could be produced as the village level with a minimum energy input. Even so, it might be more useful and efficient to use agricultural wastes as compost and fertilizer instead.

Water hyacinths, leaf protein, etc.

One of the more interesting is the idea of converting water hyacinths and other aquatic weeds to cattle feed. But, although water hyacinths are abundant in the tropics, (and pestiferous - they clog waterways) and contain protein that is high in the essential amino acid lysine (commonly lacking in cereals), their dry weight is only 5 per cent of their wet weight, which presents a tremendous obstacle even to processing them into cattle feed.

Leaf-protein

A technical difficulty is the extraction of the protein concentrate from the fiber content of the leaves, but this has been done successfully in small-scale projects and presumably could be a wider scale if the appropriate equipment were available to farmers. The leaves of forage crops such as alfalfa and sorghum produce large proportions of high-quality protein. Alfalfa yields as much as 2,400 pounds of protein per acre-more than twice as much as soybeans. The fibrous residue can be used as fodder for ruminant animals such as cattle, and the protein extract and other fractions can be fed to pigs and poultry, used as fertilizer, or made into a protein supplement for human consumption. Moreover, forages can be grown on soils too poor or hilly to support food grains.

Increasing the efficiency of agriculture crops

One of the goals of this research is to improve the protein quality of grains - that is, produce grains containing a better balance of the amino acids essential for human nutrition. This, of course, can be done fairly cheaply by enriching flour with processed amino acids (usually lysine, tryptophan, and methionine). Lysine-enriched wheat, for instance, has been shown to be beneficial both to laboratory rats and to human babies under carefully controlled conditions.

An alternative to fortified grain is to breed new varieties with more complete protein balances or other desired characteristics.

1. Use of mutant varieties

Using a naturally occurring mutant strain (known as Opaque-2 because the kernels are opaque), plant breeders have developed a strain of corn (maize) with digestible protein levels twice as high as those of ordinary corn and with adequate proportions of lysine and tryptophan. Opaque-2 corn has been shown to be nutritionally far superior in tests with farm animals and, subsequently, with malnourished Latin American children. But homozygous Opaque-2's softer, lighter kernels resulted in reduced yields, greater susceptibility to pests, and different milling characteristics. By interbreeding Opaque-2 with traditional strains, scientists at CIMMYT in Mexico have produced several new, improved protein strains suitable for various climatic conditions.

A similar programme is in progress to improve the quality of sorghum, which is the fourth most important grain in human diets, widely consumed in Asia and Africa, especially by the poor. Two strains of high-protein sorghum have been found in Ethiopia and are now the basis of a breeding program to develop a pest-resistant, high-yielding variety. Such protein-improved corn, sorghum, and other cereals (if they are developed) no doubt can make important contributions to diet improvement among the poor, who subsist mainly in such foods.

2. Use of triticale

One of the more interesting developments in agricultural research is the cross-breeding of wheat and rye to produce a completely new grain called *triticale*, which gives permise of incorporating the best characteristics of each parent species. Among these characteristics are (from wheat) high yield and high protein content for a cereal and (from rye) higher proportion of lysine, ruggedness, resistance to disease, and adaptability to unfavourable climates and soils. In some respects, triticale may ever outperform both parents. For instance, some strains rival the miracle wheats in yields.

3. Use of nitrogen fixing bacteria

Research is proceeding on developing for grains a nitrogen-fixing capacity, like that in the roots of legumes.

Some preliminary research in which corn has been inoculated with a nitrogen-fixing bacterium, *Spirillum lipoferum*, is underway, but it is too early to know whether this approach will be fruitful. Preliminary research also indicates that some species of *Rhizobium*, the genus that fixes nitrogen in symbiotic associate with legumes, may be transferable to nonleguminous crops. Another possibility is to induce the nitrogen-fixing ability by cell-culture manipulation. Microorganisms may be given genes for nitrogen-fixation by genetic-engineering techniques. It has been discovered that paddy-rice can be enhanced by 50 to 100 per cent when grown with a water fern (*Azolla*) that carries a symbiotic nitrogen-fixing alga (*Anabaena*). How to maintain the fragile water fern during the summer was once a closely guarded secret in two North Vietnamese villages; now the secret is out, and all Southeast Asia may benefit from higher rice yields without dependence on imported fertilizers.

4. Cell culture and cloning

Cell-culturing also appears to be a means whereby botanists may produce new kinds of crops. Plants that cannot be cross-fertilized might be genetically combined or specific mutations introduced into a strain by cell culture. Desired genotypes also can be screened out in early stages and rapidly propagated by cloning - a procedure now commonly used in research on several crops.

Most of these unorthodox lines of research on crops are unlikely to pay practical dividends before the end of the century, unfortunately, which is no comfort for the hungry millions of today. The exceptions are triticale and improved-protein corn, which are in relatively advanced stages of development. But the built-in momentum to population growth guarantees that the "food problem" will not go away; hence long term development projects are every bit are as essential as those projects that will help augment food supplies within a few years.

The IRRI found a traditional "floating rice", crossed it with a dwarf, and created a high-yield floating rice that can survive productively in the frequent monsoon floods of India, Bangladesh, and Southeast Asia.

New food combinations

The presscakes that remain after oil is squeezed out of soybeans, cottonseed, peanuts, and sesame seeds may be the most readily accessible, untapped source of protein for human consumption. Until recently, most presscakes were used as livestock feed or fertilizer; the rest were wasted.

Special foods have been created from conventional once in some less developed countries by combining oilseed protein concentrates with cereals and sometimes with milk. The best known of these is Incaparina, developed by INCAP (Institute of Nutrition for Central America and Panama). It is a mixture of corn and cottonseed meal, enriched with vitamins A and B. Another product is CSM formula (corn, soya, milk), a mixture of 70 per cent processed corn, 25 per cent soyaprotein concentrate and 5 per cent milk solids.

A third is Vita-Soy, a high-protein soy-based beverage that has been marketed very successfully in Hong Kong. Another product using peanut meal and soy is being distributed in India to children, and a bun fortified with milk solids has been given to school children in the Philippines.

Incaparina has been available in Central America for more than a decade, but its impact remained insignificant, inspite of determined efforts by private and commercial organizations to push its acceptance, and inspite of tremendous worldwide publicity. In 1973, the Quaker Oats Company gave up producing it. Other products seem to have been more successful, however, especially those in Asia.

A commonly wasted "by-product" of food processing is whey, the residue from cheesemaking. In 1975, although some whey was being used as animal feed and was beginning to find other uses, a great deal of it was still discarded, to become a serious pollutant in rivers and streams. Within a few years, whey's usefulness as a protein, vitamin, and mineral supplement (in place of milk, for instance, in some products), as fertilizer, as a binding agent in pills, and as a livestock feed supplement, will probably end the pollution problems. There have even been successful attempts to make wine out of it. A product combining whey with soy milk is being tested in Latin America as a weaning food for babies.

2

Cereals

The cereals are a group of annual grasses grown primarily for their large swollen grains. They provide the main concentrated carbohydrate food for the peoples of moist temperate areas and many parts of tropics. Although cereals are basically "energy" foods, they also supply a large part of the protein needs of peoples in the poorer areas of the world. They also provide feed for livestock. Altogether they provide sustenance for more than 2,000,000,000 people and their production runs to well over 1,000,000,000 tonnes annually. Over half the cultivated land of the world is devoted to cereals.

The true cereals are all members of the grass family Gramineae. The main cereals are wheat (*Triticum*), maize, corn (*Zea*) or rye (*Secale*) , rice (*Oryza*), barley (*Hordeum*). Oats (*Avena*) and man's evolution has been closely linked with the origin and cultivation of cereals. Early agriculture involved the gathering, cultivation and domestication of cereal grains and from that followed the development of modern civilizations, all of which have been dependent on one or more grain crops as the staple food. The word cereal derives from ceres - the goddess of grain of the Ancient Greeks.

All the cereals can be used for brewing although barley is probably the most important as the basic constituent of most beer. They may also be used for the production of spirits such as whisky, gin and vodka.

Cereal grains constitute a concentrated source of carbohydrate plus some protein, oil and vitamins. They can moreover be easily handled, stored and stockpiled over a long period, unlike starchy staples with a higher water content such as potatoes, yams and cassava and other vegetables which are bulky, require special handling and can only be stored for limited periods.

Structurally, the grain is a fruit (*caryopsis*) developed from a single ovuled ovary in which the fruit wall and the ovary are completely fused. In addition to the store of carbohydrate in the form of starch, the embryo (germ) and seed coat are rich in protein, oil and vitamins, but in the case of wheat these are lost in the milling for wheat flour. The major change that accompanied domestication of each of the cereals was the appearance of types with tough non-shattering heads (ears). The immediate wild progenitors, while possessing fairly large seed, had heads that shattered on maturity. Although this facilitated seed dispersal in the wild it was a definite hindrance to domestication in as much as grain had to be harvested before it was quite ripe. Conversion to non-shattering involves a very simple genetics.

RICE

Rice or *chaval, Oryza sativa* L. is the staple food of half of the world's population especially in the Oriental countries. Both rice and wheat are of major importance in the agriculture of the world, but rice culture dominates the economic and agricultural life in the South-East Asian countries. In volume of World Production, rice ranks above maize and far above all the cereals except wheat. World produces about 450 million tonnes of rice per year, rice is cultivated in about 74 countries in an area of about 145 million hectares of land. There are about 1,20,000 morphological varieties of rice in the world today (Fig. 2.1).

Rice accounts for about 40 per cent of the total food grain production in India. Rice production is about 60-80 million tonnes per year; area about 40 million ha. There are about 5,000 morphological varieties of rice in India.

Oryza belongs to the tribe Oryzae of the family Poaceae (Gramineae). *Oryza* is closely related to the genus *Leersia*. There are 27 species of *Oryza* and only 2 species (*sativa* and *glaberrima*) are under cultivation. *Oryza sativa* L. is an annual or perennial grass without a rhizome; grain is 6-14 mm long, kernel is white, but red, purple or brown pigmentation is present in some varieties.

Kato (1910) divided the cultivated rices into *indica* and *japonica* groups, based on hybrid sterility. Recent work shows that *indica* group includes another group *javanica* (indo-japonica) consisting of *bulu* rices which exhibit intermediate characters between *japonica* and *indica* rices.

Origin of cultivated rice

There are two cultivated rice species, *Oryza sativa* L. and O. *glaberrima* Steud. *Oryza sativa* seems to have derived from the perennial type of *O. perennis* distributed in tropical Asia. The wild species contains a large amount of genetic variation in its populations. The initial step of domestication may have been caused by "cultivation pressure", particularly by that due to seeding. *O. glaberrima* grown in West Africa seems to have been derived from *O. breviligulata* in parallel to the evolution of *O. sativa*.

All these species are known to have the same genome, A (n = 12). *O. perennis* is distributed throughout the humid tropics comprising various geographical forms. Its Asian form may be the progenitor of *O. sativa*.

A continuous array of integrades from perennial (*balunga*) to annual (*spontanea* or *fatua*) types is found among strains of Asian *perennis*. So called *O. sativa* var. *spontanea* can be considered to be the annual type of *O. perennis*.

Indica and Japonica types

Varieties of *O. sativa* are divided into the *Indica* and *Japonica* types. The Indica and the Japonica types differ in various characteristics, namely phenol reaction (Indica-positive; Japonica-negative), length of apiculus hair (Indica < Japonica), potassium chlorate resistance (I < J), drought resistance (I > J), low-temperature resistance (I < J), temperature response in seed germination and vegetative growth (I > J), temperature response in floral initiation (I < J) and so on.

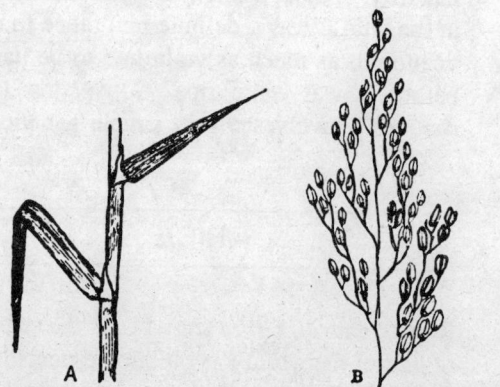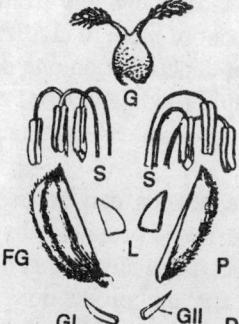

Fig. 2.1. Rice (*Oryza sativa*). A, portion of a branch with sheathing leaves and ligules; B, a panicle of spikelets; C, 1-flowered spikelet (note the glumes and stamens); D, spikelet dissected out—GI, first empty glume; GII, second empty glume, FG, flowering glume (Lemma) ; P, palea; L, lodicules; S, stamens; and G, gynoecium.

Observations of materials from the Jeypore Tract (India) have shown that the Indica and Japonica types are monophyletic.

Polyphyletic origin

The views regarding the origin of rice can be grouped into two classes, (a) *polyphyletic* origin and (b) *monophyletic* origin. According to the first view the present day rice varieties have originated from several species. According to the latter, a single species has given rise to all the varieties of cultivated rices, *O. sativa* as well as *O. glaberrima*. Many of the modern rice workers believe that the origin of cultivated rices is monophyletic.

Originally the cultivation of *O. sativa* was limited to South-East Asia. The cultivation of *O. glaberrima* is still confined to tropical West Africa. These two regions are separated by a vast geographical region which shows no evidence of growing rice in the past. Egypt was a seat of ancient civilization and from the archaeological findings it appears that they had no knowledge of rice. Nor there is evidence of any contact among the people of these two regions. This led to the assumption that *O. sativa* and *O. glaberrima* have evolved independently in their respective regions. It was concluded that *O. sativa* has evolved from *O. fatua* in Asia and *O. glaberrima* from *O. stapffi* and *O. breviligulata* in Africa as these two wild species are the close relatives of the cultivated species in their respective regions.

O. fatua resembles *O. sativa* in morphological characters except in awning and premature shedding of spikelets, grows in the same region as that of *O. sativa*, and crosses freely with *O. sativa* giving fertile hybrids. Its grains are collected by poor people and used for human consumption. Like *O. sativa*, *O. fatua* is diploid and the hybrid with the former forms bivalents in the meiotic metaphase of microsporocytes. Watt (1892), Roschevicz (1931), Ramiah and Ghose (1951) and Chatterjee (1951) believe in this view.

However, the variation present in *O. fatua* was found inadequate by many workers to account for all the variation present in *O. sativa* and hence the role of other wild species has also been proposed.

Watt (1892) supposes that *O. officinalis* played a role in the evolution of the present day rice varieties through hybridization. According to him, the umbellate panicle, long, naked peduncle, hairy glumes and hard sub-woody root-stock are the characters of *O. officinalis* present in many varieties of *O. sativa*. However, this view has never gained much support. Richharia (1960) rejects the role of *O. officinalis* in the origin of *O. sativa* on several grounds, but Sastry *et al.* (1960) again point out at the inadequacy of these grounds and claim that the role of *O. officinalis* in the origin of *O. sativa* cannot be finally rejected under the present state of knowledge.

Earlier workers believed that small-grained varieties of *O. sativa* might have come from *O. minuta*. Similarly the saline-resistant varieties of *O. sativa* were supposed to have come from *O. coarctata*. The role of these two species is, however, not probable as they are tetraploid (4n = 48) whereas *O. sativa* is a diploid.

According to Porteres (1956) *O. breviligulata* is the progenitor of *O. glaberrima*. *O. breviligulata* is a diploid species with long, red awns, hairy spikelets, and shattering character of the grains.

Monophyletic origin

Chatterjee (1948) grouped the former species, *O. cubensis* (America), *O. barthii* (Africa) and perennial wild rice of Asia as merely varieties of a single species, *O. perennis*. Thus *O. perennis* is distributed in Asia, Africa and America. Ghose *et al.* (1961) pointed out that it is the most primitive species in the genus as it has the primitive features including diploid chromosome number, wide distribution, perennial habit, tall growth, large leaves, large ligule, long stamens, anthers adapted for outcrossing, long spikelet, marked trichomes and awns, pigmentation on several plant parts, etc. It is gregarious and the seeds are actually collected by poor people for consumption. Moreover, the characters of *O. sativa* and *O. glaberrima* are parallel to a large extent and the variation in the two species is strikingly parallel, which is difficult to explain unless a common ancestor has been responsible for their origin.

Sampath and Rao (1961) postulated that *O. perennis* has given rise to *O. sativa* in Asia and *O. glaberrima* in Africa. *O. perennis* is a diploid species like *O. sativa* and forms partially fertile hybrids with *O. sativa* and *O. glaberrima*. They postulated that *O. fatua* is not the progenitor of *O. sativa* but is the progeny and originates in nature due to natural hybridization between *O. sativa* and *O. perennis*. The reasons advanced by them are as follows :

1. The seeds of *O. spontanea* collected from a single plant often give a segregating progeny.

2. *O. spontanea* segregates for the characters of *O. sativa* and *O. perennis*.

3. The progeny of the artificial hybrid between *O. sativa* and *O. perennis* resembles the natural population of *O. spontanea*.

4. The naturally occurring *O. spontanea* shows different degrees of sterility as is expected in the progeny of an interspecific hybrid.

5. A collection of *O. spontanea* crosses easily with the varieties of *O. sativa* of its own locality but crosses with difficulty with *O. spontanea* collected from a widely different locality.

Similarly, they postulate that *O. stapfii* is a naturally occurring hybrid between *O. perennis* and *O. glaberrima*. *O. breviligulata* is thought to be a stabilized form of such hybrid so as to claim the status of an independent species.

In the recent years, this view has received support from several India, Japanese and American rice workers.

Rice cultivation

Rice is mostly grown in the monsoon seasons from June to December in the tropics north of the equator and from November to April south of the equator. In Sri Lanka, the east coast of India, the Philippines, West Malaysia, Indonesia, and East Pakistan, rice is also grown in the so-called off-seasons, though to a much smaller and limited extent. Monsoon varieties have the longest maturity period, from 160 to 200 days, while the limited number of varieties grown in the off-season have shorter maturity periods, from 90 to 130 days. Most of the latter varieties have little sensitivity to photoperiod changes, e.g. the aus and boro varieties of East Pakistan; the Kuruvai and Kars of India.

There are two types of rice cultivation:

(i) *Upland rice* (dry paddy cultivation) practised on hill tops, hill sides or other areas where neither irrigation nor any device to hold rain water are available. This is also known as *terrace cultivation*. Seeds are sown by broadcast or by drilling in plough furrows.

(ii) *Floating rice* : In some low-lying areas in Bengal, Assam, Orissa, Tamil Nadu and Kerala, after sowing the seed, due to monsoon rains water level rises to a level of 5-6 feet. Rice grows rapidly keeping pace with the water level. Harvesting of crop is done in standing water by boats.

Rice grows in tropical regions where the annual rainfall ranges from 120-150 cm; otherwise heavy irrigation is essential. The crop requires constant temperature 20-25°C; warm and bright sun necessary when the crop matures. Rice is grown under diverse soil conditions over a wide range of pH.

In India there are two cropping seasons : *Kharif* season (main season; July-December), *rabi* season (off season; January-April).

There are two methods of sowing seeds : dry system - broadcasting, dibbling and drilling; wet system - seedlings raised in a nursery are transplanted; or sprouted seed directly sown in a puddled and levelled field.

Seed beds 1 m wide, 8 m long and 10 cm high are prepared and the beds are covered with polythene sheets. Sand or ash upto a thickness of 2-5 cm is applied on the beds. Seeds are water soaked for 12 hrs. and kept for germination for one day. The pre-germinated seeds are sown in the beds. Seed rate is one kg per sq.m. The beds are covered with gunny bags for the first 3 days and kept moist by spring water thrice a day 20-28 days old seedlings are ready for transplantation.

After transplanting seedlings in the field, maintenance of optimum water level is very essential. About 20 cartloads of farmyard manure, 100 kg/ha super phosphate, 100 kg/ha of green manure is given to the field as basal dressing before transplanting the seedlings. 200 kg/ha ammonium sulphate is given as top dressing. Weeding (removal of weed plants) is necessary.

Harvesting

Right stage for harvesting rice is when the ear is nearly ripe and the straw has just turned yellow. If harvesting is delayed, loss in grain occurs due to shedding. The crop is cut and allowed to dry in the field for 3-4 days and then threshing is done by cattle or Japanese pedal thresher. After threshing, the paddy is winnowed to separate the chaff.

Rice breeding in India

In India rice breeding started in 1911 in East Bengal (now East Pakistan). The Central Rice Research Institute, Cuttack, was started by ICAR in 1946. At the various rice experimental stations 430 improved varieties have been evolved, 27 by hybridization. Rice is grown in the country under widely varying conditions with maturation periods ranging from 90 to 200 days. It is grown in three main seasons; each season has its own set of varieties. Some of the most outstanding of the existing *improved varieties* are MTU 1, MTU 15, and HR 19 of Andhra Pradesh; Chinsura 7 of West Bengal; Kolamba strains of Bombay; hybrids 2 and 18 of Madhya Pradesh; GEB 24, CO 2, CO 25, CO 26, and ASDI of Madras; T 141 and SR 26B of Orissa; Basumati 370 of Punjab and T 136 of Uttar Pradesh. The variety GEB 24 was obtained as a spontaneous mutant in a traditional variety, Konamani. It proved to be a useful variety which spread far from its native habitat and contributed towards the development of several varieties.

Objectives of rice breeding

Early maturing ranging from 100-130 days; small, dark green leaves with erect growth, short and sturdy culms which are resistant to lodging; moderately firm threshability; seed dormancy; good milling percentage; good cooking behaviour; resistance to blast and other diseases.

The Chinese *dee - geo - woo - gen* variety has short plant stature; insensitive to photoperiod; dark green and erect leaves; non-lodging plant type; high nitrogen responsive; high yielding. These characters are chosen as the main objectives of rice breeding, for evolving new varieties. But rice quality of *japonica* rices is glutinous. Non-glutinous grain is preferred by consumers.

IR-8 is high yielding variety from International Rice Research Institute, Philippines, evolved from a cross between Peta x Dee - geo - woo - gen.

Bala is a high yielding variety from India, evolved from a cross N - 22 × Taichung Native - 1; 100 days maturity; plants are short statured and high yielding.

Jaya is high yielding variety from India, evolved from a cross between TN-1 × T-141; 130 days durations; high yielding; coarse grained.

IR 20 is a high yielding variety from IRRI, Philippines evolved from a cross of IR 262 - 24-3 × TKM-6; 135 days duration; short statured plants; slender, transparent grains.

Rice grain

The rice fruit is a *caryopsis*, a one-seeded fruit. The grain is provided with a brownish wall made up of seed coat and the non-separable pericarp fused together the lemma. The embryo lies on the ventral side, next to the lemma. The remaining part of the caryopsis is the endosperm. The endosperm is enclosed by the aleurone layer lying beneath the tegmen. The white starchy endosperm consists of starch granules embedded in a proteinaceous matrix.

Rice embryo

Embryo is small and situated at one end of the endosperm. Embryo consists of (i) one shield-shaped cotyledon known as scutellum and (ii) a short axis, the upper part representing the plumule surrounded by minute leaf-sheath or coleoptile and the lower radicle protected by root cap and surrounded by root-sheath or coleorhiza. A small projected structure known as epiblast, is seen opposite to scutellum (Fig. 2.2).

Rice is a major source of food protein in Asia and other countries where the daily intake of rice is high. Its value as a protein source is enhanced by its *high lysine content* relative to other cereal grains. The main limitation of rice as a protein source is its low protein content (6 to 8) per cent. Protein content of some brown rices from Korea (Chok-jye-bi-chal, Chow-sung, Santo, Crythroceros Korn), Hungary (Omirt 39), Japan (Rikuto Norin 20) ranges from 15 to 16 per cent.

Biochemical composition of rice grain

In the *glutinous* (*waxy*) varieties of rice the starch is only amylopectin and stains reddish-brown with iodine test. In the *non-glutinous* (*non-waxy*) varieties the starch contains amylose and amylopectin and stains dark blue with iodine test.

Rice is a poor source of fat and milling removes most of the fat. Rice is washed before cooking and there is loss of vitamins-thiamine, riboflavin and niacin.

Milled rice has about 9 per cent protein and 0.3 per cent fat (compared to 2 per cent in husked rice); Carbohydrate about 77-79 per cent.

Cytogenetics

The diploid (2n) chromosome number of rice is 24. There are 12 linkage groups. The linkage groups have been determined genetically by hybridization and segregation of genes.

Parboiled rice (sela chaval)

Parboiling is an ancient method of processing paddy. The process includes four different stages :

1. *Soaking* : Paddy is soaked in water for 1-2 days and the paddy absorbs water.
2. *Steaming* : The soaked paddy is steamed at atmospheric pressure for about 30 minutes. During steaming, thiamine and other water-woluble nutrients diffuse through the grain.

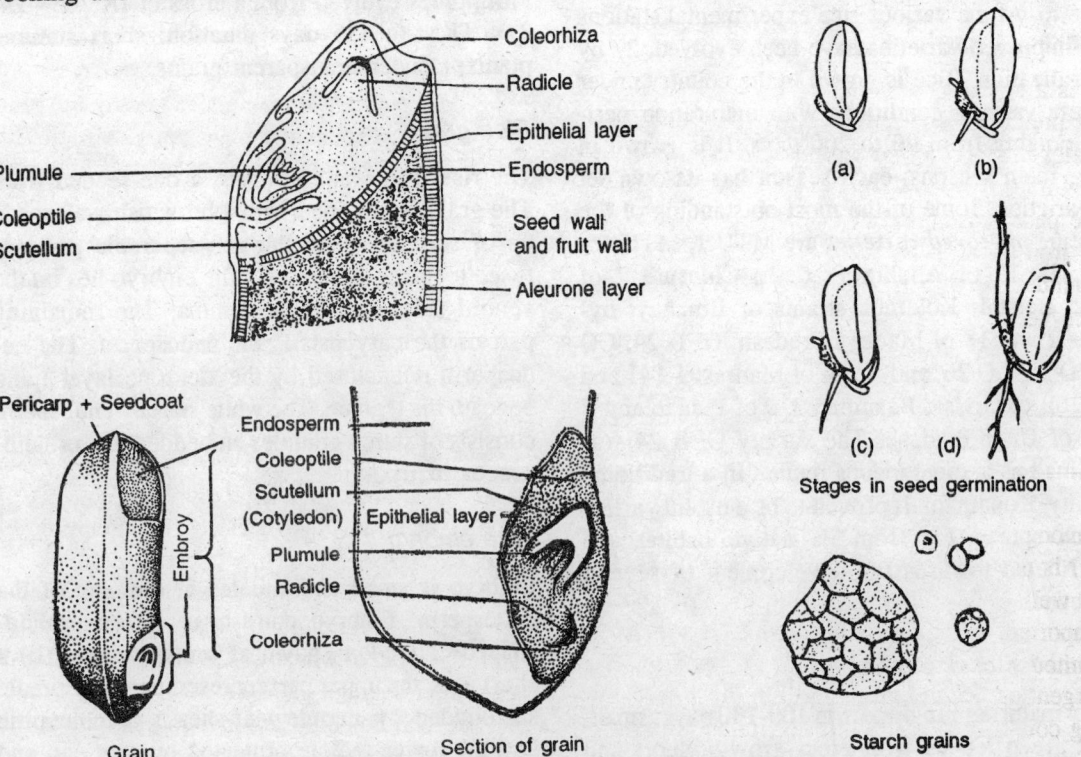

Fig. 2.2. Rice. *Oryza sativa* grain and germination.

3. *Drying* : After steaming, the paddy must be dried in direct sunlight.

4. *Milling* : Parboiled paddy after being dried is pounded or milled in the same way as raw rice. Parboiling toughens the grain and reduce the percentage of 'breakages' during milling in addition to its nutritional value.

Parched rice (murmura)

Little rice is thrown into the heated sand which is in a boiling iron pan; rapid stirring is necessary, rice cracks and swells, the rice is then sieved for separating it from sand.

Parched paddy (kheel)

It is prepared by drying the paddy in earthern pots kept in sun light. Then the paddy is moistened by adding hot water and immediately the water decanted. Then the jars are kept over-night in an inverted position. Next morning the moistened paddy is dried in hot sun and then parched in hot sand in the same way as parched rice preparation.

Flaked rice (chura)

It is a type of parboiled rice made flat and thin by pressure. It is prepared by soaking paddy in water for 2-3 days and then boiled in water for a few minutes. Then, the water is drained off, and the paddy is heated in a shallow earthern vessel or iron pan till the husk bursts open. Then it is pounded by a wooden pestle which flattens the rice kernel and removes the husk.

WHEAT

Among the world's crops, wheat (*Triticum aestivum* L.) is pre-eminent both in regard to its antiquity as well as its importance as a food of mankind. Important wheat growing countries are Russia, United States, France, India, Canada, Italy and Argentina. India ranks fourth in the wheat-growing countries of the world. In India, bread wheat (*Triticum aestivum* L.) is grown throughout the country; Jammu and Kashmir, Himachal Pradesh, Punjab, Haryana, Bihar, Orissa, Assam and other eastern states grow only bread wheat. Both bread wheat and macaroni wheat (*T. durum Desf.*) are grown in Uttar Pradesh, Rajasthan, Madhya Pradesh, Gujarat, Maharashtra, Karnataka and Andhra Pradesh. Emmer wheat (*T. dicoccum Schubl.*) is grown to a limited extent in Maharashtra, Gujarat, Tamil Nadu and Andhra Pradesh.

Wheat species are among the earliest plants to have been cultivated by man. A species called emmer, *Triticum dicoccum*, was found by archaeologists to have been cultivated at Jarmo, in Kurdistan, around 5000 BC. It is likely that ancient wheat species originated in this area from artificial crosses between two wild grass species.

Wheat is the world's leading cereal crop cultivated over area of about 229 million hectares with a production of about 493 million tonnes.

In India, wheat is the main cereal *Rabi* crop. In respect of area and production, it occupies the second position, next to rice.

India, stands fourth among the major wheat growing countries of the world both with regard to area and production (23 million hectares and 42 million tonnes respectively) and contributes about 9 per cent of the total wheat production of the world. However, the average wheat yield in India is relatively low as compared to many countries of the world.

One interesting fact about world wheat production is that the higher yields per unit area are obtained in countries (Netherlands, Brazil, France, etc.), with relatively smaller area under wheat cultivation and not in the major wheat production countries, which proportionately, cover larger areas.

The average yield of wheat in India was about 1800 kg/ha but very high yields have been obtained by individual farmers by adopting improved agro-techniques. For example, in the National Demonstrations conducted by the Indian Council of Agricultural Research (ICAR), the average yield of wheat was 3715 kg/ha against the All India average of about 1698 kg/ha.

The US, France, Canada and Australia are the major wheat exporting countries.

Major wheat importing countries are Egypt, Brazil, China, Japan, Italy, Poland and USSR.

CIMMYT

Centro International De Mejoramiento De Maizy Trigo, The International Maize and Wheat Improvement Centre, Mexico is responsible for collecting promising wheat germplasms all over the world, breeding high yielding varieties and making them available in different countries.

Cytogenetics

Schulz (1913) divided all wheats into three groups on the basis of their morphological differences. The validity of this grouping was confirmed from serological evidence (Zade, 1914) and evidence from rust reaction (Vavilov, 1914). However, Sakamura (1918) and Sax (1922) independently found that the three groups of Schulz had different chromosome numbers and that they formed a polyploid series on the basis of 7. All wheats were either diploid, with 14 somatic chromosomes, tetraploid with 28 chromosomes, or hexaploid with 42 somatic chromosomes.

The division of wheats into three polyploid groups was a significant discovery but it was Kihara (1924), Gaines and Aase (1930) independently proposed that the cultivated wheats had the genome formulae : *diploid species AA, tetraploid species AABB* and *hexaploid species AABBDD.* In general, hybrids within the groups have complete or nearly complete chromosome pairing and are highly fertile, while hybrids between groups have 7 or more univalents and are highly sterile. This hypothesis is referred to as the *ABD hypothesis*; the A genome of diploid wheats form fewer than expected 7 pairs in hybrids with the hexaploids. Furthermore the tetraploid *timopheevi* also shows variable pairing and complete or nearly complete sterility with most of the tetraploid species; but, *timopheevi* does show nearly normal pairing with one form of the wild tetraploid *dicoccoides* (Sachs, 1953). Sachs believes that all of the tetraploid species could have been derived from an original 14 paired prototype. Zohary and Feldman (1962) have further suggested that there was extensive

hybridization between amphiploids and their ancestral forms, which would lead to modification of the diploid parent genomes present in modern amphiploids. Kihara found many exceptions to the pairing expected on the basis of the ABD hypothesis.

The source of the A genome was recognized as the wild *monococcum.* A species of *Aegilops* was the source of the D genome. D genome of hexaploid wheats could have come from *Triticum aegilops* (*Aegilops squarrosa*) Kihara (1944). The sterile hybrids between the tetraploid *T. dicoccoides* and *T. aegilops* resembled the hexaploid only a step toward a satisfactory system of classification. Vavilov began the Herculian task of describing the species and varieties within the groups and determining their relationships. Bowden (1960) has revised the classification of wheat and its relatives. He merged the genus *Aegilops* with the genus *Triticum* and greatly reduced the number of species. Kihara in Japan, Sax in United States and Thompson in Canada studied the cytology of interspecific hybrids of wheat.

Winge (1917) suggests that polyploid series, such as that of diploid, tetraploid and hexaploid wheats with gametic numbers 7, 14 and 21 in an arithmatic series, could arise by doubling of the chromosomes in a sterile hybrid. If two distinct diploid species each contribute 7 non-homologous chromosomes to an interspecific hybrid, it will be sterile because the 14 unpaired chromosomes are distributed at random during meiosis. If on the other hand the chromosomes of the sterile hybrid become doubled there will be 14 pairs and the resulting tetraploid (amphiploid) can go through a normal meiosis and give fully fertile and constant progeny. This process may again be repeated by the hybridization of a tetraploid with an unrelated diploid to give a fertile hexaploid, which will contain three sets of chromosomes each from a different diploid species.

If the tetraploid wheat species arose by chromosome doubling in sterile hybrids, this should be evident from a study of chromosomes pairing in hybrids between the double hybrid and its ancestral diploids AA and BB, the hybrids AAB and ABB will have 7 paired and 7 unpaired chromosomes at meiosis. The amphiploid that arose through doubling the chromosome number of the sterile hybrid was fertile and produced fertile hybrids with the natural *spelta*. Mc Fadden and Sears (1946) suggested that *spelta* may have been the prototype of hexaploid wheats. But Kihara and Tanka (1959) have shown that the hybrid between *carthalicum* (*T. persicum*) and *T. aegilops*, which produces frequent natural amphiploids is a more likely source of common hexaploid wheats. In any case *T. aegilops* is the probable source of the D genome.

Determining the source of the B genome has been the most difficult problem of all because of the limitations of the method of genome analysis together with the lack of detailed knowledge of the taxonomy of wheat and its related genera. It was not until Sarkar and Stebbins (1956) using Anderson's (1949) method of extrapolation concluded that the wild species closest to the extrapolate was *T. speltoides*. They assumed that some form of the diploid wheats was one parent and that the tetraploid wheats (AABB) would be intermediate in morphology between some form of the diploid wheats (AA) and a wild species that contributed the B genome. This conclusion however was not entirely consistent with what was known about chromosome pairing of interspecific hybrids. There was more pairing between the chromosomes of diploid wheats and *T. speltoides* than between the A and B genomes of traploid and hexaploid wheats.

Riley, Unrau and Chapman (1958) found that the chromosomes of *T. speltoides* contained two pairs of satellited chromosomes that were similar to the two a satellited chromosomes in the B genome of hexaploid wheats. Furthermore they confirmed the discovery of Sears and Okamoto (1958) that there was a gene or block of genes of the B genome of hexaploid wheats that restricts intergeneric pairing thus promoting two by two pairing in tetraploid and hexaploid wheats. This finding cleared up the apparent inconsistency that there is no pairing between chromosomes of the A and B genomes within tetraploid wheat yet the AB hybrid obtained by crossing two diploid spe-

cies shows some pairing. Consequently, there is now good evidence that *T. speltiodes* is the source of the B genome in tetraploid and hexaploid wheats.

The demonstration that the A, B and D genomes came from the wild species *T. monococcum*, *T. speltoides* and *T. aegilops* provided an explanation of Darwin's problem of how so much variability could arise in such a relatively short time. The moderate diversity of varieties within the cultivated diploid *T. monococcum* is readily explainable as the consequence of selection of favourable mutants in the wild *T. monococcum*. On the other hand the great diversity of species and varieties within both the tetraploid and hexaploid wheats can be attributed to their amphiploid origin. Without assuming any increase in mutation rate, it is not difficult to understand the present diversity. The greater number of gene recombinations at the tetraploid and hexaploid levels coupled with the slow release of recessive genes would give an opportunity for the selection of an abundance of forms under the diversity of environments throughout the total range of cultivated wheats.

Diploid wheats (Fig. 2.3)

The cultivated varieties of einkorn differ from the wild *T. monococcum* forms largely in having a tough rachis which preserves the head intact and thus permits efficient harvesting. They undoubtedly rose through selection somewhere within the range of the wild diploid. At present the latter has a wide distribution from Greece, Bulgaria, Crimea, Transcaucasia and throughout the "nuclear arc". Vavilov (1950) recognised a centre of genetic diversity of both wild and cultivated forms in eastern Turkey and Transcaucasia. Today enikorn is only cultivated in a few mountain areas all four ancestral wild species, *T. monococcum*, *T. speltoides*, *T. dicoccoides* and *T. aegilops* are found within what Helbaek (1959) called the "nuclear arc" the piedmont flanking the mountain are extending from the Zagros Mountains (= Iran-Iraq) the Taurus Mountains (Southern Turkey) to

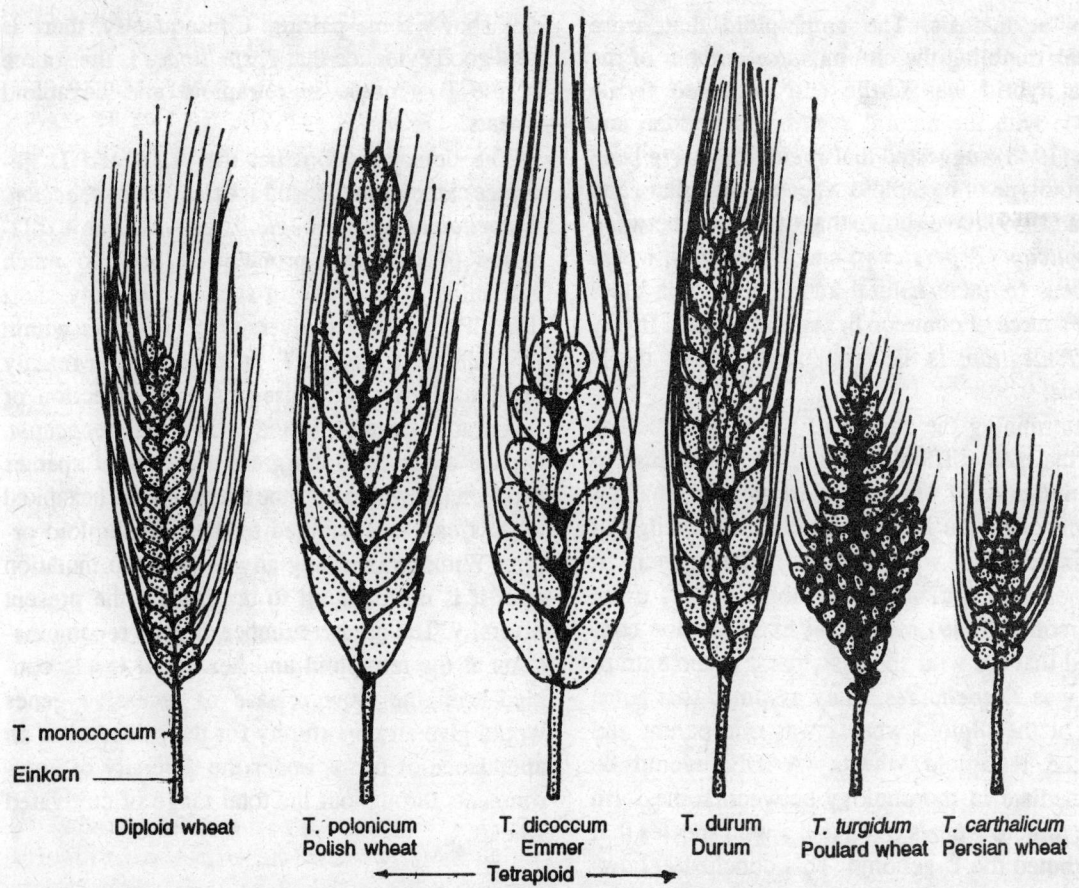

T. monococcum

Einkorn

| Diploid wheat | T. polonicum Polish wheat | T. dicoccum Emmer | T. durum Durum | T. turgidum Poulard wheat | T. carthalicum Persian wheat |

←——— Tetraploid ———→

Fig. 2.3. Species of *Triticum*.

the Galilean uplands (Israel-Transjordan). Extending from Spain along the northern Mediterranean to Asia Minor and on to Transcaucasia. Presumably these few islands are all that have survived of a once extensive cultivation in neolithic times. Einkorn together with the tetraploid emmer was cultivated from the mouth of the Danube to the mouth of the Rhine as early as the fifth millenium and from there spread throughout all of western Europe (Schiemann, 1932, Helback, 1959). Although both wild and cultivated *T. monococcum* was found at Jarmo, it apparently was not adapted to the adjacent low lands either in Iraq or Iran (Helback, 1959). Thus it migrated northward from some area within the "nuclear arc" and subsequent isolation in mountain areas was a consequence of ecological selection.

Tetraploid wheats

The question as to when and where tetraploid wheats originated took on new meaning when it was shown that all tetraploid wheat could have originated from a common 28-chromosome prototype and that prototype was undoubtedly an amphiploid derivatives of the hybrid *T. monococcum* (AA) × *T. speltoides* (BB) (Sarkar and Stebbins, '56; Riley, Unrau and Chapman, '58). The Cytogenetic evidence, however leaves unanswered the question : was the original tetraploid an ancient amphiploid that arose long before and independent of domestication or did it arise subsequent to domestication. For a long time the latter, the post-domestication origin, was favoured largely because the tough rachis, which is present in both diploid and tetraploid cultivated wheats, was thought to be derived from the cultivated diploid wheats. That is the A genome of tetraploid wheats was derived from a cultivated form of the diploid *T. monococcum.*

The present distributions of the wild tetraploid varieties *dicoccoides, timopheevii* and *tumanianaii (armeniacum)* together with their putative ancestors, *T. monococcum* and *T. speltoides,* do not yield any decisive evidence in favour of either the pre- or post-domestication hypotheses. All three are distributed throughout the Iraq-Israel "nuclear arc". Furthermore, these are distributed that their distributions have been greatly altered since domestication. Consequently the original tetraploid could have arisen any where within the mountain are either before or subsequent to domestication.

The major difficulty with the post-domestication hypothesis of the origin of the tetraploid wheats is the following one. How could so many different forms arise in the few thousand years subsequent to the origin of the cultivated diploids? Not only is there a tremendous diversity of cultivated tetraploid with the wild *dicoccoides.* While it is true that amphiploids, because of their hybrid origin are inherently more variable than their diploid ancestors, it is still short a time. The present distribution of varietal diversity within cultivated varieties of tetraploid wheats (Vavilov, '50) would speak against a post-domestication origin of the tetraploids. The cultivated tetraploids now occupy a Bikal in the USSR and so as far as Ethiopia. Within the boundary there are three centres of genetic diversity : Ethiopia shores of the Mediterranean, and Transcausasia. Only the Transcaucasian centre is within the modern range of the wild varieties *dicoccoides* and its presumed parents, *T. monococcum* and *T. speltoides.* The groups of cultivated tetraploids, *dicoccom* (emmer) and *durum* occur in all three centres, *turgidum, polonicum* and *ethiopicum* only in the Ethiopian and Mediterranean centres, while *carthalicum* (Persian), *paleocolchium, turanicum* only in the Transcaucasian centre. All of the cultivated tetraploid groups of varieties occur together with the wild variety *dicoccoides* in the Palestine-Iraq "*nuclear arc*". The origin of distinct cultivated varieties outside the range of var. *dicoccoides* would speak for an ancient origin.

The ancient origin of the wild tetraploid prototype had recently been supported by the observations and suggestions of Braidwood (1958) and Helbaek (1959). The latter reported that both wild diploid and tetraploid wheats were found together with the cultivated emmer at Jarmo. He suggested that diploid and tetraploid wheats were domesticated simultaneously and that the tough rachis was automatically selected for by agricultural operations. More grains could be harvested from those variants having a tough rachis. The cycle of harvesting, storing and sowing would automatically select genes for a tough rachis. He further suggested that as man carried agricultural operations outside the areas where wild wheat grew naturally there was a further automatic selection for forms that were adapted to the new areas. Thus, the tetraploids were adapted to farming the alluvial plains of the Euphrates-Tigris and Egypt, cultivated *monococcum* was not.

Hexaploid wheats (Fig. 2.4)

Hexaploid wheats must have originated within the area occupied by *T. aegilops (Ae. squarrosa)* if the latter is the source of the D genome as is now generally accepted. Although the present area occupied by *T. aegilops* has not been thoroughly explored as far as it is known in general it extends from the Caucasus Mountains on the West to the Tien Shan Mountains in the USSR on the East, and South to about the middle of Iran and Afghanistan. This very large area overlaps the ranges of the other basic wheat species. *T. monococcum, T. speltoides* and *T. dicoccoides* in Transcaucasia and adjacent areas of Iran and Turkey.

Vavilov ('26, '50) placed the centre of genetic diversity of hexaploid wheats, which he also considered as their centre of origin, in South Eastern Afghanistan and the adjacent mountain areas border in the upper Indus Valley. He also pointed out that the diversity of hexaploid wheats is nearly as great in Transcaucasia as in the Himalayan centre. Thus without knowledge of *T. aegilops* as the source of the D genome he placed

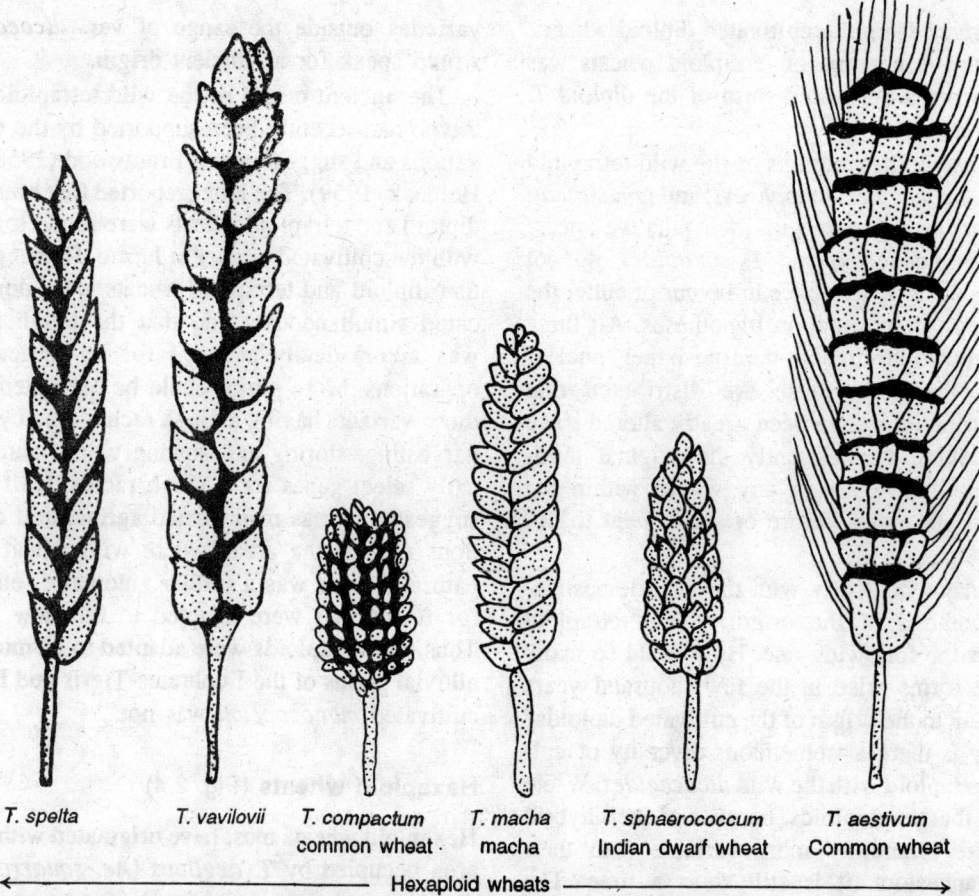

T. spelta   T. vavilovii   T. compactum   T. macha   T. sphaerococcum   T. aestivum
                                     common wheat   macha   Indian dwarf wheat   Common wheat

← ——————————————————— Hexaploid wheats ——————————————————— →

Fig. 2.4. Species of *Triticum* (contd.).

the centres of genetic diversity more or less within its range. Transcuacasia at the Western end of the range lies within the area of overlap with the other wild species while the Himalayan centre lies at the Eastern and somewhat outside the range of *T. aegilops.*

Among the varietal groups of hexaploid wheats three, *macha, vavilovii,* and *sphaerococcum,* have very restricted ranges of distribution. The first two, *macha* in Western Georgia and *vavilovii* in Turkish Armenia, are found within the area of *T. aegilops* and significantly within the area of overlap of the other basic wild species in Transcaucasia. Since they have not become widespread, it is possible that they originated within the area where they are found today. *Sphaerococcum,* which is endemic to the Indus Valley and adjacent areas,

is found wholly outside the area of *T. aegilops.* Ellerton (1939) presented convincing evidence that *sphaerococcum* varieties differ from *aestivum* by a single gene or block of genes, and that they arose as the result of a single mutation from the latter. It is very probable that *spaerccoccum* varieties also originated within the area in which they are found today. Furthermore, Mangelsdorf (1953) reported that Kernels of *sphaerococcum* have been found at the most ancient site in India, Mohenojo-Daro dated at about 2500 BC which would place the origin of *aestivum* before that date.

Up to the present no wild ancestral hexaploid has yet been recognized. Furthermore, the earliest reported occurrences of hexaploid wheats in the near East are *compactum* types that occur sporadically : somewhat before 4000 BC in Egypt, about 3000 BC in Iraq and about 2000 BC in Asia Minor, Syria and Palestine. Only after 1000 BC do they occur with an abundance. On the other hand, *compactum* wheats are abundant in the neolithic finds of Middle Europe before 3500 BC Helbaek '59 attributes this sporadic early occurrence of *compactum* in the Near East, where it most likely originated (within the area of *T. aegilops*) contemporary with its abundance in Europe as a consequence of its occurrence as a comparatively are weeds in the wheat fields of the Near East that flourished and increased when it was transported to areas of summer rainfall. He notes that imprints of *compactum* wheat were found together with those of *einkorn* (diploid) and *emmer* (Tetraploid) in the earlier *compactum* varieties could also be invoked to explain their concentration in the Himalaya centre assuming that they had been transported there from a centre at the Western and of the range of *T. aegilops*.

The *spelta* group of varieties has been the subject of considerable speculation. For a long time they were thought to be endemic to Central Europe where they were first reported in Bronze Age sites (3500-1500 BC). Further more they have never been found in prehistoric deposits outside of Europe. Since *spelta* types segregate from the cross *dicoccum* (4X) × *compactum* (6X) a number of authors (Schiemann, '51) postulated that spelta originated in the Rhine Valley at the threshold of the Bronze Age from a cross of local wheats (which included *compactum*). On the other hand McFadden and Sears (1944, '46) considered *spelta* as the prototype of all hexaploid wheats, largely because their synthetic hexploid from *dicoccoides* and *T. aegilops* resembled *spelta*. Kuckuck (1959) has reported a distinct *spelta* type that is endemic to South West Iran. Furthermore, *aestivum* and *compactum* types appeared in the later generation of hybrids between the Iranian and European *spelta* which Kuckuck considered as evidence that the Iranian *spelta* were of ancient origin and not a recent importation of European spelts.

Kihara (1959) suggested that the hexaploid prototype, which was morphologically like *aestivum*

arose in Transcaucasia. *Spelta* and *compactum* were derived from the prototype as mutants, and on crossing they produced *aestivum*, which in turn produced *sphaerococcum* by mutation. Even this hypothesis, which is consistent with the great bulk of evidence sited above is difficult to reconcile with the distribution of necrotic genes among the different wheats (Tsunewaki and Kihara, '62). Together with Helbaek's hypothesis of ecological selection, Kihara's suggestion will serve as working hypothesis. It will undoubtedly be modified and made more precise with the appearance of new evidence from archaeology and genetics.

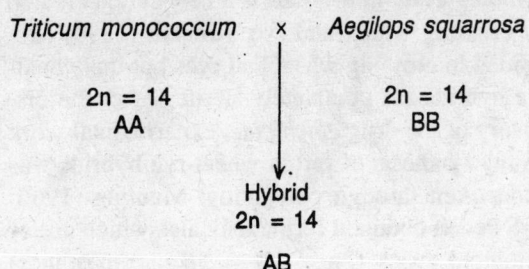

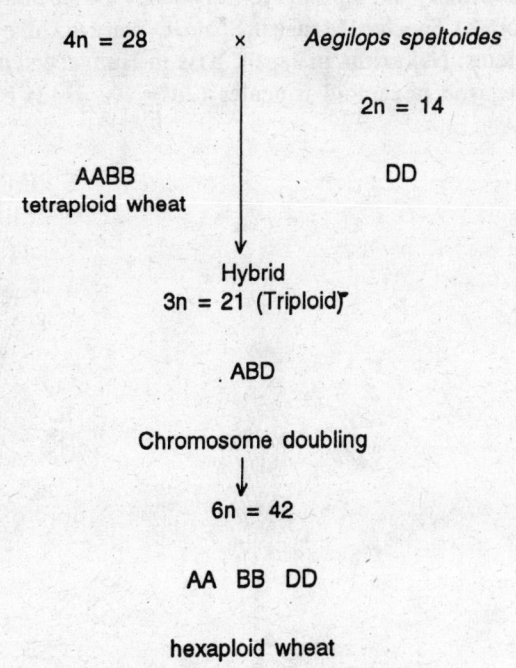

Origin of cultivated wheat

While Kihara's hypothesis assumes a single origin of hexaploid wheats, it does not exclude the possibility of multiple origins from different tetraploid ancestors and it can readily be adopted to include the possibility of multiple origins from different tetraploid ancestors, and even subsequent hybridization between different hexaploids. Since he assumes that the place of origin is in the area of overlap of the wild species, it can be adapted to pre- or post-cultivation origin.

Triticale (Fig. 2.5)

Triticale hexaploides Lart is a new cereal created by crossing wheat and rye. Rimpau (1890) succeeded in crossing wheat and ryes but the wheat-rye hybrids are completely sterile. After the discovery of the drug *colchicine*, experimental work on the synthesis of fertile wheat-rye hybrids was undertaken through polyploidy. Munting (1966) of Sweden obtained fertile triticales which are of octoploid types (2n = 8X = 56 chromosomes) synthesized from crossing common hexaploid wheat species, *Triticum aestivum* (2n = 6X = 42 chromosomes) and diploid rye (2n = 2X = 14 chromosomes). Sanchez-Monge in Spain, O'mara in United States, Nakazima in Japan, Kiss in Hungary synthesized hexaploid triticales (2n = 6X = 42) by

crossing the tetraploid wheat *Triticum durum* (2n = 4X = 28) and rye. In contrast to ocotoploid types, hexaploid triticales are more vigorous, long spikes, better floret fertility, and larger grains.

Distribution

The great wheat regions are found in temperate zones between 30-60°N and 24-40°S loamy soils with pH 8.3 are suitable.

Wheat cultivation

In India wheat is cultivated in Gangetic alluvium soil (Uttar Pradesh); Indus alluvium (Punjab); black soils (Central and South India); desert soils (Rajasthan); hilly regions of the Himalayas and Siwaliks.

About 200 kg of Ceresan treated seed is sown in field from the beginning of October to the end of November. Seed is usually sown by broadcasting or by line sowing. In North India wheat is sown by dibbling method; spacing between rows is 20-25 cm. Weed control by spraying, 2, 4-D. Adequate irrigation throughout the growing period is necessary. 40-50 kg P_2O_5, 65 kg potash, 120 kg nitrogen should be given per hectare of land. Wheat is harvested towards the middle of

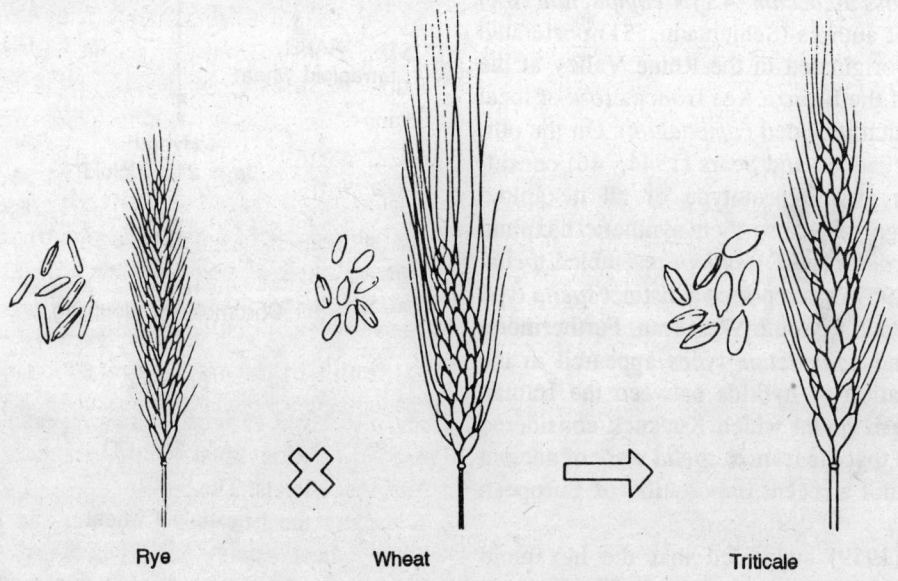

Rye   Wheat   Triticale

Fig. 2.5. *Triticale*—Hybrid by rye and wheat.

January and continues till the middle of May in various regions in India. Best time for harvesting is morning time. High yielding varieties give about 1400 kg/ha of grain.

The wheat plant

Wheat is grown both as a winter annual (winter wheat) and as a summer annual (spring wheat).

Wheat tillers freely; the root system of the mature plant proceeds from a number of sub-terranean nodes, filling the upper soil with a fine network which affords a very considerable absorbing surface. The culms are from two to four feet in height; those of most varieties grown in North America are comparatively stocky and stiff, affording straw which is valued for strawboard-making and for various other purposes.

The spikelets are nearly sessile and compactly arranged on a zig-zag axis. The broad glumes vary considerably in size and proportion in different races; they enclose from two to five florets, of which usually not more than two produce fruit. The bearded wheats produce long-awned lemmas; many varieties are beardless. The majority of wheats grown in northern latitudes are close-pollinated; durum wheat and the supposedly primitive wheat of Syria are usually cross-pollinated. Hot, arid environment appears to favour cross-pollination in all wheats (Fig. 2.6).

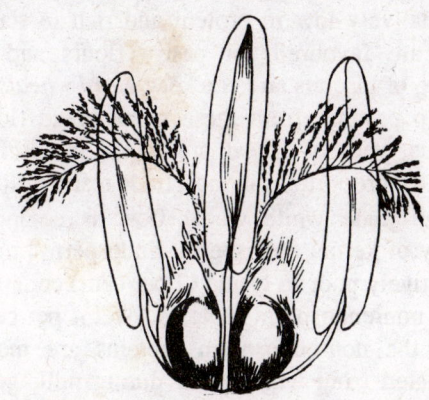

Fig. 2.6. Flower of wheat (*Triticum aestivum*).

High yielding varieties

Breeding work in India was done by B.P. Pal who was responsible for the release of several New Pusa (N.P.) varieties.

N.P. 846

Parentage Riogenero X NP 760; a tall wheat having amber, lustrous and hard grains of medium size; highly resistant to rusts; suitable for growing in North Hills Zone at lower elevations.

M.S. Swaminathan did pioneering work in mutation breeding and released the variety *Sharbati Sonora*.

Sharabati Sonora

Parentage : mutant of Sonora 64. A double dwarf wheat having amber, hard grains of good quality; suitable for late sowing; early maturing.

Contribution of N.E. Borlang : Borlang, the Nobel Prize Winner for Peace (1970) is a wheat breeder in Mexico who was responsible for the release of Norin dwarf varieties of wheat. These varieties produced *green revolution* (see chapter 1) all over the world, not to speak of Mexico.

Lerma Rojo 64 Parentage Yaqui 50-N10B X L52/Lr.2. A single dwarf wheat with wide adaptation and resistance to yellow and black rusts; grains red, soft and medium bold.

Pusa lerma

Amber mutant of Lerma Rojo 64A. Resembles Lerma Rojo in all respects except the grain colour which is amber.

Structure of the wheat grain (Fig. 2.7)

As in the fruit of most cultivated plants, there is much variation in size and shape of grain of different wheats. The second grain of the spikelet is usually the heaviest. One hundred kernels of bread wheat usually weigh between 3.5 and 4 grams.

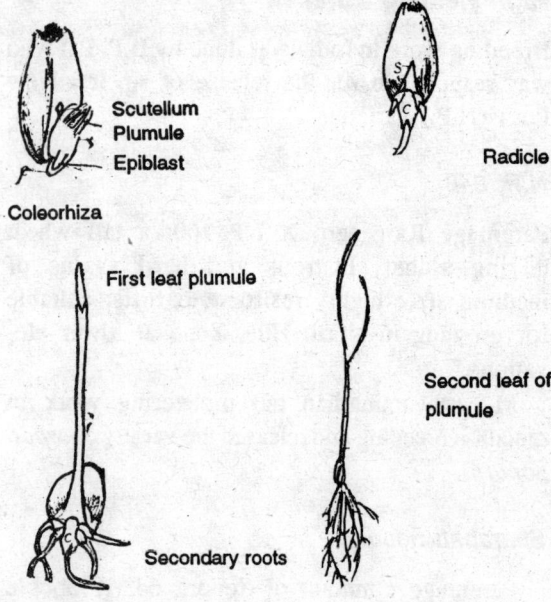

Scutellum
Plumule
Epiblast

Coleorhiza

Radicle

First leaf plumule

Second leaf of
plumule

Secondary roots

Fig. 2.7. Wheat—Stages of seed germination.

Wheat grain

Wheat grain is a caryopsis. The seed has 4 pairs seed coat (testa), embryo, nuclear layer and endosperm. The endosperm is divided into (i) aleurone layer, (ii) starch and gluten parenchyma. A short supply of amino acid *lysine* is the principal nutritional limitation of wheat protein. Middleton, Bode, and Bayles (1954) showed that the variety Atlas 66, selected from the cross Radhart X No. 11/2X Fronodoso, had more protein in its grain than other soft wheats; the protein content of seed of Atlas 66 is 17.9 per cent.

Ordinary wheat grains separate readily from the dried floral bracts; in emmer and some other forms the palea and lemma remain, embracing the kernel somewhat as in barley or oats. At the upper (stigmatic) end of each grain is a dense growth of fine hairs, the brush. A groove along the side adjacent to the palea marks the position of the placenta.

In the *section of the kernel*, the following structures may be recognized; *pericarp, seedcoats, nucellus, endosperm,* and *embryo.*

The *pericarp* consists of a few layers of hardened cells, whose structure has become much modified during the maturing of the grain and is difficult to make out. In the early stages of development there are two integuments; one of these is absorbed, and the other persists as a two-layered testa, which contains the coloring matter of the seed. The nucellus is a single-cell layer.

Pericarp, testa, and nucellus collectively make up the *bran-coats*; because of their position and toughness they are removed by the first milling processes as the familiar brown flakes of bran, to which also adhere particles of endosperm.

The bulky *endosperm* is of two portions, the aleurone layer and the starchy endosperm. The former is closely packed with the protein aggregates which give it its name. The starchy endosperm is made up of thin-walled storage parenchyma, in the cells of which are massed the characteristic disk-shaped starch grains. Proteins also occur in the starchy endosperm.

The embryo is lacking in starch, but contains protein and oil. The relative weights of the parts of the grain are stated as follows : bran coats, 8-9 per cent; aleurone layer, 3-4 per cent; starchy endosperm, 82-86 per cent, embryo, 6 per cent.

Soft wheats and hard wheats

Different wheats vary in their carbohydrate and protein content; variation is correlated with varietal characteristics and also with climatic, seasonal, and soil conditions. The so-called *soft wheats* are relatively low in protein and rich in starch; these are favoured for pastry flours and the making of biscuits and crackers. *Hard wheats* are high in protein; from these come bread flours. Various types of wheat are mixed in carefully regulated proportions in most trade marked flours.

High-grade white wheat flour is composed largely of kernel endosperm. Endosperm protein is relatively poor in lysine (2 per cent) compared to the non-endosperm proteins (over 4 per cent). Since the non-endosperm proteins are mostly eliminated from wheat flour during milling, increasing their lysine content would not change the lysine content of wheat flour much.

Milling wheat into flour decreases the protein content of the flour below that of the whole grain. The protein reduction ranges from 0.5 to 1.5 percentage point or roughly 10 per cent.

Wheat proteins

Table 2.1. High protein wheats

	Protein content (per cent) in seed
Atlas 66	17.9
Purdue 28-2-1	17.5
Triumph 64	15.2
Bezostaia	13.8
Gaines	13.3
Yorkstar	12.8

Two basic types of *storage proteins* can be distinguished in seeds : the *globulins*, which are found in the embryos of most seeds, and the *prolamines*, which occur in the endosperm of many cereals (but in oats major storage proteins are globulins) (Fig. 2.8).

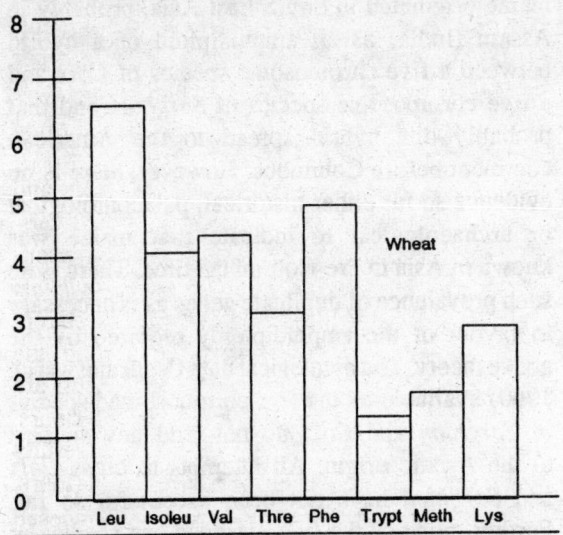

Fig. 2.8. Content of the 8 essential amino acids in wheat protein (gms per 100 gms protein).

Wheat seed prolamines

Wheat storage proteins are divided into *gliadins* and *glutenins*, which together form *glutern com-plex* responsible for the breadmaking properties of wheat. *Glutenin* is believed to impart elasticity to dough; while *gliadin* makes it viscous and provides extensibility. Typically wheat storage proteins consists of about 50 per cent *gliadin*, 10 per cent high molecular-weight *glutenin*, and 40 per cent low molecular weight glutenin sub-units.

Uses

Wheat is consumed mostly in the form of flour obtained by milling the grain. Bulgar is a par-boiled wheat product which is a traditional food item in West Asian Countries and in North Africa. Macaroni is produced from semolina (suji) which is obtained from hard wheat.

MAIZE

Maize (*Zea mays*, Poaceae) is the third most important cereal crop in India. About 90 per cent of the yield is used for human consumption. Maize is a hardy crop as compared to wheat and rice and can be cultivated under both rainfed and irrigated conditions. It is specially suited for crop rotation programmes.

Water requirement of maize is about 60 cm but it is influenced to a great extent by the season, soil type, variety and agronomic practices adop-ted.

Hybrid and composite maize is usually grown at an inter-row distance of 75 cm. Maize is a unique cereal having wide diversed soil and cli-matic conditions and is the most efficient conver-tor of sunlight into carbohydrate, among the ce-real crops. Average yield of maize is also highest - 3.0 tonnes/ha - among the cereal crops of the world. Highest yields can be achieved by growing the crop in winter. During winter season the second week of November was found to be the best time of sowing maize when weather remains warm with maximum and minimum temperature around 30°C and 20°C.

Varieties

EH-400-475, EH-400-575, EH-2380 (Ganga 9), Histarch, Deccan 101, Vijay, Amber, Opaque-2 composites, Ganga 5, 3, 101; Ganga safed-2.

Opaque-2 composites forms a series of nutritionally superior maize strains with high lysine and tryptophan content. The opaque-2 gene (O_2) increases the lysine and tryptophan contents and protein quality.

Origin

The discovery of America was the discovery of corn. Corn played very important part in the earlier civilization of the American continent.

Maydeae, which is a small tribe of the large family of grasses, has assumed its present importance on account of the single economic crop of *Zea mays*, which is one of the great contributors of food to humanity, perhaps only next to wheat and rice. The tribe Maydeae includes the following **eight** genera of which five are relatively unimportant :

1. *Coix* (Job's tears)
2. *Sclerachne*
3. *Polytoca*
4. *Chionachne*
5. *Trilobachne*

These genera are native to the region extending from India and Burma through East Indies into Australia and Polynesian islands. The three American genera are :

1. *Zea* (corn)
2. *Tripsacum* (gama grass)
3. *Euchlaena* (teosinte, closest wild relative of maize).

There are various theories regarding the origin of maize. The recent significant discovery of fossil pollen of maize from Mexico City by **Barghoorn** and others (1954) points to the existence of the plant in the valley of Mexico about 80,000 years ago, antidating the most primitive type of agriculture ever conceived in the Western Hemisphere. Several workers contributed to the elucidation of the problem of origin of maize.

Period of origin

Most authorities on corn are convinced that the place of origin of maize lies in America. Two locations in America have been suggested :

1. The high lands of Peru, Equador and Bolivia, and
2. The region of Southern Mexico and Central America. Both these regions are characterized by the occurrence of many types of corn.

An Asiatic origin for maize was also suggested by Bonafus in the last century. This theory received support from the evidence such as:

1. the existence of a type of corn with a waxy type of endosperm in Western China under conditions suggesting that it had been there probably since pre-Columbian age;
2. the existence of peculiar types of corn grown by primitive people in the inaccessible regions of the hill country of Assam and Northern Burma (Stonor and Anderson, 1949).

In 1945 Anderson proposed the theory that maize originated in South East Asia, probably in Assam (India) as an amphidiploid of a hybrid between a five chromosome species of *Coix* and a five chromosome species of *Sorghum* and that probably this hybrid spread to the American continent before Columbus. However, there is no evidence so far either historical, palaeontological or archaeological to indicate that maize was known in Asia in Pre-Columbian time. There is no such prevalence of duplicate genes as is necessary in favour of the amphidiploidy required by the above theory. The cytological data (Venkateswarlu, 1960) available on the five chromosomed species of *Sorghum* and *Coix* do not lend any support to the Asiatic origin. All attempts to cross *Coix* and *Sorghum* have not been successful so far. Further, none of the five chromosome species of *Sorghum* are known to be native to South East Asia.

Method of origin

No truly wild maize has yet been found and there is no such agreement on the method of origin as is there on the probable region of origin. Various hypotheses differ in this respect.

1. According to **Pod Corn Theory** put forward by St. Hailaire in 1829 and supported later in 1894 by Sturtevant, Pod Corn is looked upon as wild or primitive type of maize based on the conception that the kernels are enclosed in a pod in the wild progenitor of maize. It is now known that the pure pod corn differs only in one gene from the ordinary maize plant.

2. Mangelsdorf and Reeves (1939, 1959) and Mangelsdorf (1947, 1948) incorporated the above idea into their **Tripartite theory** on the origin and evolution of maize. According to this theory :

(a) Cultivated maize originated from a wild form of Pod Corn which was once and perhaps still is indigenous in the low lands of South America.

(b) The teosinte (*Euchlaena*), the closest relative of maize (Fig. 2.9), is a recent product of the natural hybridization of *Zea* and *Tripsacum* which occurred after cultivated maize has been introduced by man into Central America.

(c) The new types of maize originated directly or indirectly from this cross and exhibiting admixture with *Tripsacum* comprise the majority of Central and North American varieties.

Mangelsdorf (1958) resurrects the above view on the basis of the evidence from the cobs found in the Bat Cave in New Mexico. The cobs and kernels are of primitive variety having the characteristics of pod and pop corn. Mangelsdorf (1958) succeeded in synthesizing an ear, which resembled the primitive Bat Cave Corn and having the male and female spikelets on the same inflorescence as in *Tripsacum* by appropriate crossing between pod corn and pop corn. Mangelsdorf and Reeves (1958) stated that evolution of maize to the modern state took place as a consequence of hybridization between teosinte and corn. It is suggested that the following four principal factors operated in the evolutionary sequence :

(i) The presence of natural selection, one of the most important suppressive factors in evo-

Fig. 2.9. Teosinte (*Euchlaena mexicana*).

lution of corn during the period of general evolution.

(ii) Occurrence of mutations from the more to the less extreme forms of pod corn.

(iii) Modification of corn by contamination with teosinte.

(iv) Crossing of varieties and races resulting in new combinations of characters and a high degree of hybridity.

3. A theory of hybrid origin of corn involving *Teosinte* as one of the parents was proposed by Harshberger. In 1884, Harshberger came across a plant which he named as *Zea canina* and considered to be wild corn. Later in 1896, when he found that it was only a hybrid of teosinte and corn, he proposed that corn originated as a hybrid of teosinte and a similar but unknown wild grass. This theory was subsequently supported by Collins

(1912) who suggested that the other parent might have been a member of the Andropogeneae. This theory, however, is no longer seriously considered by any student of corn phylogeny at the present day.

4. The *teosinte mutation theory* of the origin of maize was originally advanced by Ascherson (1880) and is based on the concept that corn arose directly from teosinte by mutation. In recent times it received supported from Longley (1941) on the basis of similarities in the morphological features of the chromosomes of corn and teosinte such as total chromatin lengths, arm ratios, occurrence of knobs in the same relative positions and also pairing behaviour followed by crossing over. This is further strengthened by the genetic analysis by Langham (1940) which indicated that relatively few major gene mutations would be required to transform teosinte into a plant with most of the important characteristics of corn such as (i) a change from distichous, 2-ranked arrangement of the kernels on the ear to polystichous or many rowed condition, and (ii) a change from single to paired spikelets. Later it was found that larger number of genes than assumed by Langham would be required for transformation of teosinte into corn like plant (Mangelsdorf, 1947). Weatherwax (1950), on the basis of comparative morphological study of corn and its wild relatives came to the conclusion that teosinte is a more specialized type than a progenitor of maize would be expected to be. This, along with evidence from fossil pollen, indicates that maize is probably more ancient than teosinte thus demolishing the teosinte mutations theory.

5. Theory of origin of maize by *divergent evolution* was put forward by Montgomery in 1906 and in later years elaborated by Weatherwax (1935, 1955). According to him *Zea, Euchlaena* and *Tripsacum* stem from a common ancestor which probably gave rise also to some of the Andropogoneae. Regarding the hypothetical ancestor of corn, which he considers now to be extinct, he writes : "On one of the branches of this family tree, a little nearer to teosinte than to *Tripsacum,* there was wild plant which was to become the ancestor of corn". He visualizes that

the wild ancestor of maize was probably a perennial plant with something of the habit of the more caulescent species of *Tripsacum* with male and female spikelets borne in almost completely separated inflorescences with an incompletely eight rowed ear having the grains covered with glumes and lemmas as in most other grasses and a cob not breaking into pieces at maturity as do those of *Tripsacum* and *Teosinte.*

This theory has also not met with universal acceptance. Kemptom (1919) thinks that the fundamental differences between *Zea, Tripsacum* and *Euchlaena* have been overlooked and too much emphasis has been placed on the fact that the genera are alike in the suppression of sex-organs, while Collins (1931) considers that the theory fails to take into account the close relationship of *Zea* and *Euchlaena* is indicated by their interfertility, by their close agreement in number, size, shape and behaviour of the chromosomes and the fact that they are hosts to the same parasitic diseases. Reeves and Mangelsdorf (1959) point out that there is no evidence of an original common ancestor now probably extinct.

Cytotaxonomic and cytogenetic evidence available lends little or no support for the assumption that maize is :

(a) a hybrid of teosinte and some unknown member of Andropogoneae,

(b) an amphidiploid hybrid of an Asiatic species (like *Coix*) of Maydeae and a species of *Sorghum,*

(c) a trigeneric hybrid of pod corn, *Euchlaena* and *Tripsacum,*

(d) a result of large scale gene mutations of *Euchlaena,*

(e) a recent product of hybridization of South American *Tripsacum* with *Teosinte* as a bye product.

Some of the essential elements of a general theory of origin of cultivated maize that is consistent with relevant facts as stated by Randolph (1959) are given below :

1. The progenitor of cultivated maize was wild maize, probably an annual plant with many of the vegetative characteristics of teosinte

having numerous tiny, corn-like cobs with paired spikelets, several rows of kernels and non-corneous glumes only partly enclosing the very small teosinte-like seeds.

2. Domestication occurred in the south-western United States about 6,000 years ago, and probably also independently in Mexico and neighbouring Central America, possibly at different periods which might have been appreciably earlier on later than in the Bat Cave region of New Mexico.

3. The diversity of existing types of maize with respect to cytological, morphological and physiological characteristics, and their widespread distribution in very early times suggests that more than one species of wild corn was involved in the origin of cultivated maize.

Maize grain (Fig. 2.10 & 2.11)

Each grain consists of the following parts : *Seed-coat, Endosperm* and *Embryo*. Embryo is very small and lies in a groove of one and of the endosperm. It has a shield-shaped cohyledon, called *scutellum, plumule, coloeptile* and *radicle*. Germination of the seed is *hypogeal*.

BARLEY

Barley is the product of *Hordeum distinchum* and *H. vulgare*, each of which may be subdivided into numerous subspecies and varieties. *Hordeum* includes about 20 species, several of which such as *wall barley*, (*H. murinum*), *squirrel tail* (*H. jubatum*) are widespread annual weeds. Asiatic wild species *H. spontaneum* has been considered the progenitor of the modern barleys of cultivation.

The culms of barley, are usually less than 3 feet in height. The inflorescence is a close, heavy spike. 3 one-flowered spikelets occur at each joint of the rachis.

In 2-rowed barley (H. distinchum), the lateral

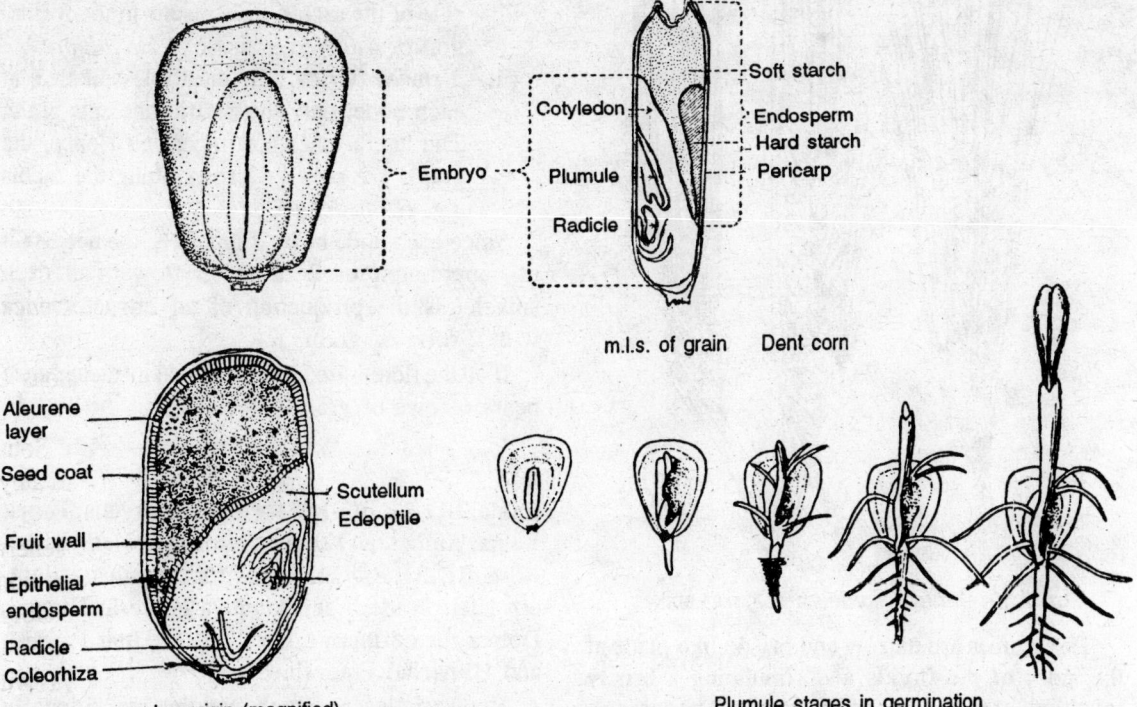

m.l.s. of grain   Dent corn

m.l.s. origin (magnified)   Plumule stages in germination

Fig. 2.10. Maize (*Zea mays*)—Grain and seed germination. *m.l.s.* = Median Longitudinal Section.

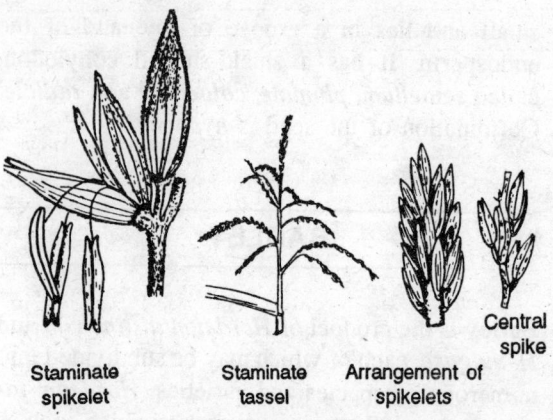

Staminate   Staminate   Arrangement of
spikelet   tassel   spikelets

Central
spike

Fig. 2.11. Maize—Spikelets.

spikelets produce no grain. In 6-rowed barley (H. vulgare), each floret is fertile; the alternating groups of spikelets form 6 equidistant and distinct rows (Fig. 2.12).

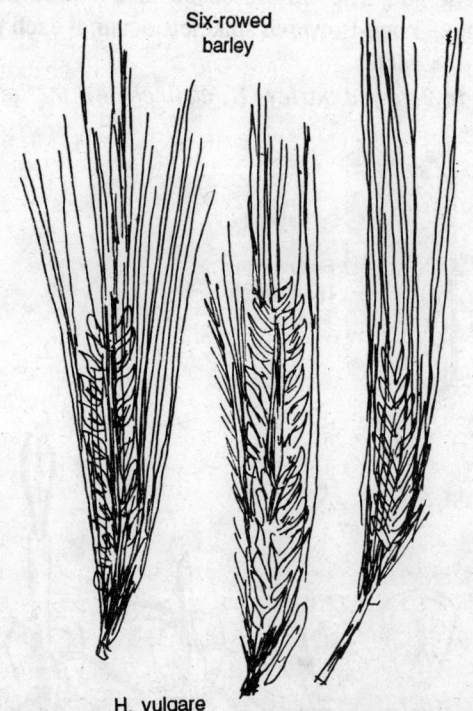

Six-rowed barley

H. vulgare

Fig. 2.12. Barley (*Hordeum vulgare*) spikes.

The glumes are narrow and bristle like place at the sides of the florets and simulating a bristly involucre at each joint of the rachis. The lemmas are broad; in most varieties, they terminate in long

awns which render barley heads among the most conspicuously *bearded* of the cereals. Which the florets usually open briefly, barley is almost invariably *cross-pollinated. The mature grain is invested by a hull consisting of the lemma and palea. the floret is also characterized by a short rachilla which remains as a kind of bristle behind the palea in the hulled barleys.*

Barley is one of the oldest of cultivated cereals. Cultivated barleys are divided into

(i) *6-rowed barleys* : All the 3 spikelets at a node are fertile. The spikelets are in 6 distinct rows and arranged at a uniform distance around the rachis. Hence, the mature grains are in 6 rows upto the axis e.g. *H. hexastichum*. 6-row forms appear in the archaeological record in 6-row types, each of the 3 florets in a cluster matures into a fruit.

(ii) *4-rowed barleys* : All spikelets at a node are fertile. The rows of grain are not at equal distance from each other on the axis. The lateral grains of one trio tend to overlap the lateral grains of the triplet on the opposite side of the rachis. This results in the formation of 4-rowed grains e.g. *H. vulgare*.

(iii) *2-rowed barley* : Of the triple spikelets at each node, only the central one sets grain. The lateral ones are imperfect. Hence, the grains are seen in 2 rows along the rachis e.g. *H. distichum*.

Since each node bears 2 spikelets, the net effect of suppression of 2 of the 3 flowers in each spikelet is the production of an infructescence with 2-rows of grain.

If all the florets are fertile, the final infructescence bears 6 rows of grain.

Varieties

Ratna, BG-25, BG-108 for Delhi, Haryana, Punja; Ratna, Amber, K-125 for Uttar Pradesh and Bihar; RS-6, RD-31, RD-11, RD-57 for Rajasthan, Southern Uttar Pradesh and Madhya Pradesh; Himani, Dolma for northern hill valleys of Uttar Pradesh and Himachal Pradesh.

The varieties are high yielding, nutritionally rich but hulled (husked).

Earlier the consumers do not prefer barley as foodgrain because 15-20 per cent of the total produce goes as husk.

Origin

No satisfactory explanation for the origin of cultivated barleys is put forth so far. Kornicke (1882) considered that *Hordeum spontaneum* was the wild prototype. Covas (1950) ruled out the possibility of a polyphyletic origin of *H. spontaneum* and the cultivated types. He considered *H. spontaneum* as the most primitive species, because it possesses two primitive characters, viz. sterile lateral florets and brittle rachis. According to him, it could have given rise to *H. agriocrithon* with six rows of spikelet and to *H. kistichon* with tough rachis, by a single gene-mutation and these two could have given rise to *H. hexastichon*.

There are two main centres of diversity of barley, one of which is located in Abyssinia and North East Africa, having the long-awned hulled forms, and the other is confined to China, Tibet and Japan, containing the hull-less, short-awned, awnless or hooded types (Vavilov, 1926) *H. agriocriton* and *H. spontaneum* are considered to be the progenitors of cultivated barleys by some workers. Vavilov considered the possibility of the genus *Hordeum* is characterized by the following characteristics :

The spikelets have one flower and there are three spikelets at each rachis node. Each floret is enclosed within two glumes. The species within this genus are often distinguished by the differential development of glumes, awns and other floral parts. In most species, the central floret alone is fertile, partially or fully fertile. There are about twenty-five species of diploid and tetraploid constitution. All the cultivated species are diploid species with 2n = 14. the tetraploid series consist of wild species only such as *H. murinum, H. bulbosum, H. jubatum, H. nodosum.*

The differences between the two cultivated species *H. vulgare,* and *H. distichum* are given below :

Hordeum vulgare	*Hordeum distichum*
1. There are six rows	There are only two rows
2. Three fertile florets, at each node	Only the central row develops normally kernels
3. Lateral kernels are slightly smaller than the central ones	Only lateral parts are reduced to sexual parts

Different authors have classified barleys in different ways. To mention a few, Linnaeus (1748) recognized six types of barleys, four species and two varieties. Jessen limited the species to one, which he termed *H. sativum.* Harlan (1918) recognized four species, *H. vulgare, H. intermedium, H. distichum* and *H. deficiens.*

There is a great deal of ambiguity in the classification of *Hordeum* sp. The use of synonyms, application of invalid names and the use of minor morphological differences in distinguishing different species (*H. vulgare, H. spontaneum,* and *H. agriocriton*) have contributed to the complexity in this respect.

The *chapati* made of husked barley is rough in colour and taste and husked barley possesses about 10 per cent crude fibre. So, husked barley is termed as inferior food grain as compared to maize, bajra and sorghum.

Karan-4, 18, 19, 163, 264, 209, 231, 152, 265 are *huskless* dwarf high yielding varieties.

Barley is eaten as a staple food grain in eastern UP, Bihar, parts of Madhya Pradesh, Rajasthan, Gurgaon and Mahendranagar districts in Haryana, and in the hill-valleys of Himalayas. In addition it is also used in bulk as animal feed.

Barley (huskless variety) gives about 4 tonnes of grain per hectare of huskless barley; 40-60 kg N, 20-25 kg P, 20-25 kg K per hectare and 2-3 light irrigations are more than sufficient for the crop. Barley varieties take 110-120 days to mature particularly in north-western India. The barley crop grows in the middle of November, matures by middle of March.

Barley flour makes good chapati. It serves food for young, old and children and also ailing persons from fever, stomach and other disorders like diabetes. It can be taken alone or with wheat in the ratio of 50 : 50 or 25 : 75 (barley : wheat).

Barley can be used for making of porridge or dahlia etc. Barley water can be made bajra grain. India exports barley to Jeddah.

Uses

Barley grain is used in the brewing of fermented beverages, beers, malt tonics, breakfast cereals etc.

Malting

In malting, the grain is steeped from 48-70 hours in water at about 55°F. During this period, the water is changed 2 or 3 times to promote aeration and check bacterial action. The steeped grain contains about 60 per cent moisture. It is spread on floors and turned and aerated for about 12 days. During this period, the enzymic processes of germination go on actively; the starch of the endosperm is changed in part into sugars and dextrins; translocation of these into the developing embryo begins. Proteins share in the enzymatic changes 3 or 4 roots appear and the shoot grows about three fourths the distance up the back of the grain. During the last 4 days on the floor, the malt is dried somewhat and the growth of the embryo stops. Drying is completed in kilns, slowly at first, and then, at about 160°F until the moisture content is reduced to 2 per cent. Sprouts are removed and may be used as feedstuff. The quality of malt is said to be improved by storage.

Malt, refers to barley which has been partially germinated and then dried.

OATS

Probably most of the cultivated forms have been derived from the weed *Avena fatua*, the *European wild oat*. Cultivated oats are *panicle oats* (*A. sativa*), *banner oats* or *Tartarian oats* (*A. byzantina*). The oat may reach a height of 4 feet; *it is easily distinguished among the cereals by the bluish-green tint of the foliage and the very characteristic inflorescence*. The culms are larger and less stiff; the leaves broader and more numerous, than those of wheat. The leaf-sheath is closed. The inflorescence is a wide and much-branched penicle. In *A. orientalis,* inflorescence is contracted and one-sided. Spikelets are large and usually long-pedicellate. The prominent, sharply acute glumes surpass the lower floret and sometimes the whole spikelet. Florets are usually three; the second is often sterile and the third almost invariably so, or, in some varieties, nearly or quite awnless. The florets open widely but briefly; they are usually self-pollinated (Fig. 2.13).

The oat grain, a caryopsis, unlike that of wheat, remains at maturity firmly invested by a hull of lemma and palea, which constitutes 25-35 per cent of the threshed product. As the hull lacks nutritive value and digestibility and adds difficulty to milling, oats with small hull content are desirable. In some species (Fig. 2.14). naked oats - the grain drops frm the hull on threshing.

The grain proper or groat, has a hairy pericarp, a double aleurone layer, a starchy endosperm lacking in gluten, and a small embryo.

Oats (*Avena sativa*, Poaceae) were a major cereal of temperate areas.

Indian oat is *A. sterilis* var. *culta* now known as *A. byzantina*. Oats are cultivated in United States, Russia, Canada, Germany, France, Britain, Poland, Sweden, Denmark, Rumania, Australia, New Zealand, South America and Mediterranean countries. In India oats are cultivated in Punjab, Haryana, Uttar Pradesh; to a limited extent in certain parts of Himachal Pradesh, Maharashtra, Gujarat, Madhya Pradesh, Orissa, Bihar and West Bengal.

Oat grows best in cool, moist climate. Cool weather is particularly important for high yield and good quality of grain. In India, seeds are sown in October or November; seed rate varies from 80-100 kg/ha. The seeds are sown broadcast or drilled. Crop is harvested during March-April.

Uses

Products from oat grain such as oatmeal, oatcakes and cookies are used as food. Oats are superior to other cereals as food for horses and excellent

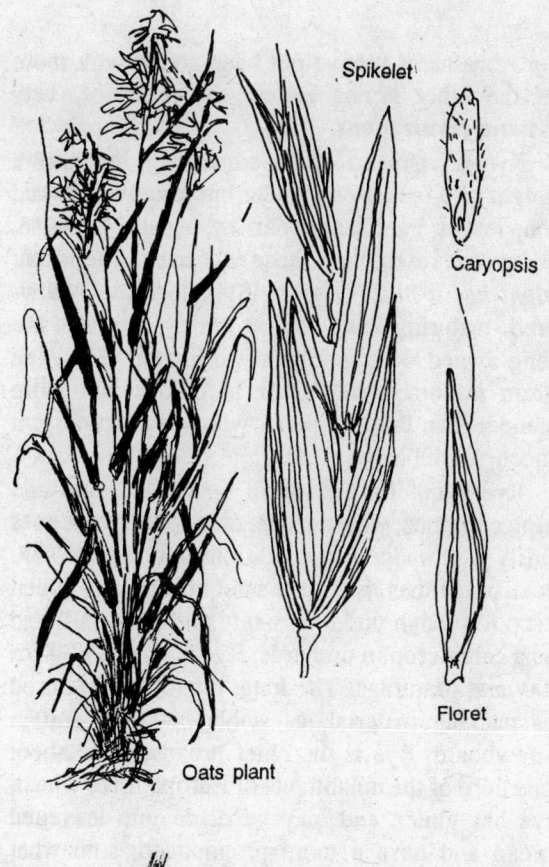

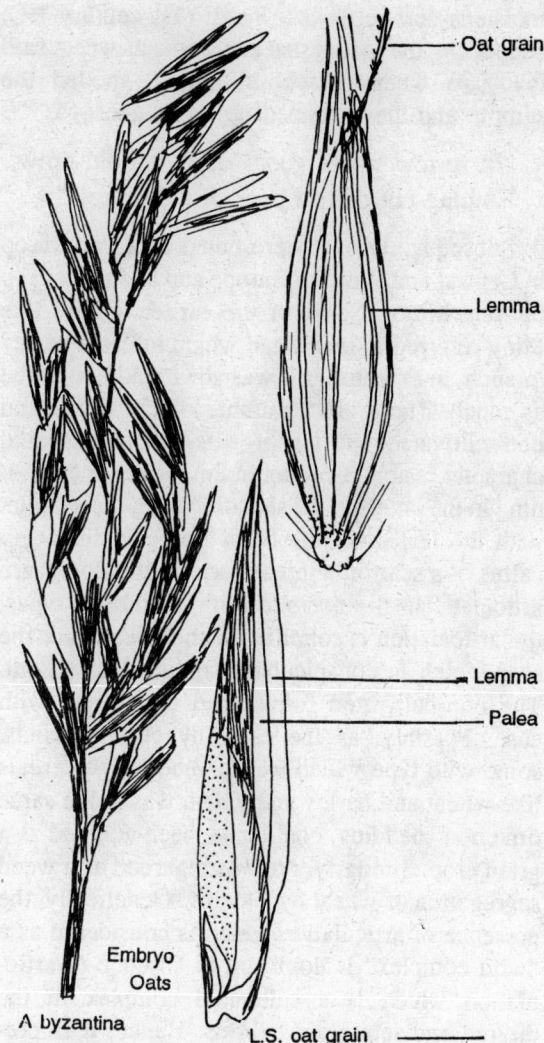

Fig. 2.14. Oats. *Avena byzantina*—spike and spike-lets.

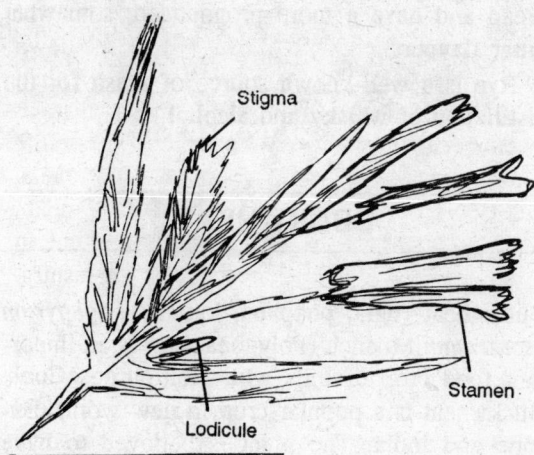

Fig. 2.13. Oats (*Avena sativa*) plant, flowers and grain.

Origin

Findlay (1956) states that oats were cultivated before 1000 BC by the cavedwellers of Switzerland.

The archaeological evidence from the lake dwellings of Switzerland indicates the grains of *A. strigosa*, sp. *strigosa* belong to *A. fatua* sub. sp. *fatua*. Historical evidence points only to its recent introduction in the Christian era till which time it was noted only as a weed. Hippocrates, the Greek philosopher, and historian (5th century BC) wrote that oats made into porridge or gruel, when eaten,

for breeding animals and young stock. Fodder oats are grown for fodder. Oat husk is a raw material for nylon manufacture, as a selective solvent in the purification of butadeine (starting point for manufacture of synthetic rubber), in refining lubricating oils, resins etc., as a source of hemicelluloses.

moistens and refreshes. Virgil (1st century BC) supporting the theory that oats came as weed, said that oats when present in barley, spoiled the sample and these "buned up fields energy".

"In furrow where good barley we did grow,
Nothing but darned poor oats did grow".

Subsequently oats were noted as a grain crop in Central and Western Europe and as fodder crop in Asia Minor. Some of the earlier writers like Pliny referred to it as weed which infested barley to such an extent that it was sown and cultivated as much. There are a number of cultivated and non-cultivated species of oats whose essential character is non-shedding in the former and shedding in the latter. In the shedding type, the spikelet with the *lemma* of the basal floret possessing a callus or a scar or, alternatively, all the florets are articulate. In the non-shedding, cultivated types, the articulation is solidified with the callus or the scar which is conspicuously reduced or absent. The non-cultivated forms shed their grain with ease. Possibly, as the early hypothesis stands, some wild type which was a weed in the cereals like wheat and barley and which was at the same time non-shedding, could have been adopted as a grain crop. Similarly, rye was regarded as a weed segregation of wheat by Vavilov. Genetically, the presence of articulation which is considered as a "wild complex" is dominant to absence of articulation which is a cultivated complex in the diploid and tetraploid hybrids. Hence, it is presumed that the cultivated type emerged as a recessive mutation. On the other hand, in hexaploids as the cultivated type is partially dominant to the wild type *fatua,* there is a suggestion that it came up as a dominant mutation.

RYE

Rye is the grain *Secale cereale,* Poaceae, an annual grass with more or less perennial tendency like its supposed ancestor *S. montanum* of the mountains of central Eurasia. Apparently rye is of more recent domestication than wheat and barley,

for remains of it have not been found with those of the other grains among the relics of early Asiatic civilizations.

Rye is vigorous grass, sometimes reaching a height of 6 feet, with slender but tough culms and long leaves. Panicles are narrow, usually 4-ranked, somewhat resembling those of barley. *The whole plant has a bluish aspect.* Spikelets are 3-flowered, maturing the 2 lowest grains. Lemmas are long awned. Rye is a cross-pollinated crop. The grain is somewhat darker in colour and more slender than that of wheat, which, structurally, it much resembles.

Rye is chiefly a grain of colder latitudes and alpine regions, where wheat fails to thrive. It goes fairly well under arid conditions and in poor soil. It is sometimes used as a sand binder, as a green crop to plough under in sandy and poor soil, and as a catch-crop in orchards. Rye is grown also for hay and pasturage. The long, tough straw is used as packing material, as stable bedding, and in strawboard. Rye is the chief breadstuff of about one third of the inhabitants of Europe. Like wheat, rye has gluten and may be made into leavened bread and have a more pronounced, somewhat bitter flavour.

Rye is a well-known source of mash for the distillation of whisky and alcohol.

BUCKWHEAT

Buckwheat (ogla, phaphua, kathu), *Fagopyrum esculentum* Moench (Polygonaceae) is an important food crop of hills with multipurpose uses. Buckwheat is a popular crop in new world, Europe and India. The plant is believed to have originated in central Asia, wherefrom possibly it has spread to countries in north America, Russia, France, Germany, United Kingdom, South Africa, India, China, Bhutan and Nepal. In India the crop is distributed in the temperate region of the Himalayas (Kashmir, Himachal Pradesh and Uttar Pradesh) in north and hilly regions of Nilgiris and Palneys in the south. Extensive cultivation of buckwheat can be seen in the Kalpa and Sangla valleys of Kinnaur district and also in the Chandra Bhaga, Bhaga, Patan and Chenab valleys of Lahaul district in Himachal Pradesh.

There are two cultivated species of buckwheat, namely *Fagopyrum esculentum* (now *Polyganum fagopyrum*) commonly known as *ogla* or *phaphra.* Another species is *F. tataricum* (Tartary buckwheat) and is known as *Kathu.* Both these species are supposed to have originated from a wild and weedy form *F. cynosum* which occur in the Himalaya abundantly (Chandel, 1980).

Common buckwheat *F. esculentum* is a dimorphic plant and needs cross fertilization by bees. It is herbaceous erect annual of about 1 m height. The stem is hollow angular with nodes swollen. The flowers are arranged in axillary, terminal densely clustered cygne. The fruit is a dry one-seeded, 3-sided nut called *achene.* The grain is brown with darker spots and lines. The endosperm is flowry, white and opaque and contains starch in the form of round or polygonal grains.

F. tataricum is similar to the common buckwheat in habit, but it tends to become ring. The stem is shorter and less fragile. Flowers are borne on axillary recesses having inconspicuous light green signals which differentiate this species from the common buckwheat (*F. esculentum*). *F. tataricum* is self-fertile and hence high yielding. Fruit is ovoid, conical with wavy outline, brownish grey or black in colour.

Cultivation

Buckwheat is usually cultivated as a rainfed crop in the hills, as it needs *moist and cool climate.* This is a short season crop maturing in about 65 to 120 days, and most suited as a *short summer crop* or *rainy season crop* on the hills. *F. tataricum* tolerates more cold as compared to *F. esculentum.* *F. tataricum* can grow well even at a very high altitude.

There are 3 types in *F. tataricum* : (i) Early types : Those maturing in about 60-65 days; (ii) Medium duration types : Those maturing in about 75-80 days; and (iii) Late types : Maturing in 80-100 days.

After preparing the field, seeds can be broadcast followed by light ploughing and planking. Seeds are drilled in furrows behind the plough 40-50 cm apart. 30 to 40 kg seed in case of *F. tataricum* (20-25 kg in case of *F. esculentum*) is required to sow one acre. In northern hills sowings are done just after the break of monsoon (June-July) while in eastern hills during August-September. In Nilgiris sowing is done during April-May as done in Lahaul Spiti Valley of Himachal Pradesh.

Buckwheat can grow on very poor soil, tolerates acidity; grow well on heavy wet soils rich in limestone; responds well to fertilizer. A mixture of 300 kg/ha N.P.K. in the ratio of 1 : 2 : 1 before sowing, ensure good yield. On an average, yield of grain is about 5-8 quintals per hectare. The yield of *F. tataricum* is more by 20-30 per cent than *F. esculentum.*

The buckwheat gives a pure white flour and the milling percentage is about 60-70. The grains have less protein as compared to cereals except maize, similarly carbohydrates are less but more than oats. The fat content is generally high while it is deficient in glutin.

Uses

In Himachal Pradesh, especially in Kulu and Sangla valleys in Kinnaur district and Lahaul Spiti valley, the flour is used for making the *chillare* a kind of bread fried with ghee.

PHOTODYNAMIC SUBSTANCES

Recent work has shown that the toxic effect of some plants is influenced by light, as they contain substances which when consumed by livestock make them sensitive to light. Such animals develop serious symptoms when they are exposed to sunlight after their feeds. Photosensitization takes place only in those animals who have no pigment in their skin and are white in colour; further it only appears if, after ingestion of sufficient quantities of the photodynamic substance, the animal is exposed to sunlight. If any one of these conditions is not fulfilled, symptoms may fail to develop. It should, however, be understood that the photodynamic substances do not always produce photosensitization. Among plants which produce photosensitization may be mentioned the buckwheat (*Fagopyrum esculentum* Moench and *F. tataricum* Gaertn.), St. John's wort (*Hypericum perforatum* Linn.), caltrop (*Tribulus terrestris* Linn.) and some species of *Trifolium, Medicago, Polygonum,* etc. Buckwheat grains and leaves produce photosensitization in pigs and other animals, but not in man. The rest of these plants are not normally eaten by human beings.

GRAIN AMARANTH

Amaranth (siriaria, seol, tulsi, kalgi, dankhar, marchae, sil) is commonly grown in Himachal Pradesh and hills of Uttar Pradesh for both grains and greens. It is cultivated along the whole length of the Himalayas, from Kashmir to Bhutan and also in the south Indian hills, Madhya Pradesh, Maharashtra and parts of Gujarat.

This species is mainly grown in central and south America. Presumably from this region it spread eastward to Europe and later to Asian continent after the great Spanish conquest around the end of 15th century. This species still occurs in semi-domesticated forms and provides rich source of vegetable protein, produces exceptionally high biomass.

Amaranthus is an annual, vigorously growing plant that, attains a height of about 2 m.

Besides *Amaranthus hypochondriacus,* other species of grain amaranth are *A. caudatus* and *A. cruentus,* which are grown mixed in different proportions with *A. hypochondriacus* in the hills. Grain amaranthus is estimated to occupy about 60 per cent of kharif land of higher hills in northwest India. In Himachal Pradesh, it is generally sown mixed with maize and French bean, whereas in higher hills of Uttar Pradesh it is grown mixed with French bean (viny types) only.

Varieties

IC 17453, IC 38047, IC 38570, IC 42254-6, IC 42257-1 are high grain yielders for general cultivation.

Amaranth is a quick growing crop. Seed should be sown in the first or second week of June just after the first monsoon shower. The depth of sowing is 2.0 to 2.5 cm, with 50 cm spacing between rows and 15 cm between plants. Seed rate of 1.5 kg/ha is used for grain yield.

The heads are cut when plant is still somewhat green and kept for sun-drying for 6-7 days. Threshing is done by beating. Usually, the harvesting is done early in the morning when the plants are somewhat wet due to night dew to avoid grain shattering in cuttings.

Uses

Amaranth is a multipurpose crop. The leaves are used as green leaf vegetable in its early growth stage when the plant is tender. The leaves are rich in, iron, protein, vitamins A and C.

The grains have 16 per cent crude protein. The protein has lysine (5 per cent), cystine (2-9 per cent), and methionine (44 per cent).

White grain

In the hills, the grains are used in various ways: white grain popped and eaten, white grain ground as flour for chapati; mixed with finger millet and ground as flour for chapati; in confectionery *laddoo*; popped grain mixed with milk and sugar; popped grain in the form of *halwa* : Black grain : Alcoholic drinks from fermented grains; used as animal feed; leaves as purgative; dried stems or considered to break the rock boulders easily with its smoke when burnt.

In India, grains are milled into flour; paw-cake like *chapatis* made from grains is a staple food in the entire Himalayan region (from Jammu and Kashmir, Himachal Pradesh, Uttar Pradesh extending eastwards). In plains of north and south India *ladoos* prepared from grains are very much liked and are named as *Ramdana*.

Medical uses

In UP hills it is considered to cure *measels* of children when they sleep over the spreaded grains. In HP hills, it is considered to cure the *foot and mouth disease of cattle* when the grains are fed to the cattle.

In HP hills, people consider that the quality of grain amaranth does not deteriorate even for storing up to 20 years and they use the grain stock for *keeping the apple fresh* for long periods (S.D. Joshi et al., 1983).

EDIBLE SEEDS

Seeds of *Coix lacryma-jobi,* of *Oryza rufipogon* and bamboo rice (seeds of *Bambusa arundinacea*) are eaten by tribals. Seeds of *Bauhinia vahlii, Entada phaseoloides, Mucana prurita, Sterculia guttata, Hodgsonia heteroclita* (cucurbit), *Cirsium lipskyi* are eaten after roasting, as such or mixed with salt. Seeds of *Canavalia, Mucana, Parkia, vigna* are eaten after boiling in soups. Seeds of *Euryale ferox, Nelumbo nucifera, Nymphaea nouchalli* are eaten raw.

The grass *Digitaria cruciata* var. *esculenta* a cold-tolerant grass is grown for food and fodder by Himalayan tribals. Seeds of *Amaranthus caudatus, Amaranthus hybridus* subsp. *cruentus* var. *paniculatus* are eaten *Pennisetum americanum, Fagopyrum cymosum* (bragina), *Glycine max, Phaseolus torusus* (gidilpi) are cultivated by tribals. A cool drink is prepared from the seeds of *Scoparia dulcis* (ghoda-tulsi) by soaking the seeds overnight.

3

Millets

A wide variety of millets are grown in India, namely *pearl millet or bajra* (*Pannisetum americanum*), *finger millet or ragi* (*Eleusine coracana*), *Italian millet* or *foxtail millet* (*Setaria italica*), *kodo millet* (*Paspalum scrobiculatum*), *common millet* or *proso millet* (*Panicum miliaceum*), *little millet* (*Panicum miliare*) and *barnyard millet* (*Echinocloa frumentacea*). At present there are 22 research centres in India, working on the improvement of millets.

The term *millets* refers to any of the small-seeded cereal and forage grasses used for food, feed or forage. *Pearl millet* or *bulrush millet* (*Pennisetum americanum*), *shama millet* (*Echinocloa colona*), *Australian millet* (*Echinocloa decompositum*), *browntop millet* (*Brachiaria ramoa*), *teff* (*Eragrostis tef*), *hungry rice* (*Digitaria iburua*), *hungry rice* (*Digitaria exilis*), *Job's tears* (*Coix lachryma jobi*) are grown in African countries, Russia, and a small extent in Europe and United States.

In India 17.5 million hectares are under millet cultivation and yield is 9.5 million metric tons/year.

PEARL MILLET

Pearl millet or bajra (*Pennisetum americanum*, Poaceae) is cultivated on area of about 11.5 million hectares mainly in the states of Rajasthan, Gujarat, Maharashtra, UP and Haryana. Yield of grain is 8 million tons per annum (Fig. 3.1).

Varieties

The exploitation of hybrid vigour has been an important facet in the development of this millet. The availability of cytoplasmic male sterile lines gave a new impetus to the development of hybrids. High yielding varieties : HB-1, HB-2, HB-3, HB-4, HB-5, HB-6, HB-7, PHB-12, BJ-104, BK-560, BD-111, BD-763, MBH-110, COH-2, CM-46, CHB-27, MH-29, MH-34, MH-36, Vijay, New Vijay, Visakha, Balaji, Nagarjuna, K-2, K-3, CO-6, HS-1, PSB-8. For forage : Pusa Napier Giant-1 (*Pennisetum purpureum x P. americanum*). NB-21, NB-5, RSB-2, T-55, K-674, K-677, L-72, L-74.

Pearl millet is the most drought and heat tolerant of all the grain crops. The millet is cultivated in the hot dry plains of Southern Asia, and the southern periphery of the African Sahara. A large number of related species of pearl millet are found wild in Africa. It furnishes an important food for the lower classes and is particularly valuable in cold weather because of its heating quality. The inflorescence is a cylindrical spike. Two types of flowers are borne in the spikelet. They are, bisexual and staminate flowers. Pearl millet is a cross pollinated crop (80 per cent). Useful part is caryopsis (endosperm).

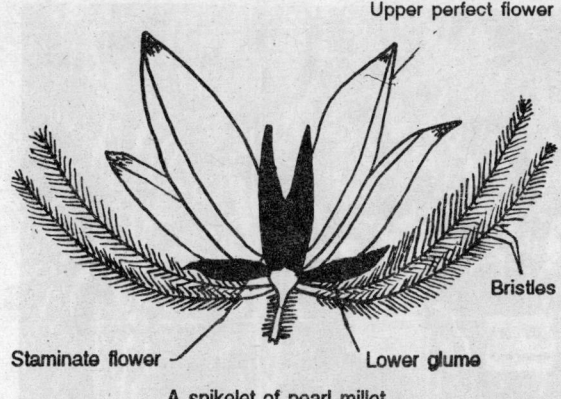

Upper perfect flower

Bristles

Staminate flower

Lower glume

A spikelet of pearl millet

Fig. 3.1. Bajra (*Pennisetum americanum*) spike and spikelet.

Cultivation

Pearl millet is grown as a pure or mixed crop and the millet is rotated with sorghum, niger, wheat in kharif crops; in the rabi crops, with pulses, as a mixed crop. Seed rate is 3-4 kg/ha in black soils, 9-10 kg/ha in other soils. Seed is broadcast, but sometimes drilled in rows 45-90 cm apart. The crop is harvested when the heads are ripe. Average yield is 770-1100 kg/ha in rainfed crop and 1100-2200 kg/ha in irrigated crop.

Uses

Bajra grain is dehusked, broken into rice, ground into flour and made into unleavened bread or the flour is cooked in water till it becomes a paste of suitable consistency.

It is considered as poor man's food. The grain is directly cooked and eaten like rice. The flour is used in making cake, chapatis, rotti, gruel etc. High quality grain is used in the manufacture of malts. Stem and leaves are used as fodder.

SORGHUM

Sorghum (jowar) *Sorghum vulgare* is native to Africa and Asia. Sorghum is the second most important cereal grown in India. Hence it is called *great millet*. Cultivated sorghums are derived from the perennial johnson grass *Sorghum halepensis*. The different varieties of sorghum are:

  (i) Sweet sorghums - grown for syrup production or for forage;

 (ii) Grain sorghums - grown for grain; grain sorghums include those varieties with relatively large palatable seeds which thresh free from glumes. They are further divided into kafir, milo, hegari, feterita, durra and shallu, based on grain characters;

(iii) Broom corn;

 (iv) Grass sorghums - for hay or pasture. Useful part in sorghum is the caryopsis (endosperm). Sorghum (*Sorghum vulgare*; Poaceae) is an annual attaining a height of 3-15 feet. The inflorescence is a loose or compact panicle, known as head. Spikelets in clusters of 2-3; one sessile and bisexual and the other, pedicaled, staminate or sterile (Fig. 3.2).

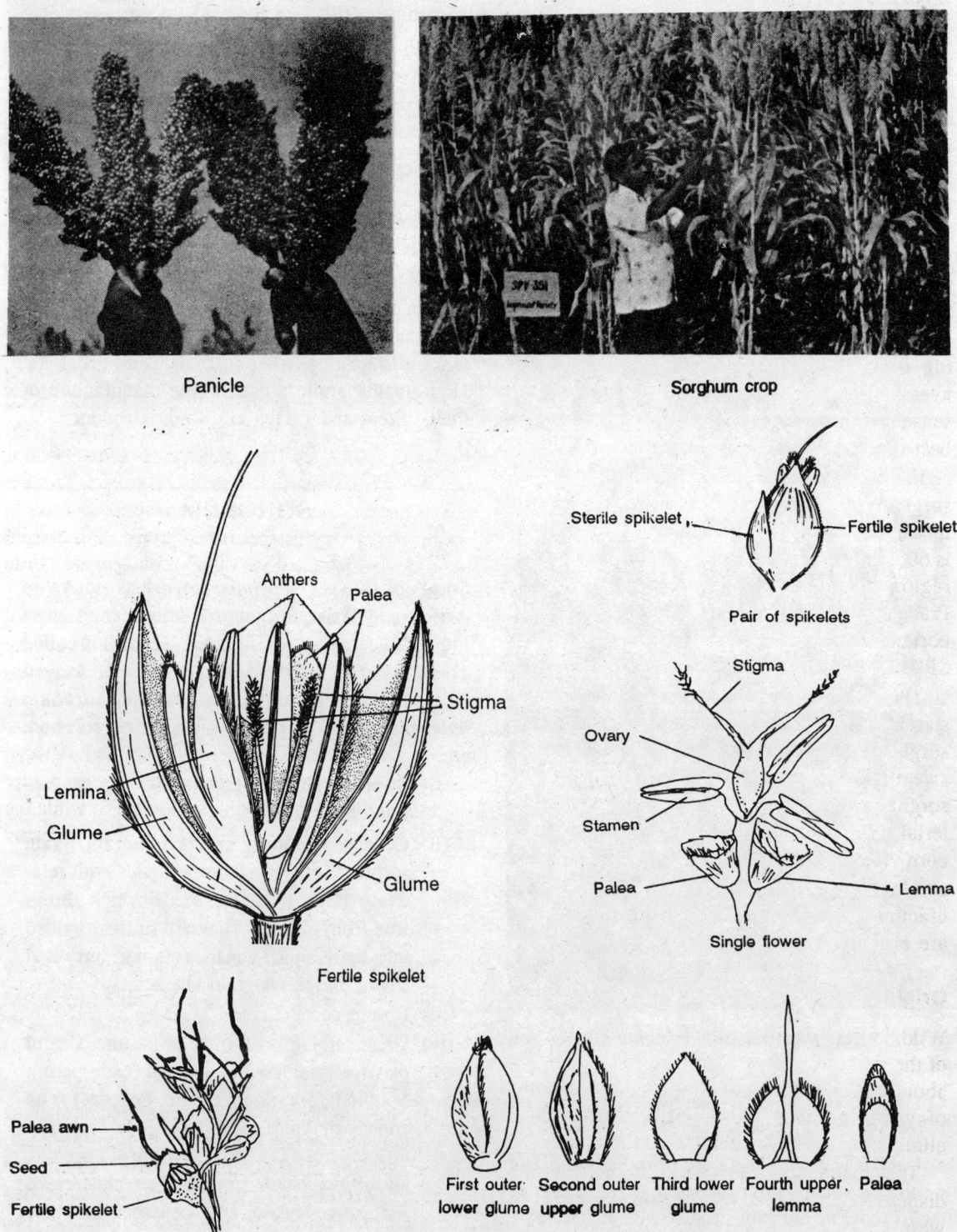

Panicle

Sorghum crop

Sterile spikelet

Fertile spikelet

Pair of spikelets

Anthers

Palea

Stigma

Lemina

Glume

Glume

Fertile spikelet

Stigma

Ovary

Stamen

Palea

Lemma

Single flower

Palea awn

Seed

Fertile spikelet

First outer lower glume

Second outer upper glume

Third lower glume

Fourth upper lemma

Palea

Fig. 3.2. Sorghum. (*Sorghum vulgare*).

Cultivation

The preparation of land with ploughs or blade harrows with least application of farm yard manure, line sowing with a seed-drill in rows 30-45 cm apart and interculturing with bullock-drawn implements is a common practice in India. A 2-year rotation (or 3 year rotation sorghum-cotton-groundnut) of sorghum-cotton is most common during *kharif*. During *rabi*, sorghum-cotton, sorghum-gram rotations are common. Mixed cropping of sorghum-red gram (arhar) is most common.

Sorghum crop requires moderate rainfall varying from 12-40 inches during the season. The average temperature required during the crop season ranges from 80°-90°F. Sorghum grows best on heavy soils (clay loams).

In India about 16 million hectares are under sorghum cultivation, yield is about 10.5 million metric tonnes of grain per annum. Average yield is 600-700 kg/hectare. Principal sorghum growing regions are Maharashtra, Karnataka, Madhya Pradesh, Andhra Pradesh and Uttar Pradesh. Important varieties are BSH-5, CSH-5, SPV-126, SPV-351, Moti etc.

The endosperm of the sorghum grain is rich in starch. This is more nutritive than rice. Sweet sorghums or sorges are used as forage crop. A sweet syrup is extracted from the stems of sweet sorghums. Grain sorghums are used as food material for making bread, chapathi, rotti etc. Broom corn variety is used in the manufacture of brooms and brushes. Alcohol, whisky, beer are also manufactured from sorghum starch. Sorghum stems are also used in paper making.

Origin

Wild Sorghums are considered to be the ancestors of the cultivated types, the changes being brought about by careful selection by man over hundreds of years, retaining the desirable characters and eliminating the undesirable ones.

The exact place of origin of Sorghum is highly disputed. De Candolle (1884), and Hooker (1917) were of opinion that Africa was the home of principal millets like *Sorghum, Pennisetum* and *Eleusine*, which later reached India. Vavilov (1935), who considered the place of greatest diversity of a crop as its place of origin, considered **Abyssinia** to be the primary centre of origin of *Sorghum*. In fact, the natural distribution of races of *Sorghum* is from Africa to India. Wild species from which cultivated *Sorghums* are supposed to have been derived, are abundant in Africa. Snowden (1936) lists 32 races of cultivated *Sorghum*, out of which 11 belong to India and South Asia, while the rest are purely African. Werth (1937) was of opinion that for the three major millets, *Sorghum, Pennisetum* and *Eleusine,* India was the place of origin from where they spread to Africa through Arabia and Abyssinia. Abyssinia was considered by him as a secondary centre of origin and distribution. Darlington and Janaki Ammal (1945) are of opinion that the centres of diversity shift from time to time depending upon several factors, and that there are not one but several centres. In India, where high standards of civilization existed for a long time the diversity in millets is very little due to intensive selection and domestication for thousands of years. A study of the *Sorghums* of India and Africa suggests that species of both the countries exhibit similar distribution of dominant and recessive genes. It could not be emphasized that any one particular set of genes were concentrated in Africa and others in India. The African *Sorghum,* as a group, shows greater vigour, bolder grains and certain amount of coarseness, while the Indian species are more palatable and probably more domesticated.

Classification

Snowden (1935, 1955) divides the genus *Sorghum* into two sections : 1. Para-Sorghum, 2. Eu-Sorghums.

Para sorghums

The para-sorghum group includes species such as *Sorghum versicolor, S. drumondii, S. purpurea, S. sericeum* and *S. dimidiatum.* Para-Sorghums are characterized by having the sheath nodes bearded and simple primary branches of the panicle. The racemes are simple. The grain is small, endosperm poorly developed, and pericarp narrow, while the inner integument occupies most of the space. The chromosome number of the species is 2n = 10. The chromosomes are characteristically larger. Snowden (1936) upheld the view that this group has played no part in the evolution of cultivated species of *Sorghum*.

Eu-sorghums

The section Eu-Sorghum is characterized by glabrous nodes, finely pubescent and not bearded, primary branches of the panicle divided, and lateral and terminal racemes. This group is again subdivided into two subsections :-

(a) Subsection *Arundinacea*

(b) Subsection *Halepensea*

Arundinacea

It has been subdivided into two series, viz. (a) *Spontanea* and (b) *Sativa*. *Spontanea* includes all the wild *Sorghums,* annual or perennial. All the cultivated *Sorghums* belong to *Sativa* series. The somatic number of *Arundinacea* has been reported through extensive studies by Huskins and Smith (1932), to be twenty. The karyotype is similar to that of the *Halepensea*. They considered these *Sorghums* as tetraploids and *Sorghum halepense* as octoploid. Snowden upholds the view that *Sorghum arundinaceum, Sorghum verticelliflorum, Sorghum aethiopicum* and *Sorghum sudanense* played an important role in the origin of the cultivated species of *Sorghum*.

The classification of the genus *Sorghum* followed by Garber (1950) and Celarier (1958) is slightly different. According to them, the genus is divided into five sub-genera : (1) *Eusorghum,* (2) *Para-sorghum*, (3) *Stiposorghum*, (4) *Chaetosorghum*, and (5) *Heterosorghum*. Whereas the section *Eu-Sorghum* of Snowden corresponds with the subgenus *Eu-Sorghum* of Garber (1950), the section *Para-Sorghum* of Snowden was divided into two subgenera *Para-sorghum* and *Stiposorghum,* mainly to include the North Australian species newly investigated by Garber. The difference is mainly cytological although minor morphological differences like better develop-

ment of awn and callus in *Stiposorghum* also occurs. The last two subgenera represent single species each and accommodate species which fit well in none of the other categories, viz. *S. intrans, S. stipoideum, S. brevicallosum, S. matarakense, S. plumosum* and *S. timorense.*

Halepensea

The Halepensea are perennials with distinct rhizomes. The somatic chromosome number is 40 although there are species with 2n = 20 as well. The chromosomes are smaller than in the Parasorghums. Hybridization of *Sorghum halepense* with the cultivated Sorghum has not met with success. *Sorghum halepense* was at one time considered to be the progenitor of the cultivated Sorghums. Snowden (1936), however, considers this improbable.

Chief types cultivated in India

Sorghum durra

It is cultivated in Tamil Nadu, Maharashtra, Madhya Pradesh, parts of Punjab and Uttar Pradesh. Varieties with semi-compact heads and yellow grain come under this group.

Sorghum cernum

It is grown in Bombay, Telangana area of Andhra Pradesh, Tamil Nadu and Madhya Pradesh. This group is characterized by its bold, white, pearly grains.

Sorghum subglabrescens

Most of the irrigated varieties belong to this group. They are chiefly grown in Tamil Nadu, Maharashtra and Madhya Pradesh.

Sorghum roxburgii

This is characterized by loose panicles grown in poor soils of Orissa, Bihar and Madhya Pradesh. This species can tolerate higher rainfall than others.

Sorghum dochna

This is grown in parts of Tamil Nadu and Maharashtra as a fodder variety with a high seed rate. The grains are enclosed in glumes.

RAGI

Ragi finger millet, African millet, coracana millet (*Eleusine coracana*; Poaceae) occupies an area of 2.7 million hectares mainly in Karnataka, Andhra Pradesh, Tamil Nadu, Orissa and Maharashtra. Total production is about 2.2 million tonnes per annum. In Karnataka 1 million hectares are under this crop and production is 1.5 million tonnes per annum.

Varieties

Early duration varieties : CO-10, R-374; Late duration varieties : KM-13, PES-110, PR-722, PR-117; Indaf-5, VL-101, Godavari (PR-202); Mid late varieties : IE-28, Kalyani, HPB-7-6, VL-204, TNAU-9, HR-919.

Ragi is cultivated both under rainfed and irrigated lands. The rainfed crop is sown in May or June during *kharif* or in September-October during *rabi*. Irrigated crop is grown throughout the year in southern states. Application of nitrogen fertilizer at 40 kg N/ha under rainfed and 160 to 200 kg/ha under irrigated conditions is necessary. The crop flowers in 60-80 days and matures in about 135 days. Yield of grain is 1000-1500 kg/ha for rainfed crop; 4000-5000 kg/ha for irrigated crop, with late maturing varieties.

Eleusine is predominantly an African genus with most of its species confined to tropical Africa. 2 groups of cultivars of ragi are :

(i) *African highland types* (tetraplod 2 n = 36) with long spikelets and long glumes, long lemmas and grains enclosed within the florets.

(ii) *Afro-Asian types* (tetraphoid 2n = 36) with short spikelets, short glumes, short lemmas, mature grains exposed distally.

In east Africa, Ethiopia, Somaliland, ragi is the principal staple food of many tribes. The millet is also used for malting and brewing.

Ragi is used as food after husking and the flour is made into cakes or porridge. When boiled with milk, the flour forms a light and pleasant meal for invalids. Ragi grain is used as a food for cage-birds and for feeding poultry. The straw is a good fodder for cattle.

MINOR MILLETS

The minor millets are cultivated in an area of about 4.5 million hectares mainly in Madhya Pradesh, Tamil Nadu, Andhra Pradesh, Uttar Pradesh, Karnataka, Orissa and Bihar. Production is about 1.93 million tonnes per annum. These millets are the staple diets of tribals and backward people.

Proso-millet

Proso-millet, *cheena,* common millet (*Panicum miliaceum*; Poaceae) is a quick growing and short duration (60-75 days) drought resistant millet crop. The millet is extensively grown as a catch crop on the marginal soils in Bihar, Uttar Pradesh, Karnataka, Andhra Pradesh and Tamil Nadu. The millet is grown mostly in soils of poor fertility - red loams, light loams and sandy loams.

Proso-millet is usually grown as a pure crop. General yield levels are 4-6 q/ha but high yielding varieties and improved agro-technology may increase the yield to 20-22 q/ha (Fig. 3.3).

The common millet is a minor millet, highly drought resistant, and quick growing with the lowest water requirement. This millet is mostly grown in the Southern States of India. It is also grown in parts of Russia, China, Japan and also the Middle East (Fig. 3.4).

Origin

Komarav (1931) considers *Panicum miliaceum* to be as ancient as wheat, from the fourth glume, which is brought about by the swelling of the

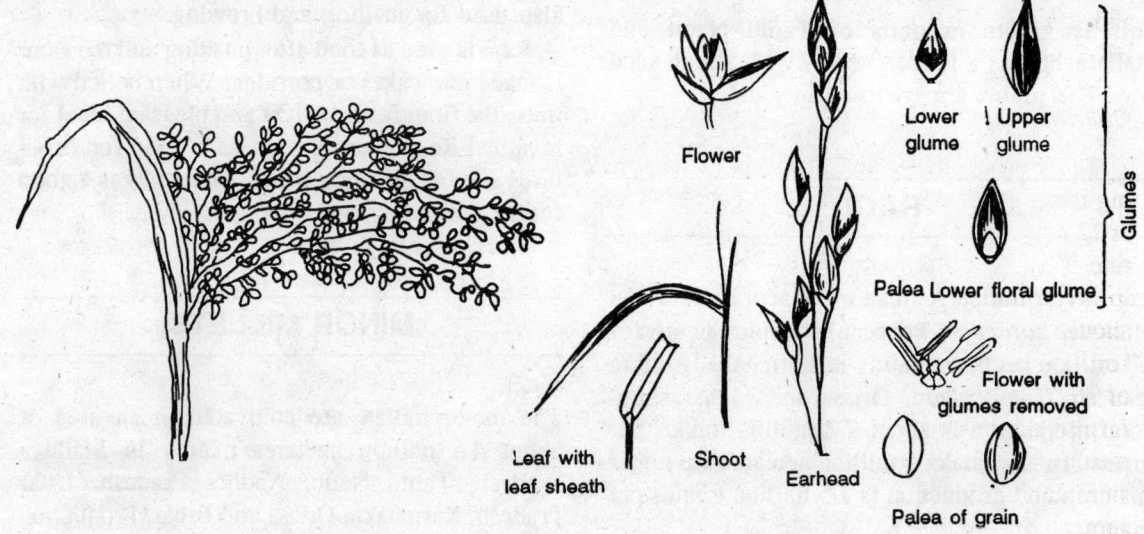

Fig. 3.3. Proso-millet *Panicum miliaceum* L.

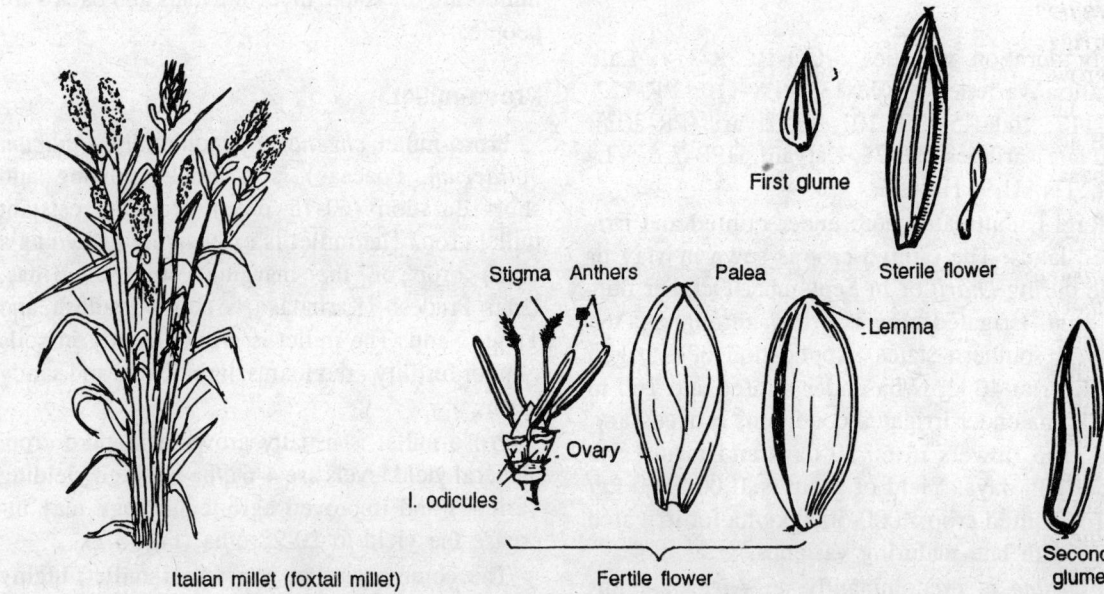

Fig. 3.4. Italian millet—*Setaria italica* plant, spike and spikelets.

lodicules. William (1899) suggests and Egyptio-Arabian region as the home of this millet. Burkill (1935) is of the view that this millet spread from the great home of nomads in Central Asia to Slavonic regions of Eastern Europe during Roman times. Werth (1937) concluded that the spread of this millet started from a broad girdle in North China through Central Asia, South Russia into Middle Europe. Brandon (1932) supposes that the Russian emigrants brought this crop into the US.

Varieties

Ram cheena, of Bihar matures in 67 days; *Shyam cheena* of Bihar matures in 66 days. **Plant height is about 60 cm. There are 6 tillers per plant. The** plants flower in 40 days BR-7 flowers in 41 days and is ready for harvest 67 days after planting. BR-9 matures in 65 days; RAU-1, RAU-2.

Cultivation

The crop is planted by broadcasting, but sowing is better. Row to row distance should be kept at 22 cm and plant to plant distance, 7.5 cm. Seed rate is 10-12 kg/ha. The seeds should be soaked in water for 24 hrs, prior to sowing. The seed should not be sown more than 4 cm deep in the soil. The crop is planted from February to middle of April. 40 kg N, 20 kg P and 20 K per hectare, fertilizer is necessary. For good results, half of the nitrogen and full doses of phosphorus and potash should be applied at the time of sowing. The remaining half dose of nitrogen (20 kg N/ha) should be given at the flowering stage i.e., 35-40 days of the crop growth, along with second irrigation (first irrigation 10-15 days of crop growth; third irrigation 50-55 days of crop growth).

Under unirrigated conditions, 20 kg N, 20 kg P and 20 kg K per hectare should be applied as basal dose, all at the time of sowing the seed.

Proso-millet crop usually matures in 60-80 days. The crop is harvested *when the plants have turned yellowish brown and 75 per cent grains in the panicle have dried up.* After maturing, the crop is not allowed to stand in the field for long, otherwise, *the seeds are prone to shattering.* The threshing is done by the help of bullocks. After threshing is done by the help of bullocks. After threshing and cleaning, the grains are dried in the sum for 3-4 days so that the moisture content in the grain cones down to 10 per cent. This is quite safe for seed storage.

Italian millet

Italian millet (foxtail millet), *Setaria italica* of Poaceae, popularly known as *Kangani, kauni* or *tanguri* is a favoured short duration crop of dryland agriculture. It is cultivated as a main crop during kharif or as a cash crop during March-April.

The millet is predominantly grown in poor soils and moisture - stress conditions in *barani* lands where no other crop could profitably be grown. Being tolerant to drought, the plant makes efficient use of available soil moisture and thrives well in areas having scanty rainfall. The crop matures in about 80-100 days (Fig. 3.5).

Setaria is grown either as a pure crop or as a component of mixed crop culture with ragi, jowar, bajra, arhar or even with the broadcast paddy in the river belts. It is suitable for cultivation right from sea-level to an elevation of 2,000 m in the foot-hills of the Himalayas.

Improved varieties

Arjuna is an improved variety of this crop. It matures in 80 days and has a yield potential of 15 q/ha. CO-3, G-1, N-1 and H-1 are dominant varieties and occupy large acreage under commercial cultivation in AP; H-1, H-2, Navane-1 were found superior for Karnataka; CO-1, CO-2, CO-3 for Tamil Nadu; Rau 1, 2, 3, 4 for Bihar.

Cultivation

The field could be prepared by one deep ploughing followed by 2-3 horrowings. There should be enough moisture in the soil at the time of planting. 8-10 kg seed is required for planting one hectare of the pure crop. The seed rate in mixed crop varies from 3-4 kg/ha. The proper time for planting, in Northern India is 20th June to 10th July. As a catch crop it could be planted during last week of March to first week of April. Row-to-row distance of 22.5 cm is optimum for high yields. Plant to plant distance within the row should be 7.5 cm. Depth of sowing should not be more than 4 cm. For higher yields 40 kg nitrogen, 20 kg phosphorus and 20 kg potassium per hectare is necessary. At sowing time 20 kg nitrogen, 20 kg phosphorus and 20 kg potassium should be given as *basal dose*. The remaining amount of 20 kg nitrogen should be applied 4 weeks after planting as *topdressing*.

Intercropping of Italian millet with black gram (urd) in the *crop ratio* 2 : 1 (base crop : inter crop) has given the highest yield. With increased package practices, yield levels at 15-20 of ha could easily be obtained.

Lower invol glume   Upper invol glume   Lower floral glume

Palea of (4)   Upper floral glume

Spikelet   Palea of (6)   Stamens and ovary

Fig. 3.5. Little millet, *Panicum miliare* L.—Spike and spikelets.

Italian millet is cultivated in India by the poorer classes in soils of very low fertility and classified under lesser millets. In India this crop is predominantly cultivated in Tamil Nadu, Maharashtra. It is cultivated in many parts of the world, viz. China, Japan, East Asia, South Africa, South East Europe, North America and Southern parts of Russia. Italian millet was one of the frive sacred plants of China in about 2700 BC.

Origin

Worth (1937) is of the opinion that China or Central Asia is the centre of origin of the cultivated species and from there fox-tail millet has spread to India and other European countries. According to Stapf the genus is divided into four sections, viz. (I) *Eu-setaria,* (II) *Panicatrix,* (III) *Ptychophylum,* (IV) *Cymbosetaria.* Under the *Eusetaria* there are 49 species out of which 7 species are listed in India. (1) *S. viridis* (2x = 18), (2) *S. italica* (2x = 18), (3) *S. gluaca* (2x = 36), (4) *S. intermedia* (2x = 36), (5) *S. verticellata* (2x = 18), (6) *S. forbesiana* and (7) *S. graciliana.* Large number of perennial species are found in Africa. Only two species *S. italica* and *S. glauca* are cultivated for grain purpose. Li *et al.* (1935) concluded that those forms with n = 9 are prim-

itive ones and those forms with higher number of chromosomes are derived ones. Further it is stated, the *S. viridis* is the ancestor of *S. italica,* both of them having a common genome "A". Further it is contended that *S. italica* may be an allopolyploid but not a simple diploid. The origin of *S. italica* from *S. viridis* seems to be beyond doubt.

LITTLE MILLET

Unlike the other millets *Panicum miliare* (Little millet) can be grown up to an elevation of 7,000 feet. It is mostly grown in the southern states of India. It is found wild in Punjab, Burma, South East Asia and to a small extent in Sri Lanka. The duration of this crop is only 75-90 days (Fig. 3.5).

Origin

There is not much evidence regarding the origin of *Panicum milliare* Lamk. According to Blatter and McCann (1935) the crop is cultivated or naturalised throughout India and Sri Lanka. Chevalier (1922) mentions that this species is cultivated only in India and Pakistan.

Pulses are important food next to cereals. Pulses are obtained from the members of Fabaceae. Pulses form the chief protein-rich food for vegetarians. Pulse crops lead symbiosis with nitrogen fixing bacteria (*Rhizobium*) which are present in their root nodules. Hence pulses are rich in proteins. Pulse crops improve soil fertility by providing it with nitrogenous compounds. Hence pulse crops are used in crop rotation (rotation crops).

Pulse seeds are non-endospermic with a large dicotyledonous embryo. Cotyledons store the reserve food. Important pulse crops are green gram, *mung* (*Phaseolus aureus*), black gram, *mashi* (*Phaseolus mungo*), Bengal gram or chickpea (*Cicer arietinum*), red gram or pigeon pea (*Cajanus cajan*), broad bean (*Vicia faba*), pea (*Pisum sativum*), soya bean (*Glycine max*), lentils (*Lens esculenta*), cowpea (*Vigna sinensis*), sweet pea (*Lathyrus odoratus*), cluster bean (*Cyamopsis tetragonoloba*), bean (*Lablab purpureus*), horse gram (*Dolichos biflorus*). "Legume" refers to the characteristic fruits of the family Fabaceae. Basically, legume is a dehiscent pod that develops from a single carpel and splits into two valves, although there are many deviations from this general structure, some fruits in the family being indehiscent and drupe-like, others transversely divided, and they may be dry or fleshy, winged or not. The name legume is also applied to those members of the family which are edible-either the pods themselves or the seeds (why they are called pulses, grain legumes, or beans), or both. The leaves are rarely eaten by man, as in the case of

species of *Pterocarpus* grown in parts of Nigeria, but several species are important fodder crops. Common temperate legumes and pulses : Scarlet, runner bean, kidney bean, haricot beans, French beans, lentil, broad bean, garden pea, asparagus pea.

Nutritionally, legumes are very important, second only to cereals as a source of human food. They are two or three times richer than cereals in protein, some are rich in oil, such as soybeans (*Glycine max*), and ground nuts (*Arachis hypogaea*), and in terms of their amino-acid composition they complement the cereals, so that a mixed diet of pulses and cereals is nutritionally well balanced and traditional in several civilizations.

Legume grains or pulses are still major components of the diet in the Indian subcontinent (especially lentil). *Lens culinaris* : Pigeon pea, *Cajanus cajan* and chickpea. *Cicer arietinum* in the Far East particularly soybean and in Latin America particularly the bean *Phaseolus vulgaris* are popular.

Only about 20 species of legumes are widely used for food, such as peas (*Pisum* species), beans (*Phaseolus* species), lentil, ground nut, cowpea (*Vigna unguiculata*), grams (*Vigna* species), mung bean (*Phaseolus aureus*) and pigeon pea (*Cajanus cajan*). Many tropical species have great potential, such as the bambara groundnut (*Voandzeia subterranea*) and the lablab hyacinth bean (*Dolichos lablab*), and efforts are being made to exploit them more fully as human food. In temperate climates, peas, beans, lentils and lupins (*Lupinus* species) are the main edible legumes. The common or garden pea (*Pisum sativum*) probably originated in the Near East and is now cultivated in most temperate regions and at high altitudes in the tropics. It is one of the four most important grain legumes and in the dried state was once a staple food of Western Europe as pea meal or split peas.

The most important grain legume in terms of world trade and production is the soybean. The leading producer is the United States, where soybean is an important cash crop and a major export. The main use of the beans is the production of protein-rich meal and oil.

Lentils are one of the oldest legume pulse crops of the New World and were involved in the origins of agriculture in the Near East along with wheat and barley. The seeds contain a high percentage of protein and are widely consumed in the Indian subcontinent, the Middle East and the Mediterranean. It has also been introduced into the New World, in Argentina, Chile and parts of the United States.

Beans derived from species of the genus *Phaseolus* are cultivated in both the Old and New Worlds. In tropical countries, the seeds are used largely as dry beans, whereas in temperate and Mediterranean countries, although there is some consumption of dry beans, cultivars have been developed for use as green vegetables such as the immature pods of *Phaseolus vulgaris* (French bean, snap bean). The dried seeds of this species are the haricot beans of commerce used in stews and in sauce as canned baked beans. *P. coccineus* (scarlet runner bean), a Middle American species, is also grown in Europe for its fleshy immature pods. The tropical species *P. acutifolius* (tepary bean) is a drought-resistant crop which is grown for its dry beans which have a high protein content.

The broad bean (field bean) *Vicia faba* is an important legume in many parts of the north temperate zone and in some subtropical areas at higher altitudes. The seeds are large and rich in protein and are consumed green and immature or ripe and dried.

Tropical legumes used for human food are many and various, but few are cultivated on a major scale. The most widely cultivated crops species are the cowpea grown as a vegetable or as a pulse throughout the tropics and sub-tropics, the ground nut, grown in warm temperate and tropical regions around the world, for vegetable oil or as an appetizer, the pigeon pea, a pulse crop grown by small farmers mainly in India, but with some production in Southeast Asia and equatorial Africa.

Pulses are the rich source of vegetable protein. India is the largest pulse growing country in the world. These crops account for 20 per cent of the acreage and 9 per cent of the production under foodgrain with about 8 per cent share in the value of output of principal crops.

Measures for improving cultivation of pulses :

(a) Introduction of pulse crops in irrigated farming system.

(b) Bringing of additional area under :
   (i) short duration varieties of *urd, moong* etc., in rice fallows by utilising the residual moisture in *rabi* season,
   (ii) in summer season with irrigation after oilseeds, sugarcane, potato, and wheat and
   (iii) in *rabi* under lentil.

(c) Inter-cropping of arhar in soyabean, *bajra,* cotton, sugarcane and groundnut both under irrigated and unirrigated conditions.

(d) Encouraging the cultivation of pulses to replace *khesari dal.*

(e) Multiplication and use of improved seeds.

(f) Adoption of plant protection measures.

(g) Use of phosphatic fertilisers and rhizobium culture.

(h) Improved post-harvest technology.

(i) Adoption of appropriate public policies including pricing and marketing of pulses.

Lens mutabilis was once a major source of protein in the Andes. But the seeds have alkaloid content.

Common warm temperate and tropical legumes and pulses are cowpea, lablab, soybean, chickpea, jack bean, buffer bean, groundnut and pea.

Legumes are additionally important in cultivation because of the association in their root nodules, with nitrogen-fixing bacteria which are able to convert free atmospheric nitrogen into nitrates. Their value as green manure which can be ploughed in to enhance the nitrogen levels in the soils is especially important in shifting cultivation in the tropics.

The mungbean is a leguminous species, or pulse crop, grown principally for its edible seeds. The pulse crops, in addition to mungbeans, include several species of legumes with edible seeds, such as garden and dry beans (*Phaseolus vulgaris* L.) lima or butterbeans (*P. lunatus* L.) broad bean (*Vicia faba* L.) garden pea (*Pisum sativum* L.), chickpea (*Cicer arietinum* L.), lentil (*Lens culinaris Medik.*), pigeon pea (*Cajanus cajan* Huth), cowpea (*Vigna unguiculata* Walp.), blackgram (*Vigna mungo* Hepper), adzuki bean (*Vigna angualris* (Willd.) ohwi and ohashi), and others of lesser importance. Mungbeans and other pulses are also referred to as grain legumes, although the latter term usually encompasses a wider range of species, including soybeans (*Glycine max* Merr.) and peanuts (*Arachis hypogeae* L.), which are grown principally as oilseed crops. But the current high market price of mungbean in the Philippines and other Southeast Asian countries is changing the image of mungbean as the "Poor man's meat"

PRODUCTION STATISTICS OF MUNGBEAN

World mungbean production is estimated to be around million mt, harvested from 3.0 million ha. This production of mungbean would be about 2% of the world production of pulses, 7% of the production of all dry beans, 18% of the production of chickpea, and about equal to the production of cowpea or lentil.

The major mungbean production area is southern and southeastern Asia, from Iran eastward through Pakistan, India, Bangladesh, Burma, Thailand, Philippines, and Indonesia. Four countries in this arc India, Burma, Thailand, and Indonesia produce almost 90% of the record world production.

Burma

Burma is the fourth ranking country in production of mungbean with 108.5 thousand ha. Among the pulses grown in Burma, mungbean ranks third, after chickpea and lime bean, occupying 14% of the area planted. Blackgram is planted on 29 thousand ha. Both mungbean and blackgram are grown after rice, often being broadcast before the rice is harvested. Mungbean is grown both for domestic use and export.

India

India is the world's leading country in production with 1.9 million ha. In India, mungbean ranks third among the pulse crops, after Chickpea and Pigeonpea. Mungbean is cultivated in all of the States of India, but largest production is in Orissa, Maharashtra, and Andhra Pradesh. Due to the diversity in local environments, the crop may be grown in any month of the year in one area or another. The major production comes from mungbean planted after rice and grown on residual moisture after the rice harvest. Multiple cropping with mungbean during the summer season (March to May) is increasing as new early maturing varieties are developed and more irrigation water becomes available. The crop is grown principally for domestic use.

SUMMER MOONG

The cultivation of moong in the summer season after the harvest of rabi crops under assured irrigation has gained popularity in the states of UP, Bihar, Rajasthan, Orissa, Punjab, Haryana, Madhya Pradesh, Gujarat, Maharashtra, West Bengal, Tamil Nadu, Andhra Pradesh, Karnataka, Assam, etc.

Indonesia

Mungbean production increased 60% between 1969 and 1976, and it now ranks third among the grain legumes grown in Indonesia. As a country Indonesia ranks third in production. Mungbean is grown for domestic use, much of being consumed in the form of "porridge".

Iran

Mungbean ranks fifth among the food legumes in Iran, with an estimated area of 25 to 30 thousand has in production. Mungbean is cultivated on the Central Plateau, and in the Southeastern and South Western areas of Iran. Mungbean is grown in small fields for domestic use, particularly among low income people, who eat it with vegetables and rice.

Japan

Production in insignificant despite a strong market demand. Use is dependent upon imports from Thailand and Burma, which are consumed mostly as bean sprouts. Blackgram is preferred to mungbean for bean sprouts in Japan.

Kenya

Mungbeans are produced for domestic use and export. Kenya is an important source of US imports.

Korea

Mungbean is a minor crop in Morea, the area planted being only 2.5% of that planted to soybeans. Mungbeans usually follow barley or wheat in the cropping system.

Malaysia

Although mungbean is generally available in local markets, traditionally they have been imported, and production is negligible in both Wes-and East Malaysia.

Middle East

There are no available estimates on mungbean production in the Middle East, although mungbean is grown in Iraq and other countries for domestic use; US imports from Turkey.

Pakistan

Mungbean ranks third among pulse crops in Pakistan after chickpea and pea, but occupies only 5% of the total area planted to pulses, compared to 75% for chickpeas. Mungbean is usually grown as a summer crop (February to June), but may be grown as a rainy season crop (July to November). The production is for domestic use.

Peru

Mungbean is a commercial crop in the Northwest Coastal area. No production statistics are available. Peru is a major source of US imports.

Phillipines

Mungbean is a favoured pulse crop in the Phillipines. Production has been relatively constant in recent years, but is insufficient to meet domestic needs. It is usually planted as a dry land crop following rice.

Taiwan

Mungbean is a minor crop in Taiwan, but demand is high with 80% of the consumption being met by imports. The mungbean is grown in an intensive multiple cropping system, usually in spring (March to May) before rice. Mungbean is a poor competitor with other crops due to the instability of yields.

Thailand

Thailand ranks second in production of mungbean after India. Mungbean is the major pulse crop in Thailand. The area planted increased from 37 to 435 thousand hectares between 1961 and 1977-78.

Other

Although seldom reported, mungbean is grown in several areas of Africa. Exports to the US in recent years have been received from Kenya, Malawi, South Africa, and other countries. The Mungbean was probably introduced to

Africa by Oriental immigrants. In the Carribean area, mungbeans are sometimes grown as garden crops.

RICE BEAN (MOTH)

Rice bean, *Vigna umbellata*. Fabaceae, an Asiatic pulse, is native to south-east Asia and is grown in Burma, Malaysia, China, South Pacific Islands, Indonesia, Philippines and India.

In India rice bean is found in the western ghats, eastern ghats and north-eastern Himalayas. Rice bean is grown as an important pulse crop in Orissa and Chhotanagpur in Bihar, where it is used as food, fodder, green manure and cover crop.

Rice bean is grown as a kharif crop during June-July and harvested in October-November. The crop has an advantage over other pulses that it is free from common pests and diseases, and storage pests. It can stand high rainfall but is susceptible to waterlogging. Since it is photo-sensitive, it can be grown only in the hills as a rainfed crop. Rice bean is suitable for cultivation as a pulse crop in north western Himalayas.

20 kg N, 40 kg P and 20 kg K would be sufficient to raise a good crop. N could be given in 2 splits, one at the sowing time and another when the crop is 20-30 days old. The distance between plant to plant is 15 cm, row to row 60 cm.

Uses

Rice bean is rich in protein content (20-24 per cent) with high methionine, tryptophan, vitamins, calcium and iron contents. The pulse is consumed alone or mixed with rice. The tender pods are cooked as vegetable, besides the green shelled seeds are also consumed like green peas or beans. The whole plant is used as forage. It can also be used as a cover crop to check soil erosion.

WINGED BEAN (CHOUGHLA SEM)

A protein rich legume, winged-bean, Goa bean, Chaudhary sem (*Psophocarpus tetragonolobus*) is suitable for cultivation in regions of the tropics and sub-tropics. In India this crop has been under cultivation in Goa, Maharashtra, Karnataka, Tamil Nadu, Kerala, Madhya Pradesh, Orissa, Bengal, Tripura and Indo-Burma border areas.

The seed resembles to some extent to that of pea. The seed coat colour varies from black, deep purple, brown, tan, light-green, yellow to white. Plants are usually creeper type, climbing to various heights from 0.5 m to < 6 m. Leaves resemble those of beans. Flowers are bluish in colour. White, red, pink and purple types also seem. Pod length varies from 5 cm in self-grown ones, to 70 cm in the extra long types. They have 4 wings and are classified as square, rectangular or flat depending on the shape of the cross section of fruit.

Winged-bean is a nutritive crop almost all parts of which are edible, have potentials of unusually high yield, high protein in seeds, pods, flowers, tender shoots and leaves as well as in *root tubers*. It is free from off-flavours and has good taste and palatability.

The immature pods are consumed as vegetable, have protein upto 3 per cent, rich in minerals and vitamins. The flowers are also consumed as vegetable, cooked as well as green and also for garnishing and decorating other dishes, have up to 5-9 per cent protein. Tender leaves are consumed as pot-herbs and have protein from 5.0 to 7.6 per cent.

Seeds of winged bean are consumed as parched or roasted similar to pea. Seed has 29.8 to 41.2 per cent crude protein.

Winged-bean is a perennial plant, capable of living and bearing for several years. The plant is adopted to warm humid high rainfall regions where cultivation of other legumes have limitations due to pests and diseases. However, the plant is sensitive to water-logging and requires a well drained soil.

Winged-bean can be sown at any time during the year. When planted during the short days i.e. winter months, most varieties flower within 8 weeks.

Most varieties grow vigorously when staked. The timing vines need support from the beginning by smooth strings etc. attached to large supports 2 m tall Y-shaped trellis is most suitable.

Winged bean is a tropical legume, immensely rich in protein and oil, offers good promise as a rich food source. Winged bean is a plant of humid and sub-humid tropics of Africa and Asia. Every bit of the plant leaves, pods, green shelled beans, tender shoots, flower inflorescence, dry grains, roots are edible and consumed in various ways. The fleshy tuberous protein-rich roots are boiled or roasted and eaten in Papua, New Guinea and Burma.

Winged bean is a ring annual : pods are produced in clusters; pods are rectangular with 4 wings; each pod has about 20 seeds (grains); seeds are smooth, shining, white, yellow, chocolate, brown, black, purple or mottled. The roots are numerous, become thick, fleshy and tuberous. Seeds contain 24 per cent protein and 18 per cent oil. More than 60 per cent of the fatty acids in the oil are unsaturated.

Tubers

The winged-bean develops vegetable tubers like those of sweet potato. These are usually consumed after boiling and peeling as afternoon snacks or made into chips. Tubers are rich in calorific value (150 calories in 100 g of edible portion). The tubers are rich source of protein (10.9 per cent).

Lablab bean

Lablab bean, hyacinth bean, Indian butter bean, *sem* (*Lablab purpureus* (L.) Sweet) is a multipurpose legume crop. The plant may be bushy, prostrate, semi-erect or winy. The leaves are trifoliate with recemose flowers having purple,

pink or white corolla. Pods are flat or inflated, linear or broad, incurred with persistent style. The pod colour varies from white, light green, dark green, light purple to dark purple. Seed colour varies from white, yellow, brown creamy, purple and black.

Depending upon the pod and grain quality and the plant growth habit, lablab bean can be classified into grain, vegetable and fodder types.

Lablab bean is used as pulse, shelled bean and green pods as vegetable. Green shelled seeds are popular in Maharashtra and north-eastern India. The dry bean is very popular as field pea bean in Karnataka, Tamil Nadu and north-eastern India. For this purpose white seeds are preferred. Dried seeds are used as wholesome palatable food, cooked and eaten. The bean is eaten while fried or boiled and salted. It is used more as dal in split form.

Tender leaves and flowers are edible.

Cultivation

Seeds is sown in the beginning of July, with a row to row distance of 75-90 cm and plant to plant 15-20 cm. Flowers and pods are produced from the month of November to February.

BAMBARA GROUNDNUT

Bambara groundnut (*Vigna subterranea*; Fabaceae) is the most popular grain legume of Africa. It is considered to be *staple protein source* from Senegal to Kenya and from Sahara to Malagassy.

The crop is adapted to climatic and edaphic conditions similar to groundnut, maize or sorghum and can be grown as groundnut, but arid or drought conditions are specially suited for its cultivation.

Africans generally grind seeds into flour and consume in various ways. The *lysine*, one of the most limiting aminoacids, is exceptionally high in bambara nut.

FABA BEAN (BAKLA, ANHURI)

Faba bean (*Vicia faba*) is a protein-rich pulse crop wish high yield and is used as *dal*, vegetable and silage.

Faba bean is cultivated in India has limited to a very small acreage to eastern UP, Punjab, Kashmir, Ladakh, Rajastha, Karnataka and Madhya Pradesh.

Vicia faba is classified into three groups, mainly based on the seed size :

(i) *Vicia faba major* : Large seeded; broad bean

(ii) *V. faba major equina* : Medium-seeded, horse bean

(iii) *V. faba minor* : Small seeded, tick bean.

It is the fourth most important pulse crop in the world after dry-ean, dry-pea and chickpea. The main faba bean growing regions in the world are West Asia and North Africa up to 4°C. In India the crop is cultivated in Kashmir at an altitude of 5,000 feet and in Kinnaur Spiti and Tibet in between 8,000 and 12,000 feet. It is also tolerant to water stress due to *higher proline accumulation* and because of this it is a more popular crop in the Middle East. Faba bean can also tolerate water logging and salinity to certain level.

Faba bean is consumed daily in one way or the other by a large part of the population in Middle East. It is used as green vegetable. Green seeds are also consumed as vegetable. Dried seeds are used asdal, flour, porched or boiled seeds, etc. Wine is also made from the seeds.

Marma bean

Marma bean (*Tylosema esculenta*) of Fabaceae, is a native of the drought-ridden semi-desert regions of the Kalahari and neighbouring sandy regions of southern Africa. Marma bean feeds the tribes in Bostwana, Namibia and the Republic of South Africa.

The plants are prostrate annual, sending long, viny stems sometimes up to 6 m long, creeping over the soil surface. The vines have soft, red-brown leaves when young, which turn leathery and grey-green with age. Golden yellow blossoms develop in mid-summer rainy season; the fruits ripen in late-autumn or early winter. The plant has large underground tubers.

The seeds contain 30-39 per cent protein and 36-43 per cent oil. The protein is rich in lysine (5 per cent) and deficient in methionine (0.7 per cent). The seeds inside are cream coloured, with firm oily fresh. The seeds are roasted for human consumption. Roasted seeds have a delicious mutty flavour.

Seed oil is edible with agreeable taste and a pleasant, nutty flavour. The meal remaining after oil extraction can be fed to cattle. The tubers are rich in protein and can be consumed as a vegetable dish.

Toxic substances in raw pulses

Kidney soyabean and black bean pulses have phytohaemagglutinic substances which have the property to agglutinate red blood cells.

Khesari dal (*Lathyrus sativus*) contain lathyrogen substances which cause *human lathyrism* - a paralytic disease (paralysis of leg muscles and death in extreme cases). The presence of β-η-oxylamino-L-alanine compound has been identified as the principal causative factor of human lathyrism.

Kidney bean is believed to contain an antagonist to vitamin E. Soybean contains trypsin inhibitors.

The ingestion of fresh raw or cooked broad beans (*Vicia faba*) is reported to cause favism in certain persons which is a disease characterized by hemolytic anaemia. Recent evidence suggests that pyramidines (divicine and isouramil) occurring naturally as β-glycosides are the causative factors of this disease.

5

Oil Seeds

The fatty oils obtained from plants are usually called *fixed oils* because they do not evaporate and become volatile. The vegetable oils contain glycerol and fatty acids like, oleic acid (palmetic acid and stearic acid are present in fats). .

Fatty oils are usually stored in the seeds, either in cotyledons or in endosperm. Most of these fatty oils are edible and hence used as *cooking oils*. The non-edible oils are converted to edible oils by a process called *hydrogenation*. Low grade oils may be used in the preparation of soaps, candles, cosmetics and in many industries as lubricants. Oils also contain little quantities of protein and carbohydrate.

Vegetable fatty oils are classified into :

1. *Drying oils* : Iodine number > 130 e.g. linseed, soyabean, tung, safflower, hemp seed soil. These absorb little oxygen but when exposed, do not form dry elastic films; oils used for foods and in soap manufacture.

2. *Semidrying oils* : Iodine number 100-130 e.g. cotton seed, sesame, sunflower, corn, croton oils. These do not absorb oxygen and when exposed, remain in liquid state. These oils remain liquid.

3. *Non-drying oils* : Iodine number < 100 e.g. groundnut, palm, olive, castor, rape, colza and almond oils.

4. Fats or tallows e.g. cocoa butter.

These absorb little oxygen but when exposed, do not form dry elastic films; oils used for foods and in soap manufacture.

Fixed oils are compounds of glycerine with certain complex organic acids - oleic, palmitic, and others - which are known as *fatty acids*. The term *oil* is used to denote a sbstance which assumes the liquid phase at ordinary temperature. *Fat* is one which is solid. Most vegetable products of the nature are classified as oils, with a few exceptions such as cacao or cocoa butter coconut oil, liquid in the tropics, is solid in temperate regions.

Linseed oil oxidizes in the atmosphere to a tough, resistant film. Although the change is not a dehydration, such oils are known as *drying oils*. Cotton-seed oil merely forms a greasy mass and become rancid on exposure; they are called *non-drying oils*.

Olive oil is derived from endosperm or cotyledon reserves of seed. Oil-yielding seeds, as a group, are known as *oil-seeds*. They contain, in addition to oil, much protein, but they are usually low in carbohydrates.

Fixed oils may be obtained by mechanical pressure, by heat, by extraction with solvents, such as carbon disulphide, petroleum fractions, and trichlorethylene, or by combination of methods. Pressure, with or without heat, is the most common procedure. It is desirable that the pressure be applied gradually but with great force. Most presses are of the hydraulic or expeller type.

The *hydraulic press* is essentially a chamber with a rigid top, whose ·floor is raised through hydraulic pressure. Materials in a hydraulic press are usually placed in sacks or wrapped in strong coarse cloth which retains the powdery residue as the oil strains through.

Expeller presses are built on the principle of the household meat-grinder; oil is forced out of the mass by a long revolving screw.

If the material being pressed is *heated,* oil is rendered more labile and larger yield is obtained. Heating, however, may injure the quality, there are often considerable price and use-differences between *cold-pressed* and *hot-pressed* oils of the same plant.

The vegetable residue or *press cake* is rich in protein and except when containing some poisonous substance as in case of castor oil cake, it may be valuable feedstuff. Press-cakes are also ground and used as nitrogenous fertilizer.

Uses of oil

Edible oils have food uses, oleomargarines and butter substitutes and in salad dressings and salad preparations. Non-food uses are in soaps, in paints and varnishes etc. Castor oil is used in lubricating.

Plant yielding fatty oils are *Schleichera trijuga* Kusum oil from seeds (Sapindaceae); *Linum usitatissimum* lin seed oil from seeds (Linaceae); *Sesamum indicum,* til oil from seeds (Pedaliaceae); *Olea europea,* olive oil from fruits (Oleaceae); *Arachis hypogea,* groundnut oil from seeds (Fabaceae); *Cocos nucifera,* coconut oil from endosperm (Arecaceae); *Ricinus communis,* castor oil from endosperm (Euphorbiaceae); *Elaeis guineensis,* palm oil from fruit (Arecaceae); *Brassica campestris,* mustard oil from seeds (Brassicaceae); *Brassica alba,* white mustard oil from seeds (Brassicaceae); *Brassica nigra,* black mustard oil from seeds (Brassicaceae); *Copernicia cerifera* wax, from young leaves (Arecaceae); *Ceroxylon andicola* wax, from leaves (Arecaceae); *Halianthus annuus,* sunflower oil (Asteraceae); *Ricinus communis,* castor oil (Euphorbiaceae); *Zea mays,* maize or oil; *Glycine max,* soybean, soya oil; cotton seed oil, *Gossypium* spp. Malvaceae; *Elaeis oleifera,* American palm oil (Arecaceae).

PALM OIL

Palm oil is obtained from the fruit pulp (palm oil) and the seed (palm kernel oil) of the oil palm, *Elaeis guineensis,* Arecaceae. The oil palm belongs to the subfamily cocoideae. It is unbranched and monoecious; grows to a height of 20-30 m and may live up to 200 years. The root system consists of primaries and secondaries. The majority of roots are found in the top 14 cm of the soil with the main concentration near the trunk and the secondary concentration 1.5-2 m from the base. Leaves are produced in spiral succession from the meristem. Crown consists of 40-50 opened leaves at a time. The number of leaves produced annually is 15-25. The leaves unfurl at the rate of about 2 per month is about two years. Inflorescence reaches the central spear stage in 2 years and a further 9 or 10 months is required for flowering and anthesis. Each flower primordium is a potential producer of male and female organs but one or the other usually remains rudimentary to produce either a male or female inflorescence. The oil palm is cross-pollinated. Pollination is affected by wind. The fruit ripens 5 to 6 months after pollination. The fruits are 2.5 cm long with an oil-rich mesocarp surrounded by a hard endocarp (Fig. 5.1).

The palm oil is extracted from the outer mesocarp of fruits. Palm oil forms the bulk part of the produce. The kernel oil is extracted from the kernel. Each fruit bunch weighs 15-30 kg; yield about 20 per cent palm oil and 2.5 per cent kernel oil. One hectare land with 135 palms gives 13 tonnes of fresh fruit bunches, which yield 2.5 tonnes of palm oil and 0.625 tonnes of kernel oil per annum.

Uses

Palm oils are used for the preparation of margarine, compound cooking fat, soaps, candles, bakery trade, glycerine production, fuel for internal combustion engine, greases and lubricants.

Varieties

Dura, Tenera are used for commercial planting.

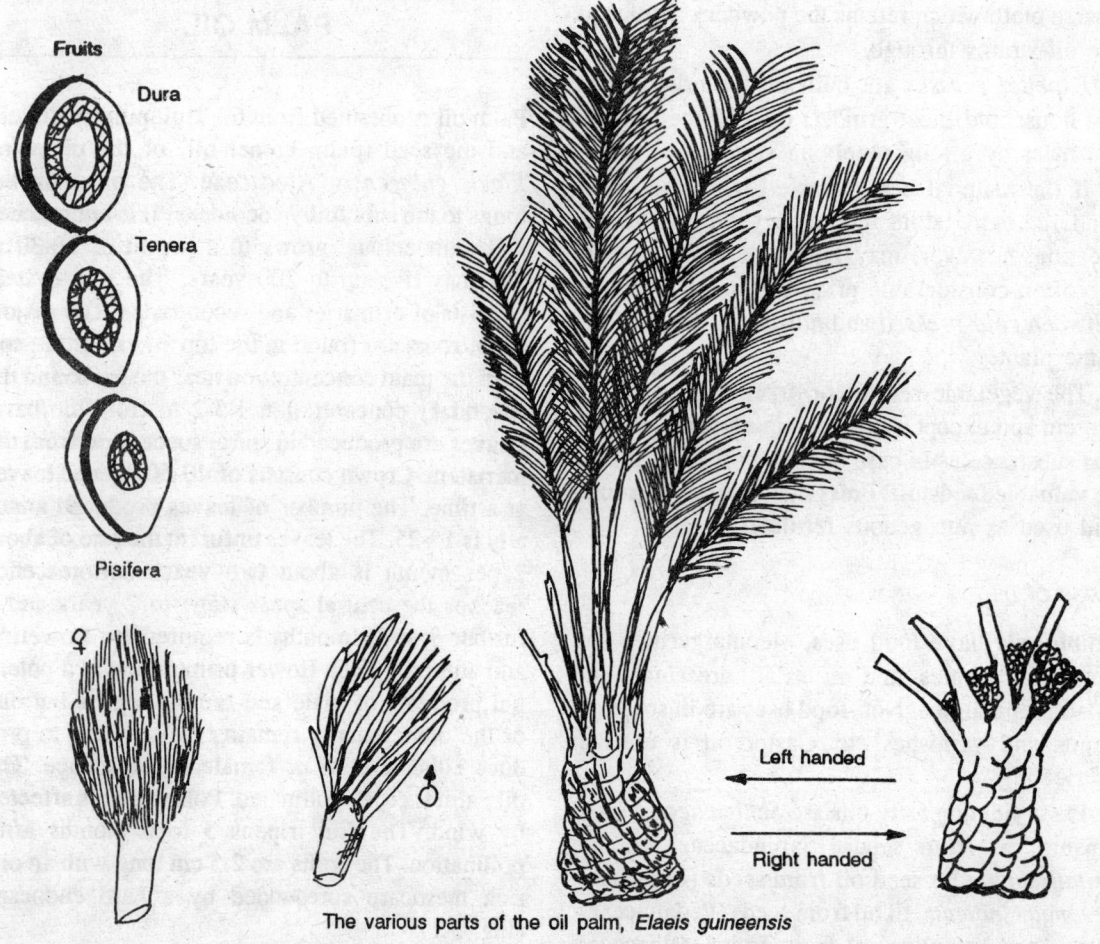

Fruits

Dura

Tenera

Pisifera

♀

♂

Left handed

Right handed

The various parts of the oil palm, *Elaeis guineensis*

Fig. 5.1. Oil palm—*Elaeis guineensis* tree, flowers and fruits.

Pisifera is an abortive type grown for purpose of breeding (to be used as the male parent). The hybrid *tenera* has gained popularity as the best variety evolved for commercial planting. It has thicker mesocarp (and high oil content) and thinner shell as compared to its mother plant *Dura*.

Cultivation

Oil palm exists in the world in a wild form. It is also cultivated in Africa, South East Asia and America. In India, oil palm is cultivated in about 6,000 hectares Yield is 2500-4000 kg oil per hectare. Oil palm can be grown successfully in Kerala, Tamil Nadu, Karnataka, Goa, Assam, Andaman and Nicobar Islands.

Oil palm requires a rainy tropical climate. A moist sub-soil is beneficial, but it should not be swampy and water logged. The palm requires well distributed rainfall 2000-3000 mm per annum. Seeds germinate at 40°C with adequate aeration and moisture. To break the dormancy, seeds are preheated in a germinator at 38-40°C, moisture content of 18 per cent, for 40 days in polythene bags. After this, the seeds are taken out, soaked in water for about 7 days and kept for germination in 500 gauge polythene bags at room temperature. Germination commences from the first week onwards and continues for about 5-6 weeks. The germinated seeds are removed and sown in small polythene bags in the prenursery.

The seedlings are transplanted to large polythene bags when they attain 4-5 leaf stage. The nursery period is 12-18 months after which they are planted in the main field.

In the field, pits are dug (0.6 x 0.6 x 0.6 m) at a suitable distance of 9 m x 9 m. The seedlings are planted in the pits after removing polythene sheets. The collar of the palm should be flush with the ground surface. The palm should not be planted too deep or on mounds. Normally no shade is required, but in case of a drought period after planting, seedlings are covered with cones of palm fronds or other materials.

Establishment of suitable cover crops, pruming of leaves, weeding in the early years include field care. Usually young palms receive a fertilizer mixture of 350 : 450 : 450 : 250 g of N, P, K and Mg. Mature palms are supplied with 700 : 700 : 1400 : 450 g N, P, K and Mg per palm per year. The fertilizers are applied in 2 equal splits during April-May and September-October.

Harvesting

Bunches of male and female inflorescence develop in the leaf axil of oil palm after 2 years in the main field. Fruit bunches ripen in about 6-8 months from the date of pollination. When *the bunches attain an orange-red colour and ripe fruits become loose on the bunch and when the bunch starts to shed a few loose fruits*, it is ready for harvesting.

In order to produce optimum yield of high quality oil, the bunches must be harvested when they are just ripe and they are taken to the factory for processing *within 24 hours*.

Extraction of oil

The harvested bunches are quartered by a chisel and *steamed in boiling water for about an hour*, to soften the fruits for pounding. The fruits are stripped off from the bunch and pounded. This pounded material is related and squeezed under the *hydraulic press*. The heavily laden material thus obtained is then boiled in a *clarification drum* where the sludge will deposit and pure oil float over the water. The oil is then drained out and stored in barrels.

Soybean oil, extracted from the seeds of *Glycine max*, an important crop; employed chiefly in cooking, in soaps and lard substitutes and as it is a demi-drying oil - as a partial replacement for linseed oil in paints and varnishes. The highly nitrogenous press-cake is about 50 per cent more valuable than the oil.

Peanut oil is used in sealed oils, in margarines, in packing sardines, etc. Like soy-bean oil.

Corn oil, a by-product of the corn-products industry, is used in cooking.

Rapeseed oil and other expressed oils from various mustards are used as food in India and elsewhere in the Orient. About 75 per cent of American use in compound lubricating oils for marined and automobile engines. Other uses are for quenching steel plate and as sanctuary oil in churches.

Sesame or *Benne oil* from *Sesamum indicum* (Pedaliaceae) is one of the staple food oils of India. It is said not to become rancid on exposure and has limited use in salad preparations, soap stock, and pharmaceuticals.

Hemp seed and *poppy seed* oils are minor commodities. The latter is used in the finest qualities of artist's paints. In the Old World it is quite largely used as an edible oil.

Expressed oil of Almonds (not to be confused with the highly poisonous oil of bitter almonds, which is a distilled product, containing hydrocyanic acid) also known as *sweet almond oil*, is a high-priced product used chiefly in pharmacy. Pressed oils of apricot and peach kernels have similar properties and are said to be sometimes mixed with it. Almond press-cake is used in cosmetics and toilet soaps.

Perilla oil from *Perilla ocymoides* of Japan, China, and northern India is a drying oil, employed in paints, varnishes, linoleum and printing inks.

Cacao or *Cocoa butter* is a vegetable fat obtained from the cacao bean (*Theobroma cacao*). It is in part a by-product of cocoa and chocolate manufacture. It is used in pharmaceutical preparations and as an aid to mixing sugar and milk into chocolate in the making of confectionery.

NIGER

Niger, *Guizotia abyssinica* (Asteraceae) known as *ramtil* is an oilseed crop having 35-50 per cent oil. The crop is said to be native of Ethiopia and other parts of tropical Africa. India, Germany, East Africa, West Indies and Zimbabwe are the important niger growing countries of the world. However, India is the chief producer of niger seed with an area of about *6 lakh hectares* and production of about *1.5 lakh tonnes of seed*. The average yield is about *240 kg/ha*. Niger is chiefly cultivated in Madhya Pradesh, Orissa, Maharashtra, Bihar, Karnataka and Andhra Pradesh.

This crop is grown on shallow, black light red or brownish loams as well as roughly and rocky laterite soil on hill tops and slopes.

Varieties

Madhya Pradesh : Ootacamund No. 5 and No. 87; *Orissa* : Ootacamund, GA-2 and GA-10; *Maharashtra* : IGP-76, Niger-13 and Ootacamund; *Karnataka* : Ootakamund, No. 71, RCP-66, No. 16 and No. 2; *Bihar* : GA-2, GA-10, No. 71, GA-5, Ghoti No. 2 and BNC-120; *Andhra Pradesh* : Gaudagude local, IGP-76 and BNC-120.

Cultivation

6 to 8 kg seed per hectare should be sown in rows 30 cm apart. The seed should be treated with thiram at the rate of 3 g/kg seed prior to sowing. Thinning is done to maintain plants 15 cm apart within the rows, about 2 weeks after sowing.

Fertilization is given a *basal dose* of 10 kg N + 20 kg P_2O_5 + 10 kg K_2O per hectare followed by *top dressing* with 10 kg N/ha about 30-35 days after sowing. Weeding is done once 15 days after

sowing and about a month of sowing 30-35 days after sowing.

Niger is a cross-pollinated plant mainly through honey bees.

Niger oil

The oil is pale yellow, odourless and edible with a pleasant nutty taste. Cold pressed niger seed oil and refined hard pressed oil in the fresh state are used for edible purposes. Lower grade oils are employed for making soap and as illuminant. On account of its comparative cheapness, niger seed oil is often used as a substitute or an adulterant for other oils. The oil is used to a limited extent as a paint oil. Niger seed cake is utilized for feeding cattle or as a manure.

Oxidative rancidity

Niger seed oil is readily susceptible to oxidative rancidity, probably due to its high content of *linoleic acid glycerides*. Refining and bleaching enhances susceptibility to oxidation. Exposure to sunlight or to diffuse light for a long period bleaches the colour of oil.

SESAMUM

Sesame *Sesamum indicum* (Pedaliaceae) (seasmum, gingelly, *til*) is an ancient crop of India. The seeds are rich in good quality edible oil (46-52 per cent) and protein (20-25 per cent). The seeds may be eaten fried, mixed with sugar or in sweetmeats. The oil, besides as a cooking medium, is also used for anointing the body, in the manufacture of perfumery oils and for medicinal purposes. The oil cake which is rich in calcium, is used as feed (Fig. 5.2).

Cultivation in India

Sesamum occupies an area of 2-4 million hectares annually (14 per cent of the total area under oilseeds) with production of about 477 thousand tonnes. The average yield per acre is 200 kg/ha.

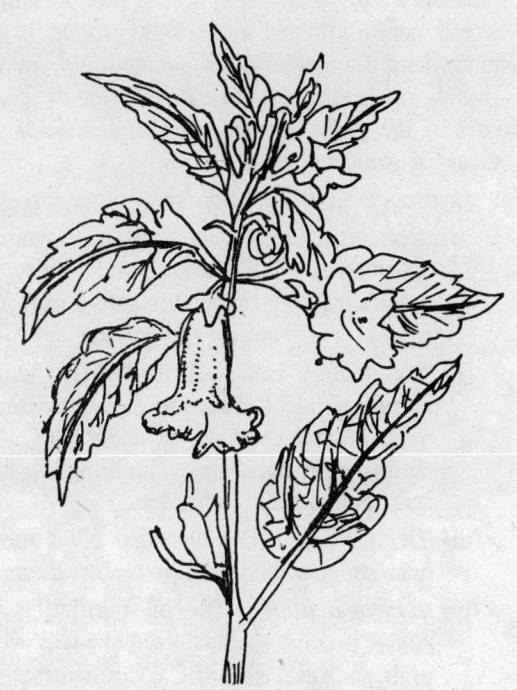

Fig. 5.2. Sesame, *Sesamum indicum.*

Varieties

T-12, T-13, T-4 (Uttar Pradesh); C-50 (Pratap); TC-25, T-13 (Rajasthan); N-32, JT-7 (Madhya Pradesh); Phule Til No. 1 (Maharashtra); T-85, Gauri, Madhavi (Andhra Pradesh); Vinayak, Kalika, Kanak (Orissa); TMV-3, TMV-4, TMV-6 (Tamil Nadu); Purva-1, Gujarat Til No. 1 (Gujarat); E-8 (Karnataka); Punjab Til 1 (Punjab); Haryana Til No. 1 (Haryana); B-67, MS Patna (Bihar); B-67, B-9, B-14 (Bengal and Assam); Kayamkulam-1, Tilottama (Kerala).

The existing varieties have yield potential upto 900-1000 kg/ha.

A *well-drained field* is pulverised and brought to five tilth by ploughing deep in summer and with the onset of monsoon 2 or 3 harrowings in crosswise direction for kharif season crop 2 or 3 harrowings are enough for semi-rabi/rabi sesame crop. In most of the states sesame is grown as rainfed kharif crop. However, in MP, Maharashtra, AP, TN and Karnataka, as a kharif as well as rabi crop; in Orissa, West Bengal, Bihar, Tamil Nadu as an irrigated summer crop.

For the kharif crop, sowings should be done in the last week of June to first week of July. Treat seeds with thiram 3 g/kg seed rate is 2.5 to 3 kg/ha for sowing; 5 tonnes of farmyard manure per hectare, is applied one month before sowing, when preparing the field for sowing. Apply 20 kg N, 20 kg P_2O_5 and 20 kg K_2O at the sowing time and 20 kg N, 30-35 days after sowing. The crop should be kept weed-free particularly during the initial 20-25 days by hand weeding and/or chemical sprays. The crop is grown without irrigation (rainfed crop).

The crop should be harvested before the plants are completely dry i.e. at the physiological maturity because this prevents shattering of capsules. The stage of physiological maturity is the one *when leaves turn yellow and start drooping but the capsules still look greenish with considerable moisture.* At this stage the crop attains the highest oil content.

Mixed cropping

Sesame is grown as a sole as well as a mixed crop. Sesame is mix-cropped in Uttar Pradesh (mixed with maize, bajra, jowar and arhar), Madhya Pradesh (maize, mung, arhar); Punjab (groundnut, cotton, maize and bajra); Rajasthan (sesamum and guar in alternate rows); Gujarat (kodo, arhar, cotton and castor); Karnataka (red-gram, castor and bajra); Bihar (maize or arhar); Tamil Nadu (cotton, redgram and bajra).

SAFFLOWER

Safflower, *kusum, karadi* (*Carthamus tinctorius* L. Asteraceae) is cultivated in Maharashtra, Karnataka, Andhra Pradesh, Tamil Nadu and Madhya Pradesh. In Maharashtra 17 lakh hectares are under oilseeds, of which safflower occupies an area of 5.15 lakh hectares. The area under this crop is Andhra Pradesh (2 lakh ha), Karnataka (1.63 lakh ha), Orissa (1.5 lakh ha), Madhya Pradesh (0.8 lakh ha) and Bihar (0.5 lakh ha).

Average yield in Maharashtra is 300 kg/ha; in United States (2021 kg/ha) and Mexico (1316 kg/ha).

Safflower thrives well on deep, well-drained and medium to heavy textured soils. It is also fairly successful in saline - alkali soils. Safflower is day natural but thermosensitive. The period September to November is more conducive to safflower growth and development. Seed rate is 5-10 kg/ha depending on the variety and place. The safflower is the ideal crop of dryland areas. The plant has deep root system, ability to withstand moisture stress.

Varieties

JSF-1, white flowered variety with about 30 per cent oil, yields about 15 Q/ha under dry conditions.

Safflower needs sufficient moisture during the early growth period of 30 days during which the plant is very succulent. After 50 to 60 days, terminal growth of plant comes to an end but branches continue to grow for sometime. After this growth period, leaves become thick and spiny having waxy coating on them which prevents transpiration and hence the water utilization is less during later growth period. Moreover, the transpiration losses through soil are also checked due to its bushy plant type.

Safflower oil

Safflower gives a good quality edible oil and it contains high per cent of linolic acid which is very useful in checking the cholesterol level in the human body.

COTTON SEED OIL

Gossypium hirsutum (Malvaceae)

After ginning of cotton, the seed is screened from refuse, put through de-lintering machines to remove the short hair, fuzz, or linters, and then hulled, after which the kernels are heated and pressed. 27 m mt of cotton seed is produced in the world, per annum and most of the product is used in the manufacture of hydrogenated fats - margarine or vanaspati. Hydrogenation converts only a part of the glycerides of unsaturated acids into those of saturated acid.

Hydrogenation of cotton seed oil involves :

(i) *Removal of the acids* : The oil is warmed and treated with *sodium hydroxide,* which neutralizes free fatty acids, salts thus formed appear as serum on the surface;

(ii) *Bleaching* : The oil from the first tank is decanted into the second. It is now treated with *animal charcoal* at about 90ºC. The animal charcoal absorbs colouring substances. Later the oil is filtered.

(iii) *Deodorising* : The bleached oil is treated with super-heated steam to destroy the smell;

(iv) *Hydrogenation* : The oil purified in the above process is transferred to a tank which is heated at 150-200ºC. *Nickel formate* is added to the oil and *hydrogen gas* is passed into under pressure. Nickel formate is reduced by hydrogen into fine nickel which act as a catalyst in hydrogenation. The process of hydrogenation is continued till a *fat of desirable consistency* is obtained. The solidified oil is then filtered to remove the catalyst.

The hydrogenated fats are many times more stable than the oils from which they are produced i.e., they are less likely to become rancid. *Rancidity* is the unpalatability of the oil or fat due to the breakdown of glycerides into few fatty acids, aldehydes, ketones etc.

In order to increase the stability of the hydrogenated fat, antioxidants like *tocopherol* (*Vitamin E*), and synergists like *phosphotides* are added.

Cotton seed oil did not find acceptance as an edible oil till a few decades back due to the presence of small quantities of *gossypol*, a toxic phenolic compound, which also contributes to the colour of the oil in the unrefined state. Methods are now available for easy removal of gossypol both from the oil and from the meal remaining after extraction of oil.

Refined cottonseed oil, which contains practically no gossypol, is pale yellow in colour, and can be used directly as cooking medium. In fact, in Egypt, cottonseed oil is the main oil for cooking purposes. Cottonseed oil is rarely used as a direct cooking medium in India and its use as an edible oil is mainly as a component of oils used for manufacture of vanaspati or hydrogenated fat.

Average recovery of oil is 12 to 13 per cent of cottonseed weight.

RICE BRAN OIL

Bran is the most important by-product of rice milling industry. Bran oil is extracted from bran immediately after milling the rice, otherwise it deteriorates rapidly. Oil is extracted by expression in hydraulic presses or extraction with solvents. Bran-oil contains about 3-9 per cent wax, which has to be removed by filtering or centrifuging. Rice bran oil is an edible oil, the oil has bettern qualities for storage due to the presence of *alpha* and *gamma-tocopherols* which are antioxidants. Bran oil is also used in preparing *vanaspathi,* soaps, oleins, stearins, textile and leather industries, flexible films, enamels etc.

SUNFLOWER

Sunflower (*Helianthus annuus*; Asteraceae) is believed to be native to South America. Russia, Rumania and Argentina are the largest producers of sunflower seeds. In recent years it has become an important crop in India. The plant produces a heterogamous head inflorescence which after fertilization is converted to an infructescence. Useful part for oil is the *cypsela fruit* (embryo).

Varieties

Moderm, EC-68414, BSH-1.

Uses

Embryo of each cypsela contains around 40 per cent oil. The oil after hydrogenation is used as edible oil. It is rich in vitamin A and D. Sunflower oil is semidrying type and used in making soaps, paints, varnishes. Seed cake is an excellent feed for poultry birds.

CASTOR OIL

Ricinus communis (Euphorbiaceae)

Castor oil is used in lubricating marine engines and for other lubricating purposes, being blended in some cases with mineral oils (Fig. 5.3).

Fig. 5.3. Castor, *Ricinus communis.*

The principal use of castor oil is in the making of Turkey red oil and other alizarins. Alizarin assistants are made by treating oils with sulphuric acid. They are used in dyeing and finishing textiles and to some extent in finishing leather.

Castor oil is somewhat employed in transparent soaps.

Castor press-cake is poisonous and is used only in mixed fertilizers.

Castor seed is chiefly raised in India.

Castor seed is crushed in India but little oil is exported. The United States and other manufacturing nations express most of their own oil from Oriental seed.

Castor oil is also used in margarine manufacture and in industry as a lubricant. Castor oil in addition to its medicinal properties, is largely used in the industry.

LINSEED OIL

The flax plant (*Linum usitatissimum*; Linaceae) is a bushy plant produced chiefly in regions of short-growing seasons of intense sunlight.

Flax seed contains about 33 per cent of oil. The seed is usually hot-crushed in hydraulics with a pressure of about 3,600 pounds per square inch. It may be also cold-pressed, either in hydraulic or expeller presses. Extraction methods, with hot light petroleum as solvent, are also practised. Linseed oil is refined by treatment with sulphuric acid. It is used either raw or boiled. In the production of boiled oil, dryers such as lead and manganese oxides are dissolved in it by means of heat, which, however, does not usually reach the boiling-point of the oil. Boiled oil dries more rapidly than raw, but the film is said to be less durable.

About 65 per cent of linseed oil consumption is in paints and varnishes and about 20 per cent in linoleum and other floor-coverings. The remainder is utilized in oil-cloth, patent leather, imitation leather, putty, printer's and foundry oils, and soft soaps. *Foots*, a mucilaginous low grade of oil, is chiefly used in soaps.

Hot-pressed linseed cake may be used in stock-feeds, but cold-pressed cake is poisonous.

In times of scarcity and high price, other oils may be substituted - generally in part rather than as a whole - for linseed oil in paint and varnishes. Among substitutes are Chinese nut oil, perilla oil, soy-bean oil, and menhaden oil, the latter obtained from fish.

Flax seed is a more important article of international trade than is linseed oil, which is generally crushed for domestic use by the large manufacturing countries. Argentina is the largest flax-seed producer, in some years contributing over half the world's supply. Russia, the United States, and India are the other largest factors in import and export of oil.

CHINESE NUT OR TUNG OIL

This is obtained from species of *Aleurites*. Japanese nut oil, from the same genus, is a less important, similar product. Oils of *Aleurites* have valuable drying properties and have largely replaced linseed oil in waterproof varnishes.

JOJOBA

Jojoba, *Simmondsia chinensis* (L.) Sch. (Buxaceae) is native to Mexico, Arizona and California. Jojoba is a hardy shrub; dioecious, long lived and evergreen; it generally attains a height of 2-3 m and during winter and spring months produces flowers and seeds. The nuts of the plant yield *liquid wax*. It is a substitute for the prized sperm whale oil, a high pressure machinery lubricant which is becoming scarce because of the ban on the killing of sperm whales from which it is used to be derived.

The female plant of jojoba produces a nut-like seed and about 50 per cent of the seed's dry weight consists of lipid stores in the form of simple wax esters that have properties similar to those of spermwhale wax. Unlike other conventional oil-seed crops that produce triglyceride, jojoba oil is an ester of average chain length of 42 carbon mono-unsaturated fattyacid and 20-24 carbon alcohols. The wax and its component fattyacid and alcohols have a wide range of industries uses, including cosmetics, pharmaceuticals, extenders for plastics, printer's ink and lubricants.

Jojoba oil is the only *unsaturated liquid wax* readily expectable in large quantities from a plant source. Jojoba seeds contain about 50 per cent liquid wax, which can be obtained in high purity by pressure technique or through solvents. The wax can be easily hydrogenated to a produce a solid, herd, white wax with important potential for use also in polishing, carbon paper etc.

Being a renewable vegetable source, jojoba has attained much importance in recent years and its cultivation has spread to Mexico, United States, Israel and some African countries.

The plant tolerates extreme desert temperatures (35°-45°C) and can thrive well on the lands inhospitable for traditional agriculture.

It grows in soil of marginal fertility, needs less water, withstand salinity and apparently has a low fertilizer requirement. Jojoba prevails from the relatively mesic environment of the temperate coast to the severe desert climate of the inland and has superior adaptation to drought, soil salinity and alkalinity and extreme temperature. It is a genuine drought-endurer, because it maintain a large leaf surface and possesses mechanism enabling it to maintain a favourable carbon and water balance even during long period of drought and high temperature. It requires no specialized cultivations and its oil can be extracted inexpensively with conventional machinery used for vegetable oil. As the species is economically potential and gaining importance throughout the world, the efforts are already under the way to introduce it in India.

MINOR OILSEEDS

Neem (*Azadirachta indica*), kusum karanja (*Pongamia glabra*), sal (*Shorea robusta*) rubber seed, mango kernel, kakum, dhupa, undi, maroti, pisa, khakan, mahor, pales, babul, gokburu, cashew kernel waste, silk cotton seed, ratan jyoti offer potential source for augmenting the supply of vegetable oils (Table 5.1).

Table 5.1. Minor oil crops

Babassu oil	Orbigyna barbosiana (O. speciosa)	Arecaceae
Ben oil	Moringa pterygosperma (M. oleifera)	Moringaceae
Cohune oil	Orbigyna cohurne	Arecaceae
Brazil nut oil	Bertholletia excelsa	Lecythidaceae
Candle nut	Aleurites molucana	Euphorbiaceae
China wood oil	Aleurites fordii	Euphorbiaceae
Japan wood oil	Aleurites cardata	Euphorbiaceae
Carapa andiroba	Carapa guianensis, C. procera	Meliaceae
Cashewnut	Anacardium occidentale	Auacardiaceae

Seeds of tobacco, ambadi (mesta), jute, watermelon, tea, silkworm pupae also offer a potential source for vegetable oil. 3,000 tonnes of tobacco seed oil is produced per year.

Mahua

Mahua butter tree (*Madhuca latifolia,* Sapotaceae) are concentrated in Madhya Pradesh, Bihar, Gujarat, Maharashtra, Orissa, Andhra Pradesh, Uttar Pradesh and Karnataka. The seed contains 35 per cent oil which is hard in nature. Total potential of mahua oil is estimated to be around 3.5 lakh tonnes. Mahua seeds yield a pale-yellow, semi-solid fat which is akin to tallow. The oil finds use in making washing soaps. As the oil is prone to develop rencidity, the oil is not used in making toilet soaps. The oil is used in wool mills for smoothening raw wool for easy spinning. The oil is used in the manufacture of candles, besides its use as a lubricant.

Suitably processed mahua fat can be used in confectionary and in chocolate-making; for edible cum cooking purposes.

Mahua cake contain saponin, a toxin, so it is unfit for use as cattle feed. Saponian prevents the decomposition and nitrification in soil, so use of mahua cake as a fertilizer is not advocated.

Mahur trees are found in Dehra Dun and on the Saharanpur Siwalika, Oudh, Bihar, Chota Nagpur, Orissa. The Central Provinces, Central India,

Gujarat, Konkan, North Canara, Southern Maharashtra country, Northern Circars, and the Deddan; largely planted in other parts and liable to run wild. It is apparently not wild in the Punjab, though as it has been a long time since it was introduced, it is apt to be taken as such Kangra it thrives on dry stony soil (Fig. 5.4).

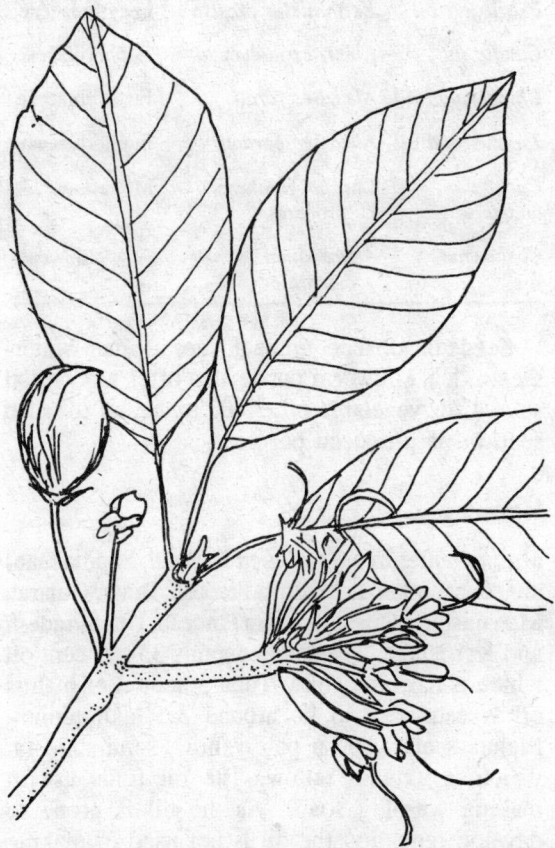

Fig. 5.4. Madhuka tree—*Madhuca indica.*

The tree is valued for its flowers, fruits, seeds, and timber. The oil extracted from the kernels of the fruit is largely used in the Central and South India for culinary and lighting purposes, and as adulterant of ghee. It is also used in the manufacture of soap and candles. The fruit is sometimes eaten, but chiefly it is the succulent flowers (corollas), which are eaten raw or cooked or made into sweetmeats. A case of dangerous vomiting with cerebral symptoms caused by eating an excessive quantity of these flowers is on record. In indigenous medicine the flowers are consid-ered as astringent, tonic and appetizing. The bark is also used as an astringent and as a tonic.

According to the Pharmacopoeia of India, "the spirit distilled from flowers has a strong smoky odour, somewhat resembling Irish Whisky, and rather a pungent foetid flavour. Which, however, disappears with age. The freshy-distilled spirit proves very deleterious, exciting gastric irritation and other unpleasant effects". "It is evidently a powerful diffusible stimulant, and when matured by age may be used as such, when breandy and other agents of the same class are not available".

After steeping the flowers in water and allow-ing them to ferment, a spirit is distilled, which is largely consumed by the inhabitants of the moun-tainous tracts of the tablelands of Central India. These flowers are considered to be a good and cheap raw material for the manufacture of alco-hol, and are now being extensively used for its production on a large scale in Bihar, Orissa, Bombay and Bengal.

KUSUM

Kusum (*Schleichera oleosa,* Sapindaceae) (Fig. 5.5) trees are found in the dry forests of Bihar, Orissa, Madhya Pradesh, Uttar Pradesh, Andhra Pradesh and West Bengal. Kusum trees serve as a host for lac insects. Seeds are collected before the onset of monsoon during May-June. Estimated potential of kusum seed oil is 6000 tonnes per year.

The oil content is 33 per cent of the weight of the seeds. Kusum oil contains certain cyanogenic compounds which emit poisonous fumes at high temperatures. The oil has therefore to be pro-cessed with extreme precaution and isolation.

Kusum oil is used in soup making, preparation of hair-dressings and in some medicines for skin diseases, rheumatism and headaches. In tribal areas, the oil is used in small quantities as a cooking medium and also as a fuel.

SAL

Sal (*Shorea robusta*) (Fig. 5.6) trees are found in abundance in Bihar, Orissa, Madhya Pradesh, West Bengal, Assam and Uttar Pradesh. There is a potential of 60 lakh tonnes of sal seeds. Sal kernels contain 12 to 13 per cent of hard oil which is greenish in colour. The oil is used in the manufacture of laundery soaps and in soap-making. *Sal fat* can be processed and used as cocoa butter extender in chocolates. Sal fat as a partial substitute for cocoa butter is being used in Japan, Switzerland, UK and Italy.

Cucurbit oil

The oils of ghia-tori, sponge gourd (*Luffa cylindrica*); kali-tori, ridge gourd (*Luffa acutangula*) and kaddu, pumpkin, (*Cucurbita moschata*) belong to the non-drying group with iodine values less than 100. The oils of chachinda, snake gourd (*Trichosanthes anguina*) and karda, bitter gourd (*Momordica charantia*) are characteristically drying oils with iodine values above 143. The chachinda

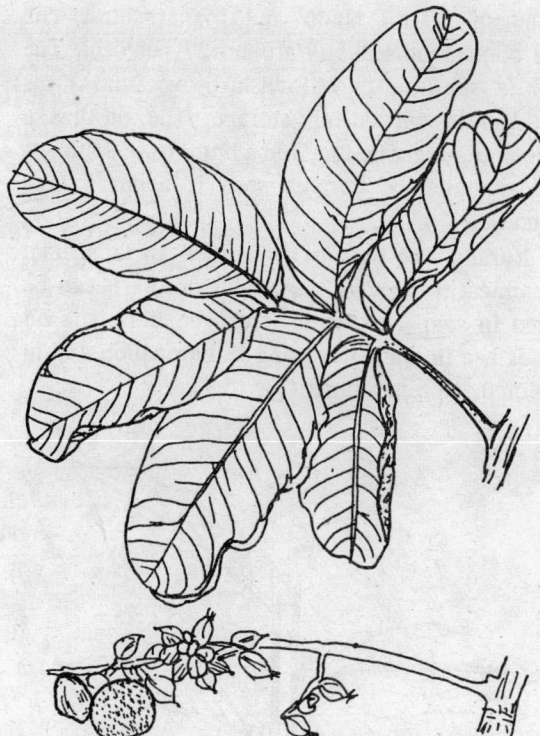

Fig. 5.5. *Schleichera trijuga* (Sapindaceae) Kusum, lac tree.

Sal *Shorea robusta*

Fig. 5.6. Sal, *Shorea robusta,* Dipterocarpaceae. Seed oil remains solid at room temperature; after processing, can substitute cocoa butter. The tree gives wood, and gugal resin.

and Karela oils could very profitably used for industrial purposes in the making of paints, varnishes and other products requiring drying oils as a base. The cucurbit kernal oil would be of particular interest to those having coronary and heart troubles, as the oils have a much higher degree of unsaturation than the conventional oils.

Karanja oil

Karanja tree *Pongania pinnata* (Fabaceae) are scattered in Andhra Pradesh, Gujarat, Madhya Pradesh, Tamil Nadu and Maharashtra. The oil content of seeds is around 27 per cent. The oil is semi-hard, yellowish brown and turns reddish brown during storage. The oil has a peculiar pungent odour and a bitter taste. Estimated, potential of karanja seed is about 5 lakh tonnes.

Karanja oil has to be refined by a special treatment to remove the colour before it can be used in soap making or hydrogenation. The oil finds use in leather tanning in lubrication and in medicinal preparations (Fig. 5.7).

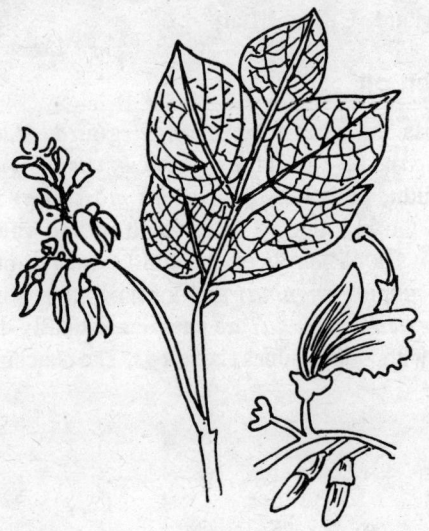

Fig. 5.7. Karanja *Pongamia pinnata* (Fabaceae).

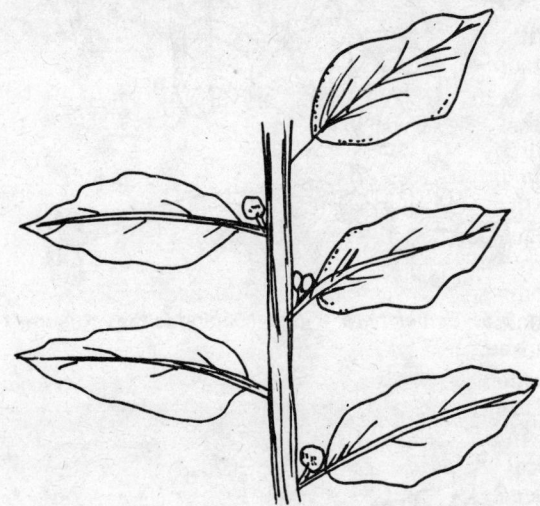

Fig. 5.8. Pilu, *Sal vadora persica*. Seeds yield a yellow non-edible oil used in local medicine and soap-making.

6

Sugar and Starch Crops

INTRODUCTION

Although sugar (in the form of sucrose) is manufactured by all green plants, it is only extracted on a commercial or substantial scale from a relatively small number of species, the most important of which are sugar cane (*Saccharum officinarum*) and sugar beat (*Beta vulgaris*). Starch is likewise produced by the vast majority of green plants, and all cereals as well as some vegetables provide us with starch. Pure starch is extracted commercially from a number of species for nutritional purposes, the most important being arrow root (*Maranta arundinacea*), cassava (*Manihot esculenta*) and sago palms (*Metroxylon rumphii, M. sagu* and other genera).

In several tropical countries, commercial and local supplies of sugar are obtained from various species of palms, including the date palm (*Phoenix dactylifera*), the sugar palm (*Arenga pinnata*) and honey palm (*Jubaca chinensis*). A number of thick-stemmed strains of *Sorghum* are cultivated for their sweet cane known as *sorgo* and may be used for chewing or for making syrup or sugar. The North American sugar maple (*Acer saccharum*) and black maple (*A. nigrum*) are sources of maple syrup.

Important *sugar crops* are sugarcane (*Saccharum officinarum*), sugar beet (*Beta vulgaris*), sugar maple (*Acer saccharum*), black maple (*Acer nigrum*), barley (germinating seeds) (*Hordeum vulgare*), sweet sorghum (sorgo), *Sorghum bicolor*, wild date palm (*Phoenix sylvestris*), palmyra palm (*Borassus flabellifer*), sago palm (*Caryota ureus*) (Fig. 6.1), coconut palm (*Cocos nucifera*), sugar palm (*Arenga pinnata*), honey palm (*Jubaca chinensis*), nipa palm (*Niga fruticans*) and manna ash (*Fraxinus ornus*), (Oleaceae).

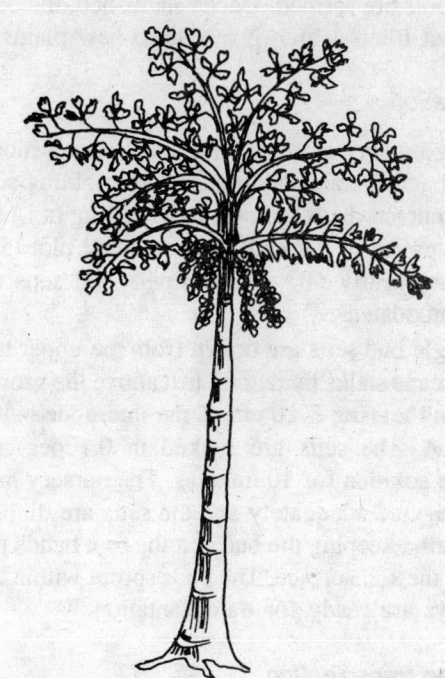

Fig. 6.1. Sago palm—*Caryota urens* tree.

SUGARCANE

Sugarcane, *Saccharum officinarum* is a tropical grass belonging to the tribe Andropogonee of Poaceae. Sugarcane had 2 geographical centres of origin, New Guinea and north India. The large barrelled tropical species *S. officinarum* probably originated from wild species of *S. robustum* in

New Guinea. The north Indian sugarcane *S. sinense* and *S. barbari* are believed to have originated by natural hybridization between the migrating forms of *S. officinarum* and wild *S. spontaneum.*

Modern sugarcane is a complex hybrid of *S. barberi, S. officinarum, S. sinense* and *S. spontaneum.* Many forms of these species interbreed, making it into a highly diverse genus.

All sugarcane is propagated vegetatively by means of *setts.* A *sett* is a part of the stem which include lateral buds and a circle of cells which give rise to a number of adventitious roots. The initial adventitious roots absorbs nutrients until new roots are formed. Once established, the cane puts out tillers which give rise to new plants.

Cultivation

Setting nursery is raised in a small area a month before actual transplanting. Before dibbling setts in the nursery beds, gamma BHC at 1 kg (a.i.) per ha is applied to the soil. In each small plot (5 m x 10 m) nearly 600 to 800 single bud setts are accommodated.

Single bud setts are drawn from the upper half of the cane stalks by cutting just above the growth ring and leaving 8-10 cm of the internode below the bud. The setts are soaked in 0.1 per cent Arctan solution for 10 minutes. The nursery beds are irrigated adequately and the setts are dibbled vertically, keeping the bud and the root bands just above the soil surface. The buds sprout within 20-30 days are ready for transplantation.

Spaced transplanting

Transplanting settlings at a distance of 90 cm between rows and 60 cm within the rows is recommended. 19,000 settlings are needed for transplanting in one hectare.

One of the means for improving productivity would be by minimising seed rate; so, spaced transplanting is useful in increasing yield.

The life cycle of a 12 month sugarcane crop can be divided into three phases : *Formative period*, first 120 days; *Grand period of growth*, from about 150th to 270th days; *Maturity phase*, from about 300th to 360th day. Nitrogen application should be completed by 3 months after planting. The plant needs N 120-250 kg/ha, P_2O_5 kg/ha, K_2O 75-135 kg/ha. N may be applied by single or split applications, as per practice.

Sugarcane is a long-duration crop. Autumn planted sugarcane is sub-tropical India yields more millable canes followed by spring. Early harvested cane as also early maturing varieties in many instances give lower tonnage because of short duration.

Varieties

Co-419 (parentage POJ 2878 x Co 290) is very popular in the tropical regions. At one time (1952-53) it occupied 96 per cent of the area in the tropical region. Now, although a number of superior location - specific varieties have been developed and recommended, the variety Co 419 still continues to be grown in tropical India.

Some of the high yielding varieties are BO-11, CO-622, CO-712, CO-658, B-14, HM-320, CO-513, CO-527, CO-321. Disease resistant varieties are CO-331, CO-213, CO-349. BO-11 is cultivated in Uttar Pradesh and Bihar. CO-740 is cultivated in Maharashtra. CO-453 is cultivated in Haryana and Punjab.

Uses

Most of the sugarcane is utilized for the manufacture of jaggery (gurs, white sugars (vaccum pam sugar) and khandsari sugar (open pan sugar). Bagasse is used as a fuel in the manufacture of sugar or gur. It is also used in the manufacture of paper.

SUGAR BEET

Sugar beet together with beet root, mangolds or mangel-wurzels and fodder beet are all cultivars of *Beta vulgaris* subspecies *vulgaris.* As a sugar crop plant, beet dates from the mid-18th century when the presence of sugar in the sap of fodder beet was noted. This led to the selection of improved strains and the beginnings of the sugarbeet extraction industry in Silesia. Sugarbeet is produced mainly in temperate climates, especially Europe, the USSR and North America, unlike cane sugar which is a crop of tropical and subtropical climates (Fig. 6.2).

Beet root

Fig. 6.2. Beet root (*Beta vulgaris*).

Talin

Thaumatococcus deniellii (Marantaceae) is a source of *talin* which is a natural sweetning agent. *Talin* is produced by the West Africa and is already in limited use in Japan. *Talin* is a sugar substitute. It is many times sweeter to the taste than sucrose, but being a protein it is coagulated on boiling and loses its sweetness as a result; thus it is for use only in foods that are not to be cooked.

Stevia leaves

Stevia (Asteraceae) is a perennial tropical plant. *Stevia* leaves contain substances which are several Lundred times sweeter than saccharins : 100 kg of stevia is equivalent to 600-700 kg sugar. The sugar can well be used to make soft drinks and tea. A technology for growing *Stevia* in Georgia (USSR) and producing a sugar substitute from it for low calory and dietetic dishes has been developed.

Starch crops

Starch is obtained commercially from the stem tubers of potato (*Solanum tuberosum*) and from the root tubers of cassava (*Manihot esculenta*). In addition, a number of aroids are important as food crops in the tropics because of their edible rhizomes which contain large amounts of starch, such as the taro (*Colocasia esculenta*), dasheen (*C. esculenta* var. *globifera*) and tanier (*Xanthosoma* species) (f. Araceae). They rarely enter commerce except through local markets although their potential importance is great since their starch is easily digested and suitable for children. Several yams (*Dioscorea* species) are a source of starch as are various palms, notably the sago and sugar palms.

LESSER YAM (*DIOSCOREA ESCULENTA*)

Dioscorea alata (Dioscoreaceae), yam bean, sweet potato, elephant foot yam, cassava, xanthosoma, arrowroot, coleus, winged bean, potato are the tuber crops of today. In Kerala cassava is grown on a large scale, followed by Tamil Nadu and Andhra Pradesh. Sweet potato is the most important crop of Bihar.

Important *starch crops* are potato (*Solanum tuberosum*), cassava (*Manihot esculenta*), arrowroot (*Maranta arundinacea*), Queensland arrowroot (*Canna edulis*), taro (*Colocasia esculenta*), giant taro (*Alocasia macrorrhiza*), dasheen (*Colocasia esculenta* var. *globifera*, giant swam taro (*Cyrtosperma chamissonis*), cocoyam (*Xanthosma atrovirens*, *X. sagittifolium*, *X.*

violaceum), East Indian arrowroot (*Curcuma angustifolia*), Fijian arrowroot (*Tacca bontopetaloides*), greater asiatic yam (*Dioscorea alata*), white guinea yam (*D. rotundata*), yellow guinea yam (*D. cayenensis*), air potato (*D. bulbifera*), cush-cush yam (*D. trifida*), sagopalm (*Metroxylon rumphii*), gonuti palm (*Arenga pinnata*), American cabbage palm (*Oreodoxa oleracea*), kaffir bread (*Encephalartos caffer*), bread tree (*Encephalartos altensteinii*), queen sago (*Cycas circinalis*), Japanese sago palm (*Cycas revoluta*), maize (*Zea mays*), wheat (*Triticum species*), rice (*Oryza sativa*).

Encephalartos is the second largest genus of Cycads distributed in tropical and southern Africa. Meal can be prepared from the *starch-rich pith tissues of the stem* of a number of species, including *E. caffer* and *E. altenstenii*. *Cycas* is a genus of palm-like gymnosperms widely distributed from Madagascar to northern Australia, Polynesia and Japan. The pith of *C. circinalis* and *C. revoluta* yields a type of sago.

Edible flour is prepared from the starchy tubers of *Codonopsis ovata*, in the Himalayas. Tubers of *Hitchinia caulina*, *Dioscorea hispida* are processed to extract starchy contents. The dried tubers of *Cyperus bulbosus* are pounded into flour and baked into bread at even cooked as pudding. Rhizones of *Limnanthemum*, *Sagittaria* and *Nelumbo* are eaten cooked.

The tubers of *Dioscorea, Alocasia, Colocasia, Vigna, Moghania, Ceropegia, Alpinia, Curcuma, Zingiber* are consumed in various ways. The tubers of *Vigna capensis, Moghania tuberosa, Moghania vestita, Eriosema chinense, Peucedanum dhana* var. *dalzellii* are eaten raw. Tubers of *Decalepis, Coleus* and *Curcuma* are pickled. Tubers of *Asparagus racemosus* var. *javanicus* are candied. Preserves are prepared from the rhizomes of *Costus speciosus*. Seeds of *Alpinia galanga*, leaves of *Murraya Koenigii* and rhizomes of *Zingiber* species are used as indirect foodstuffs as favourable additions to dishes in curries etc. mainly.

Tubers of *Cyperus rotundus* are eaten in Rajasthan. *Cyperus bulbosus* tubers are eaten in the peninsular India. The powder of the bulb is used with flour of bajra or *jowar*. Tubers of *Curcum a angustifolia, Codonopsis ovata, Aponogeton nataus, Aponogeton crispum, Phaseolus adenanthus*, fusiform roots of *Borassus flabellifer*, tubers of *Costus speciosus, Balanites aegyptiaca, Hyphaene thebaica, Plesmonium margaritiferum, Scirpus grossus* are eaten in various parts of India.

Alocasia macrorhiza, Amorphophallus campanulatus, Colocasia esculenta, Curcuma amada, Cucurma zedoaria, Tacca leontopetaloides, Tulipa stelleta, Zingiber zerumbet, Scirpus lacustris, Dioscorea esculenta, Dioscorea alata are cultivated on a small scale.

POTATO

Potato (*Solanum tuberosum*, Solanaceae) has high yielding potential per unit area and time and because of its high nutritional value, it has attained a great importance in the national economy. Potato is a labour-intensive crop and provides useful employment in rural and urban areas.

The cultivated potato and several less well known tuber-bearing *Solanum* are natives of South America. Today, potatoes are grown throughout temperate parts of the world and in upland tropical areas. Planting is almost exclusively by seed tubers or by planting portions of the tubers which contain dormant buds; in potato, seeds are almost always nonfertile, so pieces of tubers are used for planting (Fig. 6.3).

For optimum yield potato requires 120 kg N, 80 kg of P_2O_5 and 100 kg of potash in the plains of India. Half of the N and entire quantity of P & K is to be applied at the time of planting and rest half of N at earthing. In hills, 100 kg N, 100 kg P_2O_5 and 100 kg potash should be used. However, in fields where exhaustive crops like maize and paddy were taken in kharif, the dose of N should be applied at 150 kg/ha. Application of formyard manure @ 30 m.t./ha in furrows in combination with 120 kg N gave the highest yield (Fig. 6.4).

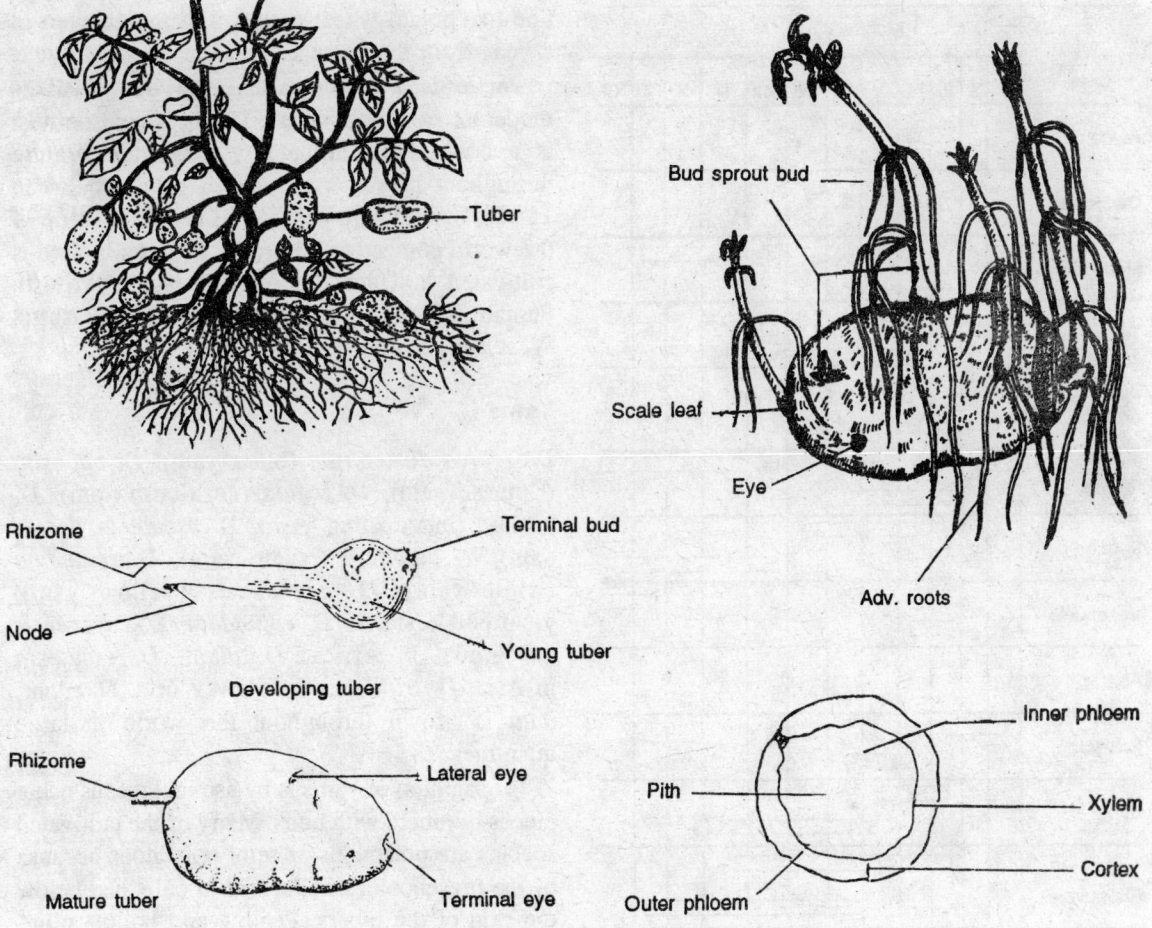

Fig. 6.3. Potato (*Solanum tuberosum*) plant, tuber etc.

Dormancy breaking

Potato tubers harvested in autumn season remain dormant for 2-3 months and are not suitable for immediate planting in winter. Dormany of tubers can be broken by the use of ethylene chlorohydrine thiourea and gibberellic acid.

Varieties

Kufri Chandramukhi : Early bulking food tuber characters; an ideal variety for multiple cropping; suitable for planting in hills, plains and plateau areas.

Kufri Alankar : Late blight immune and rapid **bulking**; an ideal variety for late planting after **paddy**; in north-western plains.

Kufri Sheetman : *Frost tolerant* main crop variety; north-western plains.

Kufri Chamatkar : Good cropper; *white round tubers*, main crop variety; north plains.

Kufri Sindhuri : Good cropper, *red tubers,* matures late; north central and eastern plains.

Kufri Dewa : Good cropper, *pink splesh tuber, frost tolerant,* matures late; suitable for central plains and tarai region.

Kufri Jyoti : Late blight resistant, *ward immune,* relatively early; suitable for hills and West Bengal, Karnataka, Gujarat.

Kufri Muthu : Late blight resistant, round tubers; Nilgiri hills.

F-5242 and F-3977 are wart-resistant hybrids suitable for cultivation in Darjeeling etc. Weed

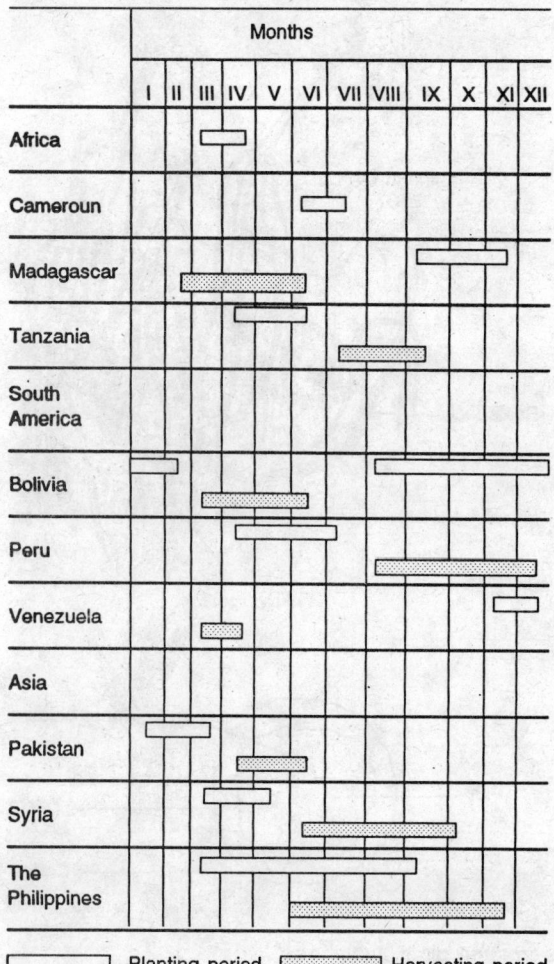

Fig. 6.4. Terms of planting and harvesting potato in various countries (FAO, 1978).

control : Pre-emergence application of Lasso @ 3 litre/ha, and post-emergence application of Stam F-34 @ 2.5 litre/ha and Gramoxone @ 1.5 litre/ha can control the weeds effectively.

Frequent light irrigations required for good yields in plains. In hills, a few irrigations during the drought period of April-June can substantially increase the tuber yield.

The yield of early autumn crop in plains can be increased by the use of *mulch* such as rice husk, paddy straw etc., during germination in plains and pine needle in hills.

Kufri chandramukhi variety fits well in a rota-tion like potato-wheat-maize, 3 different crops in a year from the same field.

Potato is a cool season crop that tolerates moderate frost conditions. The best environment is a cool, moist climate; average temperature throughout the growing season ranges between 15°-18°C. About 90 per cent of the potato crop of the world comes from Europe. In India, potato is cultivated in Uttar Pradesh, Himachal Pradesh, Punjab, Madhya Pradesh, Bihar, Maharashtra, West Bengal etc.

Yams

Dioscorea dumetorum (bitter yam), *D. opposita* (Chinese yam), *D. trifida* (cush-cush yam), *D. hispida* (intoxicating yam), *D. esculenta* (lesser yam), *D. bulbifera* (potato yam), *D. rotundata* (white yam), *D. cayenensis* (yellow yam), yield edible starch. *D. rotundata, D. cayenensis* are grown in Africa; *D. alata, D. esculenta* in Asia, *D. trifida* in the New World. *D. rotundata* is grown throughout the world in large quantities.

Propagation of yams is by asexual means using pieces of tubers with buds. Many of the cultivated species are poisonous or semi-poisonous because of the presence of oxalic acid in cells just below the skin of the tubers. Peeling and boiling elim-inates most of the potential problems caused by the crystals.

Tapioca (Cassava)

Tapioca (*Manihot esculenta,* Euphorbiaceae) is a tuber crop. Tubers are rich in carbohydrates and serve as a subsidiary food for the poorer classes and also as a vegetable. Starch, sago, semolina and flour are also made out of these tubers.

Ecology

A crop of the tropics and sub-tropics, it can be grown up to an elevation of 800 m above mean sea level. Open sunny situations are best suited. The crop is resistant to drought to a certain extent. A deep, rich and friable soil is desirable for good formation of tubers. Popular varieties are M-4, H-165, H-226 and H-97 (Fig. 6.5).

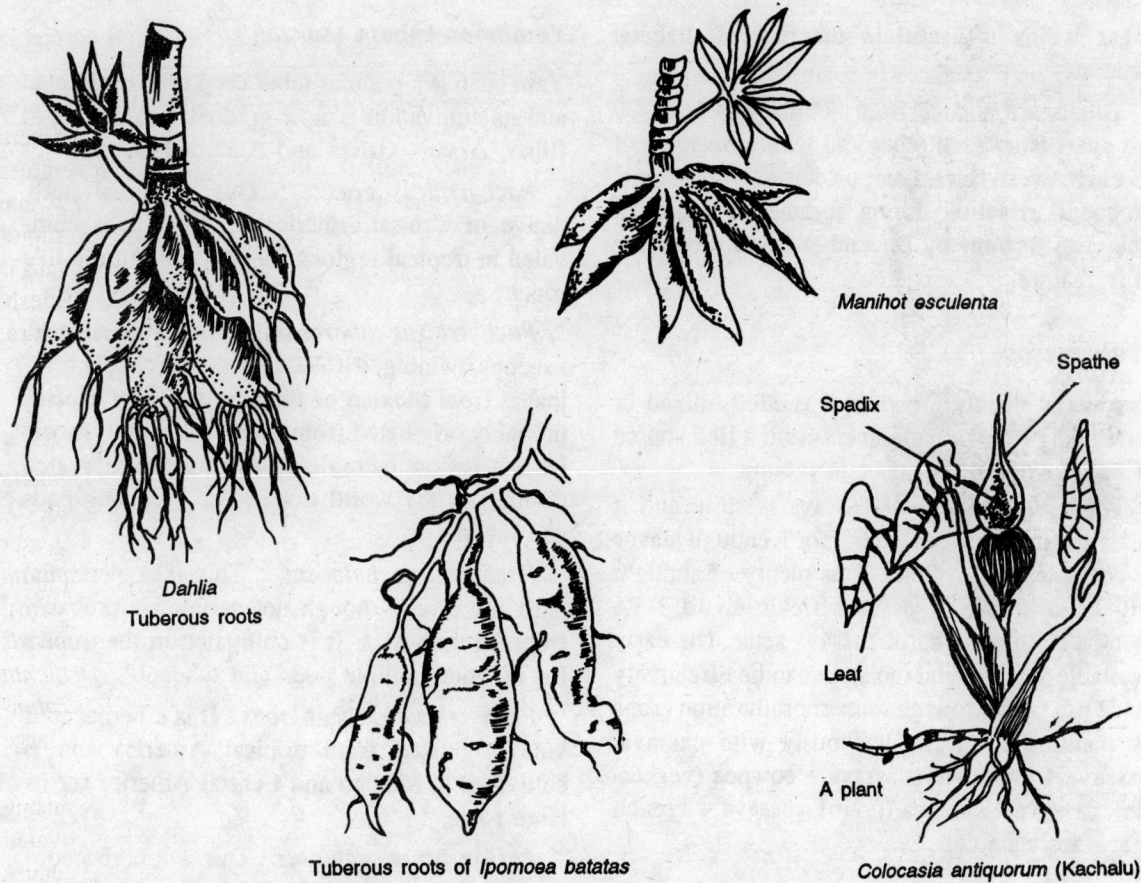

Manihot esculenta

Spathe

Spadix

Dahlia
Tuberous roots

Leaf

A plant

Tuberous roots of *Ipomoea batatas*

Colocasia antiquorum (Kachalu)

Fig. 6.5. Tuberous plants—*Dahlia, Manihot, Colocacia* and *Ipomoea* species.

Cultural practises

It is propagated by *stem cuttings*. The crop though perennial is treated as an annual. Setts of 15 to 22 cm length are obtained from the stems of the previous crop which are stored in shade for three months. The top tender portions are rejected. The setts are planted in a *slanting position* burying the greater part of them. A spacing of 45 cm either way or 60 x 45 cm is adopted. 12,500 to 18,000 setts are required per hectare. The crop is at its best during June and December-January.

A basal dose of 10 to 15 tonnes of compost or farmyard manure is applied and then a top dressing of 125 kg per hectare of ammonium sulphate.

Ratalu

Dioscorea alata L. (*Dioscorea esculenta*) are native to the region extending from India to Malaya. These plants are grown for the fleshy roots which are edible.

The upper portion of the main root and side roots are planted about 60 cm apart in a deeply dug, well manured field. Planting can also be done in shallow trenches. Frequent irrigation is required during summer.

Jerusalem artichoke (*Helianthus tuberosus* L. *Compositae*)

The plant is native to North America. Its **tuberous** roots are used as a vegetable. The tubers contain *inulin*. Inulin on hydrolysis yields levulose, a sugar which is useful in the diet of diabetic patients.

Tubers are planted from February to May, 30 cm apart from each other and with a distance of 75 cm between rows. Later on earthing up is done. Frequent irrigation during summer is required. The crop matures by December when tubers are fully mature.

Intercropping

Cassava is mainly grown as a rainfed, mixed or rotational crop. Though tubers form a rich source of carbohydrates, but poor in protein.

The initial growth of cassava is slow and it takes considerable time to put forth enough leaves to cover the entire field. Thus plenty of sunlight will be available in cassava fields up to 3-3½ months from planting of cassava setts. The early available sunlight and moisture can be effectively made use of by growing some short duration crops by planting them simultaneously with cassava. Cassava + groundnut, cassava + cowpea (vegetable), cassava + cowpea (grain), cassava + French beans are suitable.

Tubers eaten by tribals

Alocasia macrorrhiza, Alpinia galanga, Alpinia spiciosa, Asparagus racemosus, Bupleurum falcatum, Colocasia esculenta, Costus speciosus, Curcuma zedoaria, Dioscorea bulbifera, D. hispida, D. puba, D. versicolor, D. oppositifolia, D. pentaphylla, Eleocharis dulcis, Eriosema chinense, Eulophia campestris, Flemingia vestita, Houttuynia cordata, Lasia spinosa, Nelumbo nucifera, Nymphoides indicum, Oxalis martiana, Polygonatum multiflorum, Polygonatum verticillatum, Polygonum bistosta, Sagittaria sagittifolia, Tacca lentopetaloides, Vigna vexillata, Zingiber zerumbet.

Tubers of *Flemingia vestita* (soph-long) are eaten row by Himalayan tribals. It is grown as a rainfed crop in jhums in Khasi and Jaintia hills.

Yam bean tubers (*sakalu*)

Yam bean is a popular tuber crop of West Bengal and its cultivation is now gradually extending to Bihar, Assam, Orissa and Andhra Pradesh.

Pachyrrhizus erosus : This perennial herb, native of Central America or Mexico, is cultivated in tropical regions for its young turnip-like roots.

Pachyrrhizus tuberosus : This perennial, herbaceous, twining, trifoliate species probably originated from Mexico or twining, trifoliate species probably originated from Mexico or South America but is now naturalized and widely cultivated throughout the world tropics for its young pods and tubers.

Pachyrrhizus bulbosus : This is a perennial herb. Its origin, though not certain, is probably from tropical Asia. It is cultivated in the tropics for its young edible pods and tubers.

Pachyrrhizus palmatilobus : It is a herbaceous vine originated from tropical America and is cultivated in Mexico and Central America for its large roots.

Pachyrrhizus angulatus : This is a herbaceous vine originated from Malaya. It is not known in a wild state but is cultivated more or less widely throughout the tropics for its tuberous roots.

The crop is sown in the end of June or beginning of July. The seed can be sown on flat soil or in ridges. Ridge sowing ensures better tuber yields. The seeds can be sown in rows spaced at 50 cm with the help of seed drill or by dropping the seeds in furrows opened by *bukkhar*. Plant to plant distance of 15 cm is preferable. In ridge sowing, the seeds are dibbled on top of the ridge or on both sides of the ridge at 15 cm spacing. The crop does not require any irrigation during the rainy season. The crop should be kept weed-free as far as possible to have maximized production.

The harvesting of the crop depends on the purpose for which the crop is grown. *Picking of green pods* for vegetable purpose is done from time to time when pods are young and tender. For *fodder*, the crop is harvested at 50 per cent flowering stage to ensure high yields and better nutrition. The *tubers* are dug when the plants start showing signs of drying. Harvesting of *grain* is done when the crop attains maturity after 6-7 months of sowing.

Uses

The tubers are a substitute for yams and are a good source of pure white starch which is palatable and is used for custard and puddings. The tubers are also used as vegetable when cooked. They can also be eaten raw as *salad* and can be used for making pickles and *chutneys*. A tasty *kheer* can be prepared from tuber gratings when boiled in milk.

Young tender pods are eaten as vegetable when cooked.

7

Economic Botany of Flowers

Flowers have been grown in India since early days and their use forms a part of heritage and culture. The world flower consumption today is around 8750 million of which Federal Republic of Germany's share works out to 1500 million - 17 per cent. Flowers suited for export are rose, *Gladiolus,* carnation, *Chrysanthemum* and orchids (Fig. 7.1).

Orchids command the highest demand in the international market. Through orchids made India their home for millions of year, the country's performance as an exporter has been abysmally poor. Even a small country like Thailand is far ahead of India in orchid exports. In India, orchids mainly grow in Himachal Pradesh, North Bengal (Mirik, Kalimpong, Siliguri, Darjeeling and Kurseong), Meghalaya, Arunachal Pradesh and Sikkim. India exports orchids to United States, Netherland, Australia and Switzerland. Orchid exports from India may get a major boost by following tissue culture, biotechnology and growing of orchids in cymbidiums. The Botanical survey of India, forest department of West Bengal, Orchid research laboratory of Arunachal Pradesh and Tata Energy Research Laboratory, Haryana have established that the Indian orchids are the best in the world (especially the Himalayan orchids).

The Government has sanctioned various assistance like air carriage concessions, cold storage facility, speedy clearance from plant quarantine, fumigation centre, fax benefits, Roses, spray carnations, gladioli and ruscus greens are impor-

tant items of the flower business today which is a $ 20 thousand million industry. There are about 1,500 varieties of roses in red, orange, pink, yellow, mauve and white, with long stems and short, large blossoms, small blossoms and mini-multiples. Black tulips, *Agapanthus*, Zephyranthus, chrysanthemums, daisies, carnations and freesias are most popular in Europe. Spain, France and Italy export carnations to northern Europe. Israel, the world's third biggest flower exporter and in 1990 shipped over 800 million stems to Europe, earning $ 135 million. Columbia is today the world's second largest exporter and export roses, daisies, alstroemerias, chrysanthemums and carnations. Thailand, Singapore, Taiwan, Australia and New Zealand export tropical flowers to Europe, United States and Japan.

Flowers are a perishable product. Pre-cooling techniques can lower the temperature of newly picked and packed flowers to 2°C in 30 minutes by pushing cool, humid air through holes cut in the ends of the boxes of flowers. Cold-storage warehouses and refrigerated trailor trucks can hold the temperature constant.

Today the Dutch control over 65 per cent of the world export trade. The Dutch grow their flowers in artificial substances like foam, rock wool, perlite - any inert material that can take water and air.

India exports Rs. 2.5 crores worth (1990) flowers, roses, gladioli, carnations, chrysanthemums, golden rod, amaryllis and jasmine to the Gulf countries, Germany, France, England and Singapore.

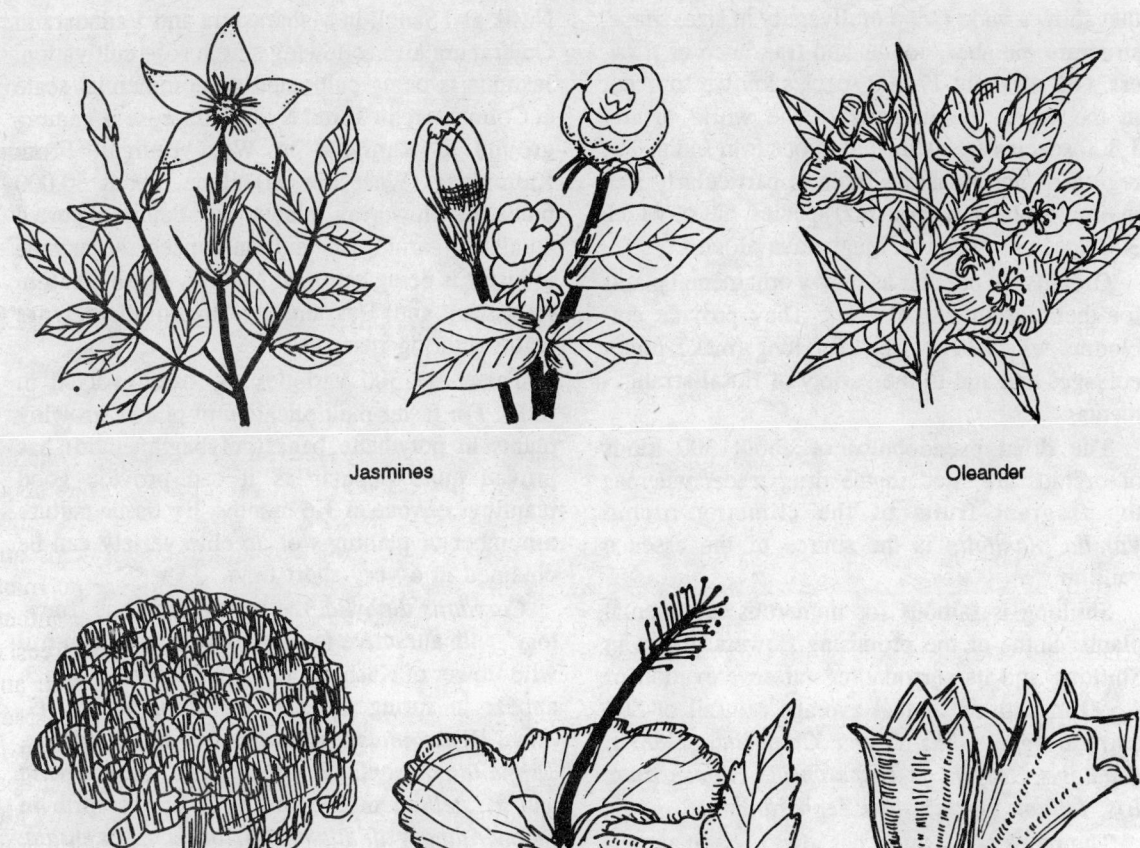

Jasmines

Oleander

Chrysanthemum

Shoe flower

Artabotrys

Fig. 7.1. *Flowers*—Jasmins, *chrysanthemum, Hibiscus,* and *Artabotrys.*

Australia exports *Gypsophila* (baby's breath) kangaroo paw (a velvety, russet brown bloom on a slender long stalk) to Sydney and Japan. The Japanese prefer orange and yellow varieties of this flower. These plants do well in a sandy red soil of central Australia and do not require great quantities of water. The current flower rate value is between 150 and 300 million Australian dollars.

Today there are almost 300 chrysanthemum varieties. World production is nearly 4000 million stems a year and top producers are Japan, the Netherlands, Colombia and Italy.

The flowers of the orchid *Dendrobium maccarthial* are considered to be the lucky flowers in Sri Lanka, and used as a temple offering at a Buddhist festival during May. The flowers are, therefore, locally called 'Wesak mala' (May flower). The fruits of some species of the orchid *Vanilla* are the source of deliciously-scented Vanilla beans. In Mauritius, tasty "Laham tea" is prepared from the dried leaves of the orchid *Angraecum fragrans.*

Orchids are a unique group of flowering plants that show a wide range of diversity in size, shape, structure, number, colour and fragrance of flowers. Out of about 17,000 species known to occur in the warm humid parts of the world, nearly 1,300 species are estimated to occur in India. The region of north-eastern India is particularly rich in orchids, having about 600 species, out of which 300 species occur in Meghalaya alone.

Orchids are popular as showy ornamental plants for their fascinating flowers. They provide cut-blooms which keep fresh for long, make pretty corsages and add to the variety of floral arrangements.

The dried pseudobulbs of about 300 kinds of orchids are used in the drug trade, whereas the fragrant fruits of the climbing orchid *Vanilla planifolia* is the source of the essence vanillin.

Shillong is famous for numerous ornamental plants. Some of the promising flowers grown in Shillong and its surrounding (at an elevation of 1,500 m with an annual average rainfall of 800 cm) are *Agapanthus, Canna, Crocosinia, Dahlia, Gladiolus, Hedychium, Hemerocallis, Hippeastrum, Iris, Lilium, Tigridia* and *Zephyranthes.*

The name of Kashmir has always been associated with beauty and meadows of wild flowers. *Anemone biflora, Colchicum luteum, Corydalis diphylla* and *Sternbergia fischeriana* are the earliest harbingers of spring as they push out their blooms through snow in February.

Under the All India Coordinated Floriculture Improvement Project a centre was established at Shillong in 1973. Since then, work was undertaken mainly on orchids and bulbous plants like *Iris* and *Lilium.* 50 native species of orchids from north-eastern region mostly from the genera like *Aerides, Arachnanthe, Arundina, Coelogyne, Cymbidium, Dendrobium, Renanthera, Rhynchostylis, Vanda, Paphiopedilum* and *Phiaus* are commercial ornamental plants.

Flowers have a vast export market. At Pune, roses are being grown in place of grape vineyards, not just for aesthetic pleasure but for strong economic considerations. Cut flowers are being sent to the nearby Bombay market and are exported to the adjacent Middle East countries. Nasik and Sangli in Maharashtra and Vadodara in Gujarat are also following suit in rose cultivation. Jasmine is being cultivated on commercial scale at Coimbatore in Tamil Nadu. Tuberose is gaining ground at Ranighat in West Bengal. From Kalimpong (West Bengal) alone, about 50,000 bulbs of flowering plants are being exported annually, earning 45 million rupees. A jasmine industry is being set up at Madras. Haldi Ghati in Rajasthan and Hassain in Aligarh district are manufacturing rose oil.

There are 300 varieties of roses evolved in India. For faster multiplication of plants, growing plants in polythene bags (polybag method) has proved quite popular as it can provide good plantings of rose in 3-6 months. By tissue culture a number of plantings of an elite variety can be obtained in a very short time.

Corydalis diphylla, the purple tuberous 'fumitory' with attractive fern-like leaves is a beautiful wild flower of Kashmir. Other wild flowers which appear in spring are *Gageas kashmirensis, G-lutea, Hyacinthus orientalis, Narcissus tazetta, Tulipa aitchinsonia, T. clusiana* var. *cashmiriana* and *T. stellata* in March. April brings forth in bloom *Allium griffithianum, A. loratum, Iris ensata, I. Kashimiriana alba, I. kharput, Notholirion thomsonianum,* and *Tulija lanata. Tulija lanata* is the largest tulip in the world, which measures 22 cm across. In May bloom *Eremurus himalaicus, Fritillaria roylei, Iris kashmiriana purpurea, Paeonia emodi. Iris aurea, I. notha* and *Lithium polyphyllum* blown in June.

The seeds and corns of *Colchicum autumnale* is an old well known remedy in Ayurvedic and Unani medicine for curing gout and rheumatism *Gagea kashmirensis, star-of-Bethlehem* is a beautiful wild flower. *Stenbergia fischeriana,* the 'winter daffodil' or the Biblical 'lily of the field' displays its large, golden flowers soon as the snow melts.

Iris kashmiriana alba has highly scented cream-coloured flowers. *I. kashmiriana purpurea* has dark purple flowers. These irises possess medicinal value and are sources of commercial 'orris root'.

Notholirion thomsonianum grows wild in Kashmir from 750 m to 2100 m altitude. The plant has funnel shaped sweet scented, pale rose or rose purple flowers with recurved types and yellow anthers.

Annuals suitable as cut flowers

Ageratum houstonianum, Anchusa capensis, Angelonia grandiflora, Arctotis sp., *Browallia elata, Cacalia* sp., *Calendula officinalis, Callistephus chinensis, Celosia plumosa, Centaurea cyanus, Centaurea moschata, Chrysanthemum, Clarkia elegans, Cleome spinosa, Digitalis purpurea, Felicia bergeriana, Gypsophila, Gaillardia, Gerbera jamesonii, Helipterum roseum, Helianthus, Iberis, larkspur, Linum, Linaria, Lupins,* marigold, *Nasturtium, Petunia,* Phlox, *Rudbeckia, Scabiosa,* sweet pea, *Salvia,* statics, tulip, *Verbena, Viola, Zephyranthes, Zinnia, Angelonia* are scented.

Winter flowering annuals

Sowing September-October in plains or March to April in hills. *Acroclinium roseum, Althaea, Alyssum, Antirrhinum,* asters, *Brachycome iberidifolia, Calceolaria pinnata, Calendula,* candytuft, carnation, *Centauria imperialis, Clarkia elegans, Delphinium, Godetia, Helichrysum,* sweet pea, *Linaria, Linum grandiflorum,* lupins, *Nemesia strumosa,* pansy, petunia, poppy, shirley, *Rudbeckia bicolor, Salvia, Statice sinuata, Cosmos* etc.

Summer and monsoon flowering annuals

Sowing October to November in hills; April to May in plains. *Amaranthus,* balsam, *Celosia cristata, Cosmos bipinnata, Gaillardia grandiflora, Gomphrena globosa,* sunflower, marigold, *Kochia, Zinnia* etc.

Summer flowering bulbous plants

Cooperanthes; Copperia drummondii, C. oberwettii, C. pedunculata; Haemanthus multiflorus, Hymenocallis speciosa, Narcissus, Spreklia formosissima, Nerine flexuosa, Zephyranthes carinata, Alocasia, Iris tectorum, Moraea iridioides, *Watsonia roses, Chlorophytum, Dianella intermedia Hemerocallis, Kniphofia folises,* water lily, *Alpinia sanderae, Cyclamem, Amaryllis, Begonia, Acidanthera bicolor.*

Winter flowering bulbous plants

Nymphaea, Acidenthera, Dahlia and *Canna.*

Spring flowering bulbous plants

Amaryllis, Eucharis, Hippeastrum, Leucojum aestivum, Narcissus, Phaedranassa schizantha, Vallota hybrida, Zantedeschia aethiopica, Dahlia, Dicaryum pendulum, Freesia, Gladiolus, Ixia, Iris, Sparaxis glandiflora, Watsonia, Agapanthus africanus, Allium, Asparagus sprengeri, Chlorophytum, Fritillaria, Hyacinthus orientalis, H. azureus; Lilium bulbiferum, L. longiflorum; Muscari, Ornithogalum, Oxalis, Anemone, Ranunculus. Rainy season flowering bulbous plants: Hedychium flavium, Alpinia sanderae, Mirabilis jalapa, water lilies, *Hemerocallis flava, Gloriosa superba, Caladium, Zephyranthes carinata, Alocasia, Polianthes tuberosa, Crinum.*

Fragrant flowers of bulbous plants

Amaryllis belladonna; Crinum americanum, C. deflxum, C. latifolium, C. longifolium, Eucharis grandiflora, Hymenocallis speciosa, H. calathina, Narcissus jancuilla, N. poeticus, N. pseudo-narcissus, Polianthes tuberosa, Iris kashmiriana, Ixia odorata, Moraea iridioides, Canvallaria majalis, Hemerocallis flava, Hosta plantaginea, Hyacinthus orientalis, Lilium candidum, L. wallichianum, Ornithogalum aravicum, Tulbaghia natalensis, Nymphaea lotus, Alpinia sanderae, Hedychium coronarium, Kaempferia rotunda. Shrubs with attractive flowers and foliage : *Acalypha wilkesvana.*

Fragrant shrubs

Artabotrys uncinatus, Azalea, *Brunfelsia latifolia, Cassia artemisioides, Cestrum nocturnum, Cytisus scoparius, Dombeya natalensis, Gardenia jasminoides,* **Hamiltonia suaveolens**, *Jasminium* *sambac, Nerium indicum, Osmanthus fragrans, Rondeletia odorata.*

Climbers with scented flowers

Aganosma caryophylla, Chenomorpha macrophylla, Hiptage madablota, Jasminum grandiflorum, J. officinale, Trachelospermum jasminoides.

Fragrant trees

Alstonia scholaris, Anthocephalus indicus, Gardenia latifolia, Magnolia grandiflora, Murraya exotica, Mimusops elengi, Tabernaemontana divaricata.

Globose, flowering cacti

Borzicactus, Echinocereus, Gymnocalycium, Lobivia, Echinopsis, Mammillaria, Coryphantha, Notocactus, Parodia.

Edible flowers

Mahua (*Madhuca indica*) flowers are consumed as staple foods especially during summer or monsoon by some tribals. Flower buds of *Bauhinia* are edible. Flowers of *Bauhinia racemosa, Boswellia serrata, Cassia fistula,* young buds of *Ficus* spp, *Musa ornata,* flowers of *Mussaenda frondosa,* leafy buds of *Phoenix* flowers of *Pterocarpus marsupium, Wrightia tinctoria* are used as famine foods by some tribals. Flower buds of *Bauhinia variegata, Madhuca indica, Periploca aphylla, Capparis decidua, Calligonum polygonoides, Indigofera desua, Polygonum runcinatum, Cardamine hirsuta,* sweet calyx of *Astragalus multicaps,* are eaten by Himalayan tribals. The scarlet flowers of *Rhododendron arboreum* are used in preparing jams and cold drinks. The scarlet flowers of *Woodfordia fruticosa* are used in preparing cold drinks. *Calligonum polygonoides* of Polygonaceae is a shrub of arid regions; found in Rajasthan and adjoining tract. The flowers of this plant are cooked in oil and eaten. Flowers are also made into edible bread. The flowers and buds of *Capparis decidua,* a leafless hardy arid zone bush, are eaten as vegetable or preserved as pickle. The flower buds of *Capparis spinosa* are used as condiment. The flowers of *Holostemma annularis* are eaten as vegetable. The sweet succulent corolla of the flowers of *Madhuca indica* are eaten raw or cooked and even made into sweet meats; also brewed into a local beer, much consumed by the tribals. The inflorescence of phonochoria hastaefolia is eaten. The flower buds of *Orthanthera viminea* (a leafless shrub of Asdepiadiaceae, growing in Rajasthan and Himalayan foot hills are eaten as a vegetable. The sweet flower buds of *Periploca aphylla* of Asclepiadaceae are eaten raw or cooked as vegetable.

Breeding of flowers

Thomas Fairchild of London, apparently made the first artificial hybrid in 1717, when he pollinated a *carnation* (*Dianthus caryophylla*) with *sweet william* (*D. barbatus*). Certain plants produce large number of seeds in each seed pod, e.g. *Antirrhinum, Petunia* etc., unless you can utilise a large area with the seedlings that will be raised, do not pollinate more than a flower or two. With the double flowers of such plants like *carnation* (*Dianthus caryophyllus*) or *Indian pinks* (*Dianthus heddewigii*) where the seed capsule is present, reduce the number of petals to a minimum as well as removing the anthers of the seed parent.

To decide when a bud has to be emasculated you must know when the flower opens. There are some that open in the early morning and with the first warm rays of the sun the authers burst and the pollen is available. Such flowers must naturally have the anthers removed at least 12 hours, previously, it is safer to say 24.

Other flowers open at dusk, or as night falls, the pollen of some, e.g. *Cooperia* (Amaryllidaceae), has been out of its anthers several hours previously. Here again 24 hours before the flower opens is a safer time to fix.

In *Althaea* the anthers are massed together on a short hollow shaft which must be removed very carefully to avoid damaging the number of thread like stigmas inside.

In Asteraceae, e.g. *Cosmos*, when a disc floret has to be emasculated the chimney like group of anthers has to be gripped at the tip and pulled away with a sharp jerk. This requires plenty of practice otherwise the anthers break, or the stigma comes away at the same time.

Emasculation of the disc florets of the Asteraceae is difficult, but in addition, as the disc florets open in succession, a circle each day, it may mean 3 or 4 days of hard work in emasculation. Use glue, gum arabic, durofix etc. to prevent the unwanted anthers from scattering their pollen.

If the ovules of a *Petunia* have been fertilised by the application of pollen, it cannot be affected by further applications.

In *Hibiscus* with 5 stigmas, each responsible for possibly a 12 or more ovules in its parts of the capsule, one could apply 4 or more varieties of pollen to each stigmatic point, the seeds of the one capsule thereby producing at least 20 different combinations.

Mutations can be induced in ornamental plants the following methods are used - extremes of temperature at the time a flower is forming, puncturing the buds with a needle; drought at flowering time, severe pruning, vibration, electric shocks, irradiation of seeds and plants with X-rays and other rays, the use of plant hormones, and weak doses of various poisons. The chief method, however, is the use of colchicine, a poison extracted from the corns of *Colchicum autumnale*, the *autumn crocus*. this poison is also obtained from the tubers of *Gloriosa superba*, the *glory lily*.

A good example of mutation by pruning is *Bougainvillea*, mary palmer, a bicolor variety that is so popular. *Shirley poppy* is a mutant from the scarlet field poppy, *Papaver rhoeas*.

Often a bud mutation takes place in a hybrid plant which, if propagated vegetatively, will breed time. This is particularly noticed in *Chrysanthemum*, roses, carnations and, among fruits in citrus and apple.

Bi-generic crosses unite genera which are closely allied, chiefly among the orchids. *Althaea x Malva*, *Zephyranthes x Habranthus* etc. hybrids are made.

Seasonal flowers of India

Holly hock, gul-e-khera, Althaea rosea Malvaceae, a winter annual. *Madwort, sweet alyssum, Alyssum maritimum*, Brassicaceae, a winter annual. *Love lies bleeding, Amaranthus caudatus*, Amaranthaceae, grows during rainy season. *Angelonia grandiflora*, Scrophulariaceae, a perennial, though normally grown during winter. *Snapdragon, Antirrhinum majus*, is a winter annual. *Columbine, Aquilegia chrysanth-skinneri*, Raunnculaceae, a perennial with red or yellow flowers, sown March onwards. *Blue-eyed daisy, Arctotis stoechadifolia* var. *grandis*, Asteraceae, grows well during winter. *English daisy, gul-e-ashrafi, Bellis perennis*, Asteraceae, grows during winter. *Browallia, Browallia elata*, Solanaceae, a hardy herbs, grown from seeds in June-July or October-November. *Calandrina speciosa*, Portulacaceae, grown during the rains or mild winter. *Potmarigold, Sadberg, Calendula officinalis*, Asteraceae, grown best during winter, seeds sown during October-November, also when rainfall not heavy, plants June-July. *China aster, Callistephus chinensis*, Asteraceae, grows best during winter, sown from September-October, but also during rainy season with less rainfall, sown in June. *Bell flower, Campanula* spp., Campanulaceae, winter annual, but also during rainy season with moderate rainfall. *Cockscomb, Celosia cristata*, Amaranthaceae, grown during summer which is not too hot and during winter which is not too cold. *Coruflower, Centaurea cyanus*, Asteraceae, winter annual, but also during monsoon at places, with 15 to 25 inches rainfall. *Sweet sultan, Centaurea moschata*, Asteraceae, winter annual, seeds sown in late October or early November. *Wall flower, Cheiranthus cheiri*, Brassicaceae, winter biennial. *Gul-e-daudi, Chrysanthenum carinatum*, Asteraceae, winter annual but grows with moderate success in places with light rainfall and cool monsoon winter.

Painted daisy, Chrysanthemum coccineum, Asteraceae, seeds sown in February-March. *Gul-e-daudi, Chrysanthemum morifolium*, winter ornamental, flowers during December-January. *Clarkia elegans, Onagraceae*, winter annual, seeds sown in September-October. *Spider flower, Cleome spinosa,* Capparidaceae, tall shrub-like annual, grown in rainy or summer season. *Coleus blumei,* Lamiaceae, shade-loving perennial, with pretty colourful leaves, winter ornamental, but thrives in summer and in the rainy season provided the climate is not too hot or rains too heavy. *Tickseed, Coreopsis stillmani,* Asteraceae, winter annual with feathery foliage with daisy-like yellow flower heads. Tickseed, *Coreopsis tinctoria* is a hardy annual, grows throughout the year except in intense cold; foliage green, feathery; flower heads daisy-like velvety, bright coloured orange yellow, bronze or crimson. *Cosmos bipinnatus,* Asteraceae, hardy annual or biennial, grows at its best in the monsoon; flower heads in shades of pink or pure white. *Chinese forget-me-not, Cynoglosum amabile,* Boraginaceae, hardy dwarf plant, produces tiny waxy flowers in blue, white and pink colours, grows best during winter. *Larkspur, Delphinium ajacis,* Ranunculaceae, hardy annual herb, with single or double flowers in many colours, such as blue, purple, pink and white, grows best during winter; seeds sown in November directly in rows in beds. *Sweet William, Dianthus barbatus,* Caryophyllaceae, herbaceous winter annual, seeds sown in September-October. *Carnation, Dianthus caryophyllus,* Caryophyllaceae, herbaceous winter annual, seeds sown in October. *Pink, Dianthus chinensis,* herbaceous winter annual grows from October onwards in plains; grown as summer annual in plains. *Foxglove, Digitalis purpurea,* Scrophulariaceae, hardy winter annual grown in the hills, seeds sown in February-March, produces flowers almost after a year. *Dahlia* spp, Asteraceae, winter annual, flowers large, single or double of various hues and size. *Cape marigold, Dimorphotheca aurantiaca,* Asteraceae, hard winter annual.

Erysimum perofskianum, Brassicaceae, a winter annual herb, often mistaken for wall flower; flowers resemble those of mustard. *Californian poppy, Eschscholzia californica* Papaveraceae, hardy winter annual, seeds sown in October-November. *Blue daisy, Felicia bergeriana, Asteraceae,* a winter annual green from seeds as well as cuttings, flowers small skyblue in colour. *Gaillardia pulchella,* Asteraceae, hardy annual grown throughout the year; seeds sown June to October or February for propagation in winter or summer. Flowers bright yellow, purple, cream yellow, orange, scarlet copper or bronze colour. *G. grandiflora* is more suitable as perennial. *Transval Daisy, Gerbera Jamesonii,* Asteraceae, seeds sown in June. *Gilia caitata,* Polemoniaceae, hardy winter annual, seeds sown in September-October. *Oenothera witneyi,* Onagraceae, a winter annual, flower resembles miniature hollyhock, seeds sown in September-October. *Globe Amaranth, Gomphrena globosa,* Amaranthaceae, a hardy annual, grown during rainy season, but preferably in the monsoon; seeds sown in May or June. *Gypsophila elegans,* Caryophyllaceae, is a winter annual herb, grows well in cold weather, flowers white. *Sunflower, Helianthus annuus,* Asteraceae, a hardy annual, grows all the year round, but does well in the rainy and summer season as compared to winter. *Everlasting flower, Helichrysum bracteatum,* Asteraceae, a hardy cold weather annual herb, grows well from October onwards. Flower heads dry on account of chaff-like involucre, do not decay for a long time. *Heliotropium peruvianum,* Boraginaceae, a winter annual with tiny, sweet scented tubular flowers, purple pink or white. *Immortelles, Helipterum roseum* Asteraceae, a winter annual, the flower heads retain their colour and shape even after getting dried. *Candytuft, Chandul, Iberis amara, I. umbellata, I. odorata,* Brassicaceae a hardy annual of dwarf growth habit; grows in winter and rainy season also; flowers usually white, but of pink and lilac colour also. *Balsam, Impatiens balsamina, Balsaminaceae,* moisture loving herb best cultivated during rains, seeds are sown in May-June or July. *Morning glory, Ipomoea purpurea,* Convolvulaceae, an annual turiner, growing mostly in monsoon season; seeds are sown in June-July, after the first showers; flowers large, bell shaped, white, pink, scarlet, blue or purple. *Burning bush, Kochia scoparia,* Chenopodiaceae, a hardy annual, grows best in rainy or summer season; old plant has a dull fireball-like appearance; nature flowers are dark red. *Sweet pea, Lathyrus odoratus,* Fabaceae, an annual climbing herb, flowers during winter; seeds sown during

November-December. *Sea-Lavender, Limonium bondwellii, Plumba*ginaceae, hardy annual, grows in cool weather on sea coasts, seeds sown in October-November, the plant retains the colour even though dried. *Toad flax, Linaria bipartita,* Scrophulariaceae, a small hardy shrub of dwarf growth habit, grows best during winter. *Flowering flax, Linum grandiflorum,* Linaceae, a hardy annual, grows best during winter; flowers smell, cup shaped, dark red in colour. *Lobelia erinus,* Campanulaceae, a dwarf bushy annual grows best during cold weather, seeds are sown in October-November, flowers tiny, tubers with 5 lobes and of deep or light blue colour. *Gili, Mathiola incana,* Brassicaceae, a winter annual, flowers single or double ranging from rose, lilac, red, white and yellow to various shades. *Fig marigold, Mesembryanthemum tricolor.* Aizoaceae, a dwarf trailing herb, grown in water when air is dry, flowers daisy-like deep pink, pink or white in colour. *Champa, Michelia champaca,* Magnoliaceae, a cultivated plant, flowers sweet, fragrant. *Monkey flower, Mimulus gracilis,* Scrophulariaceae, an annual herb flowering during winter; needs a high attitude for producing ideal flowering specimens. *Gulabashi, 4 o'clock plant, Mirabilis jalapa,* Nyctaginaceae, a perennial herb grow during monsoon; flowers white, pink, red and pale-yellow.

Flowers in worship

On New Moon Day, duringt the month of Ashada (June-July) flowers and leaves of the following are offered at Bheema Amavasya Puja : Flowers of *dronapushpam (Leucas cephalotes)*; *girikarnika (Clitoria ternatea var : alba) Karavira (Nerium indica)*; *Kamala Kuranti (Barleria prionitis), malati (Echites caryophyllata) mallika (Jasminum angustifolium), savanthika (Chrysanthemum indicum).* Leaves of *aswatha (Ficus religiosa), bilva (Aegle marmelos), bhringaraja (Eclipta prostrata), maachi (Artemesia vulgaris), tulasi (Ocimum sanctum).*

Mangala Gowri Puja

Performed during the month of Shravana (July-August). Flowers of *jaji (Jasminum grandiflorum) Ketaki (Pandanus tectorius), Kamala* mallika *(Jasminum sambac), sevanthika (Chrysanthemum indicum)* are offered. Leaves of *bilva (Aegle marmelos), Kasturika (Acacia farnesiana), maachi (Artemesia vulgaris), maruga (Sansevieria roxburghiana), tulasi (Ocimum sanctum)* are offered.

Vara Mahalakshmi Puja

Performed on the last Friday, prior to the full moon, during the month of Shravana (July-August). Flowers of *jaji (Jasminum grandiflorum) mallika (Jasminum angustifolium); nilotpala (Monochoria vaginalis); padma (Nelumbo nucifera); pooja (Achyranthes aspera), punnaga (Calophyllum inophyllum); sevanthika (Chrysanthemum indicum)* are offered. Leaves of *bilva (Aegle marmelos); maachi (Artemesia vulgaris); maruga (Sansevieria roxburghiana); sevanthika (Chrysanthemum indica), tulasi (Ocimum sanctum, vishnukranti (Evolvulus alsinoides)* are offered.

Ganesh Chaturdhi

On the fourth day of the month of Bhadrapad (August-September), Vinayaka Puja is performed. Flowers of *atasi champakam (Michelia champaca); jaji (Jasminum grandiflorum); girikarnika (Clitorea ternatea var. alber), kalhara (Gloriosa superba), karavira (Nerium indicum); ketaki (Pandanus tectorius); Koond (Jasminum pubescens); malati (Echites caryophyllata), parijat (Nyctanthes arbor-tristis)* are offered. Leaves of *apamarga (Achyranthes aspera), arka (Calotropis gigantea), aswatha (Ficus religiosa); badari (Zizyphus); bilva (Aegle marmelos); bhringaraja (Eclipta prostrata), brihati (Solanum xanthocarpum); dadima (Punica granatum); devadaru (Cedrus deodara); gandalika (Murraya koenigi); jaji (Jasminum grandiflorum), karavira (Nerium indicum); maachi (Artemesia vulgaris), maruvaka (Marjorana hortensis* syn. *Origanum marjoranum); shami (Ceratonia siliqua); sindhuvarma (Vitex negundo); Tulasi (Ocimum sanctum); Cynodon dactylon* are offered.

Worship of Lord Shiva

Flowers : *Bauhinia tomentosa; Bauhinia acuminata; Calotropis gigantea; Cassia fistula;* purple form of *Datura fastuosa; Vishnukranthi (Evolvulus alsinoides); Ervatamia coronaria; dronapushpam (Leucas aspera); Nerium indicum* white flowers; *Stereospermum suaveoleus.*

Leaves : Bilva (*Aegle marmelos*); *Santalum album, Ocimum basilicum; Vitex negundo*

Worship of Lord Vishnu

Flowers : *Acacia farnesiana; Bauhinia variegata; Cananga odorata;* blue flowers of *Clitoria ternatea; Guettardia speciosa; Hiptage benghalensis* syn - *H. madhablota); Jasminum sambac;* red flowers of *Nelumbo nucifera; parijatha (Nyctanthes arbor-tristis); padiri (Stereospermum suaveolens);* nilotpala (*Monochoria vaginalis*).

Leaves : *Dhavanam (Artemesia abrotanum); Calophyllum inophyllum; Evolvulus alsinoides; Ocimum sanctum.*

Worship of Goddess Lakshmi

Flowers : *Nelumbo nucifera, Nyctanthes arbor-tristis.*

Worship of Goddess Saraswati : Flowers of *Nelumbo nucifera* var. *alba;* Leaves of *Cynodon dactylon.*

Worship of Goddess Kaali

Flowers of *Hibiscus*, without the staminal column; *Ixora coccinea (virushi).*

Worship of Lord Muruga

Flowers : A variety of *Cassia, Ixora coccinea, Nerium indicum* (single white).

Lord Narasimha is worshipped with flowers of *Nerium indicum*, single red.

Worship of Sun God

Flowers of *Calotropis gigantea* (white); foliage of *Cynodon dactylon.*

There are a number of flowers that can be used in the worship of all gods and these are *jasminum, Michelia champaca (champakam), Mimusops elengi, Pergularia odoratissima, Punica granatum.*

Prohibited flowers

Flowers of the following plants, while apparently sacred, are prohibited for use in worship. These are *ketaki (Pandanus tectorius),* rose flowers, marigold (*Tagetes* sp.). Lord Vishnu should never be offered flowers of *Cassia fistula, Couroupita guianensis (nagalingam* except at Siva linga Puja), *Nelumbo nucifera*, white petals only, are used in *Sahasranaama puja.*

On certain auspicious occasions the banana complete with leaves and their sheath bases will be used to decorate the gate or entrance to the *pandal* where the ceremony is to take place. The addition of a *ghara*, topped with a green coconut, will often be seen.

For application on the forehead (*tilak*), saffron paste (dried anthers of *Crocus sativus*) is used. Ladies smear their feet with turmeric paste.

The Nagalingam or Sivalingam tree

The cannon ball tree *Couroupita guianensis* (f. Lecithydaceae) which was introduced to India from its habitat of Guiana and South America, is sacred to the God Siva because of the shape of the flower.

The stigma protrudes, like a lingam, from a bed of sterile anthers which are continued on a wide strap over the stigma, there the anthers become fertile and the strap bends over like the expanded hood of a mag.

The following trees are generally worshipped by various tribes and cults of India : am (*Mangifera indica*), asok (*Saraca indica*), beel (*Aegle marmelos*), champa (*Michelia champaca*), bor (*Ficus bengalensis*), doomur (*Ficus glomerata*), kadam (*Anthocephalus cadamba*), Kanak champa

(Pterospermum acerifolium), mulseri (Mimusops elengi), peepul (Ficus religiosa), piyal (Buchanania lanzan), rudraksha (Elaeocarpus sphaericeus syn. *E. ganitrus),* nischinda *(Vitex negundo),* and *vahedra (Terminalia belerica).*

This prevented the cutting down of some valuable trees for fuel purposes. Definite sin, however, attaches to any one responsible for the destruction of *bad, bor, narial (Cocos nucifera)* and *peepul;* on the other hand great merit is acquired by those who plant *pakier (Ficus lacor* syn. *F. infectoria)* and *tal (Borassus flabellifar).* In Upper India *shisham (Dalbergia Sissoo),* and in South India *amla* are also added.

At every puja, a burnt offering of 5 twigs must first be made, these are called *panch pollah,* and comprise pieces of *bor,* beal, am, pakur and *peepul.* In Keroda the selection is somewhat altered, peepul and *bor* remain but the other 3 are *palas (Butea monosperma), deodar (Cedrus deodara)* and *chandan (Santalum album).*

Many tribal societies revere a particular tree which they have preserved for time immemorial. The Mundas and the Santhal of Bihar worship *mahua* trees *(Bassia latifolia)* and Kadamba *(Anthocephalus cadamba);* the Bhuiyas and the Gonds of Madhya Pradesh regard palash *(Butea monosperma)* as sacred, the tribals of Rajasthan have reverence for the khejri tree *(Prosopis cineraria);* the tribals of Orissa and Bihar worship imli tree *(Tamarindus indica)* and mango *(Mangifera indica)* during weddings.

8

Essential Oils

The *essential oils* or *volatile oils* are the odorous principles which are generally responsible for the odour of plants in which they occur. They usually occur as such in plants, but in some cases they are found in a state of combination as glucosides from which they may be liberated by the interaction of enzymes, as is known to occur in some members of Brassicaceae. The essential oils differ from the fixed oils by being volatile in steam. They are generally mixtures of different chemical compounds which may include hydrocarbons known as *terpenes* and *sesquiterpenes, open-chain alcohols and aldehydes, aromatic alcohols* of the camphor series and their ketones, *aromatic alcohols* of the benzene series and their aldehydes and ketones, sesquiterpene alcohols, *phenols* and their derivatives, *esters* of different alcohols, and *sulphur* compounds.

The *fixed oils* are generally compounds of glycerol with different kinds of fatty acids containing sterols and other substances dissolved in them. They are greasy liquids occurring quite commonly in the seeds of plants. When heated, they decompose giving off acrid acrolein vapours. They are insoluble in water or glycerine, sparingly soluble in alcohol, and freely soluble in ether; chloroform, benzene, carbon disulphide, etc. They generally have laxative properties. Some of the fixed oils, however, have a drastic purgative action, e.g., those obtained from *Croton tiglium* Linn., *C. oblongifolius* Roxb., *Jatropha curcas* Linn., *Ricinus communis* Linn., etc.

Essential oils are a class of vegetable oils which are made up of complex mixtures of volatile organic chemicals. Essential oils occur in a wide variety of plants-herbs, trees and shrubs belonging to gymnosperms and angiosperms (Mytaceae, Lamiaceae Apiaceae, Lauraceae, Asteraceae).

Perfume oils

The art of perfumery is as ancient as recorded history. Ancient Rome used to import *myrrh* from Persia, spikenard from India, saffron from Spain, sandal wood from China, thyme from Algeria and jasmine from Arabia. Perfume oils are obtained from *Geranium (Pelargonium* spp.); *bergamot* (*Citrus aurantium* var. *bergamia*), *attar of roses* (*Rosa damascena, R. centifolia, R. moschata*), ylang-ylang (*Cananga odorata, C. latifolia*), tube-rose (*Polianthes tuberosa*), citronella (*Cymbopogon nardus*), Lemongrass (*Cymbopogon citratus*), patchouli (*Pogostemon patchouly*), khus-khus (*Vetiveria zizanioides*), scented boronia (*Boronia megastigma*), lavender (*Lavendula latifolia, L. angustifolia*), mimosa (*Acacia dealbata*),*jasmine* (*Jasminum officinale, J. miloticum, J. odoratissimum*) (see Fig. 8.1), *gardenia* (*Gardenia florida*), narcissus, *Narcissus poeticus*, violets (*Viola* sp.).

Wood oils

Cade (*Juniperus oxycedrus*), camphor (*Cinnamomum camphora*), cidar wood (*Juniperus virginiana*), eucalyptus (*Eucalyptus dives, E. globulus*), san-

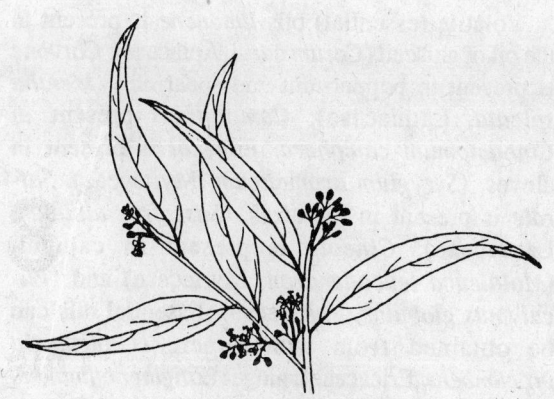

Fig. 8.1. *Eucalyptus* plant.

dalwood (*Santalum albidum*), pine needle (*Pinus sylvestris*).

Flavouring oils

Peppermint (Mentha piperita), spearmint (Mentha spicata), star anise (Illicium verum), celery (Apium graveolens), cloves (Syzygium aromaticum), lemon (Citrus lemon), orange (Citrus aurantium, C. sinensis), lime (Citrus aurantifolia), caraway (Carum carvi), nutmeg (Myristica fragrans), thyme (Thymus vulgaris, T. capitatus), aniseed (Pimpinella anisum).

Essential oils

Essential oils are liquid components of plant cells that are volatile and as a rule have a pleasant fragrance. Chemically their principal constituents are a group of complex substances known as terpenes and their compounds or derivatives. They generally occur in special cells or glands where they accumulate possibly as waste products. This is typical characteristic of many of the families noted for their content of essential oils, for example Pinaceae, Labiatae, Myrtaceae, Rutaceae and Umbelliferae.

The plants bearing essential oils are distributed widely in the vegetable kingdom while certain families, such as Labiatae, Rutaceae, Umbelliferae, Myrtaceae, Lauraceae, Piperaceae, and Coniferae are specially rich in such plants. Among the plants containing essential oils with toxic constituents may be mentioned species of the genera *Artemi-*

sia, Ruta, Mentha, Petroselinum, Chenopodium, Myristica, Eucalyptus, (Fig. 8.1) *Goultheria, Juniperus, Pinus, Eupatorium, Anemone, Ranunculus, Caltha, Prunus, Allium, Brassica* and other crucifers, *Piper, Ferula* etc.

Essential oils are substances with multiple effects. They exert an antiseptic action in that they check the growth of germs. Well known, for instance, are the marked germicidal properties of thymol, a constituent of the essential oils of garden thyme (*Thymus vulgaris*) and wild thyme (*T. serphyllum*). These are extensively used in mouth washes and gargles. Antiseptic properties are also exhibited by the essential oil of garlic (*Allium sativum*) and onion (*Allium cepa*). These are well established remedies for bronchitis and the common cold. The action of essential oils in the treatment of mycoses, scabies and other skin diseases caused by parasites, is also well known. Anethol, a constituent of the oil of anise (*Anisum vulgare*), has a marked odour that repels certain insects. The anthelminthic action of certain volatile oils is also well established.

Some essential oils irritate the skin, and in many cases cause inflammation and even swelling. Administered in therapeutic doses, however, they cause a marked increase in the blood supply and many are, therefore, used as a constituent of anti-rheumatic liniments. The oil of black mustard (*Brassica nigra*), for instance, is used in this way. Also significant are the properties of reducing inflammation and counteracting fever of certain volatile oils contained, for example, in chamomile (*Matricaria chamomilla*), milfoil (*Achillea millefolium*) and peppermint (*Mentha piperita*).

Individual constituents of essential oils, such as eugenol, camphor and menthol, have an anaesthetic effect. Others, like camphor and borneol, affect heart muscle and the circulatory system, producing a stimulation. Essential oils also affect the central nervous system and are used as stimulants, sedatives and even narcotics. Plants containing these types of essential oils include valerian (*Valeriana officinalis*) and common balm (*Melissa officinalis*).

The smooth muscles are influenced by a number of essential oils. Most of these exert a favourable action on the digestive system - both the liver and gall bladder as well as the stomach and intestines. The effects of oil of peppermint (*Mentha piperita*), anise (*Anisum vulgare*), and caraway (*Carum carvi*) are well known. Their constituents stimulate the flow of gastric secretions, alleviate digestive disorders, stimulate the appetite, and the secretion of bile.

Some essential oils have an undesirable effect on the womb in that they increase the blood supply to this organ and may cause miscarriage in pregnancy. Particularly dangerous are certain constituents of essential oils such as aptol and myristicine. On the other hand many of these components have excellent diuretic properties and also exert an antiseptic action in the urinary passages, such as the oil of restharrow (*Oninis spinosa*) and parsley (*Petroselinum crispum*).

Other oils have a favourable effect on the upper respiratory passages and have a wide medicinal use in this connection, for example, the oils of garden and wild thyme (*Thymus vulgaris* and *T. serpyllum*).

Many of the essential oils are important components of flavouring which enhance the taste and aroma of food and increase the enjoyment of a meal. The favourable effect on the digestive process is an additional benefit. Fresh herbs, or rather their roots, leaves, flowers and fruits, are sometimes used but generally they are employed in dried form, whole or crushed. For culinary purposes, however, fresh herbs always have a better aroma. If dried material is used it should be freshly ground to a powder to conserve as many of the volatile principles as possible. When possible herbs should be added to a dish towards the end of the cooking time and the vessel should have a tight-fitting lid to help retain the flavour and aroma.

The food industry - the producers of beverages, both alcoholic and non-alcoholic, are included in this grouping - also use volatile oils extensively as flavouring and perfuming agents, and they are employed in the same way by the confectionery and tobacco trades. Today essential oils are generally isolated from the fresh or dried plants on an industrial scale.

Volatile (essential) oils *limonene* is present in the oil of aniseed (*Carum carvi*, Apiaceae). *Carvone* is present in peppermint and spearmint (*Mentha spicata*, Lamiaceae). *Camphor* is present in *Cinnamomum camphora*. *Eugenol* is present in cloves, (*Syzygium aromaticum*, Myrtaceae). *Safrole* is present in sassafras, (*Sassafras albidum*, Lauraceae). *Cineole* is present in cajuput, (*Malaleuca leucodendron*, Myrtaceae) and (*Eucalyptus globulus*, Myrtaceae). Essential oils can be obtained from wintergreen, (*Gaultheria procumbens*, Ericaceae); ginger (*Zingiber officinale*, Zingiberaceae); mustard (*Brassica* and *Sinapis* spp., Brassicaceae).

Davana oil

Davana oil is obtained from *Artemisia pallens*; davana oil has davanone, a sesquiterpene ketone, linalool, dihydro-a-linalool, terpinin-4-ol, nordavanone (C_{11} terpenoid) and davnofurans.

Agar wood oil

The agar wood tree grows sparsely in the thick forests of Himalayan foot hills of Manipur, Mizoram and Arunachal Pradesh. Its value increases the moment a fungus poisons its heart wood, turning its wood into yellowish black. The wood is exported to Gulf countries. An oleoresin develops in agar wood (aloe-wood), *Aquilaria agallocha*, as a result of fungal infection.

Camphor

The camphor tree, *Cinnamomum camphora* Nees and Eberm; Borneo camphor tree, *Dryobalanops aromatica* Gaertn.f.; camphor tulsi, *Ocimum kilimandscharicum* Guerke are good sources of camphor (Fig. 8.2).

The camphor tree is tall (20 m) with shiny, dark, evergreen leaves. The tree is native to China, Japan and Formosa. The tree thrives well on hilly slopes upto 1800 m. elevation. Japan, Formosa have virtual monopoly of the camphor industry. Camphor is an essential oil used as an incense, in sweet meats, in medicine, photographic films, smokeless gun powder etc.

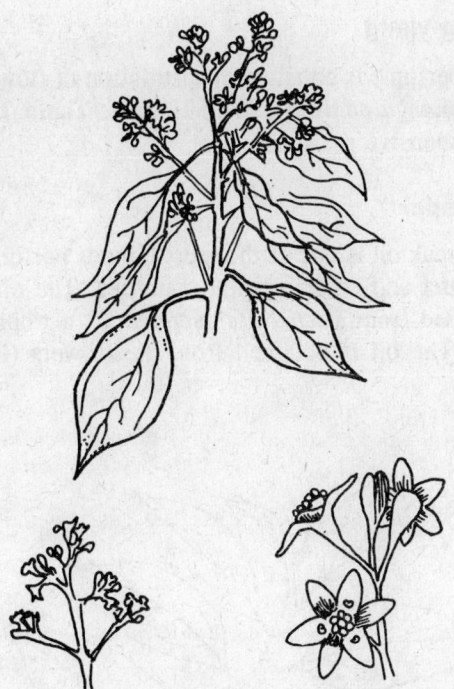

Fig. 8.2. Camphor tree—*Cinnamomum.*

Distillation of wood yields *Whole Camphor Oil* which contains 25 per cent camphor oil, 8 per cent safrole, and the remaining light camphor oil. Distillation of leaves and twigs yields camphor oil (40 per cent) but no safrole. Safrole has a higher market value than camphor and light camphor oil.

Roots of *Sassafran albumin*, leaves of *Doryphora sassafran* contain *safrole.*

In recent years, the camphor herb or camphor tulsi. *Ocimum kilimandscharicum* is widely cultivated for obtaining camphor. The plant is perennial. The plant begins to yield camphor (from the leaves) with in 4-6 months after planting. The leaves can be cut number of times each year. The leaves after being cut can be dried and stored even for an year without loss of camphor. Another advantage of this crop is that cattle, sheep and goats do not eat these plants.

Vetiver oil

A grass oil obtained by distillation of the roots of *Vetiveria zizanioides* (L.) Nash. The oil is used as a perfume and as a fixer of violet odors.

Sandal wood oil

The oil is obtained by distillation from the heart wood and roots of *Santalum album* L. The tree grows wild in India and some parts of South Eastern Asia; the tree is also cultivated. The oil consists of a-and-b-santanol. The oil is used as a perfume, in coaps, in medicine, as a fixative etc. (Fig. 8.3).

A twig of the sandal wood tree

Fig. 8.3. Sandalwood tree—*Santalum album.*

Lemon grass oil

In India lemonrass oil is obtained from *Cymbopogon flexuosus* Stapf and *C. citratus* Stapf. The grass is cultivated in Central America, Madagascar, Indo-China, Comori Islands etc. The oil is reddish yellow in colour with strong odour and taste of lemon. The oil contains *micrin.* The oil is used in perfumery, toilet soaps, manufacture of **vitamin A,** synthetic violet etc.

Citronella oil

The oil is obtained by distillation from the **grass** *Cymbopogon nardus* (L.) Rendle. The grass is chiefly cultivated in Java. Citronella oil is **a pale** yellow oil used for inexpensive soaps, perfumes, insect repellents etc. The oil has 80-90 per cent geraniol.

Cymbopogon martinii (Roxb) yields ginger grass oil (palmarosa oil) which contains large amounts of geraniol.

Geranium oil

The oil is distilled from the leaves of *Pelargonium* spp. *P. capitatum* (L.) Ait., *P. graveolens* L. Herit, *P. odoratissimum* (L.) Ait. etc., Geranium oil is used as a substitute of *otto of roses*, perfumes, soaps etc.

Rosemary oil

Rosemary oil is distilled from the fresh flowering tops of the rosemary plant, *Rosemarinus officinalis* L. The oil has borneol. The oil is used in the preparation of the perfume *eau de cologne* and toilet soaps.

Patchouly oil

The oil is distilled from the fresh flowering tops and leaves of *Pogostemon perilloides* (L.) Manof. The oil is dark brown in colour and has a strong odour similar to that of sandal wood. The oil is used in perfumes, soaps, hair tonics, and for scenting tobacco and kashmeri woolen shawls.

Peppermint oil

The oil is distilled from the leaves and shoots of *Mentha piperita* L. The oil is used in confectionary, pharmaceutical and perfumery trades. Menthol which is used in the treatment of colds and in tooth pastes, is obtained from the plant. The oil has menthol and carvone.

Thymol

Thymol is obtained from *Thymus vulgaris*. The oil obtained by distillation of leaves is the source of thymol. Thymol is used in soaps, perfumery, mouth wash lotions, dentrifices and as an anthelmintic.

Ylang-ylang

The perfume is obtained by distillation of flowers of *Cananga odorata* (Lam.) Hook & Thom. It is an expensive perfume.

Champak

Champak oil is one of the most famous perfumes of India and other oriental countries. The oil is obtained from *Michelia champaca* L., a tropical tree. The oil is obtained from the flowers (Fig. 8.4).

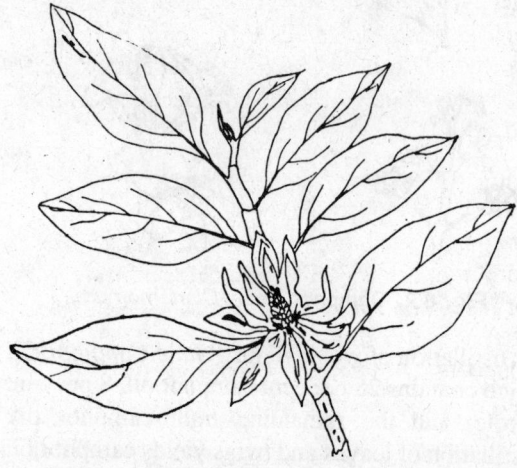

Fig. 8.4. Champak—*Michelia champaka.*

Jasmine

Jasmine oil is obtained from the flowers of *Jasminum officinale* L. var. *grandiflorum* (L.) Kobaski.

Oil of Neroli

Oil of Neroli or orange blossom oil is obtained from the flowers of *Citrus aurantium* L., by distillation. Italy, Spain and Portugal are important countries of production. The oil has linalol, limonene, geraniol, various terpenes - dipentene, pinene and camphene. The oil is an ingredient in synthetic perfumes and *eau de cologne*. Orange blossom oil is different from orange oil which is obtained from the rind of sour orange. Petitgrin oil is obtained from the leaves and twigs of *Citrus*.

Attar or otto of roses

Attar or otto of roses is obtained by distillation of the oil obtained from the flowers of the damask rose (*Rosa damascena*) and to a lesser extent from the cabbage rose (*Rosa centifolia*) and the musk rose (*Rosa moschata*), which are grown for this purpose in Bulgaria, Italy, France, North Africa and India. Otto of roses is one of the expensive perfume oils.

Lavander oil

Lavander oil is obtained from the flower tops of *Lavndula angustifolia* (= *L. officinalis*) and *L. latifolia*, the plant is native to southern Alps in Europe; lavander is grown in France, Spain and England (Fig. 8.5).

Palmarosa grass

Cymbopogon martini Stapf. var. *motia*, Poaceae is an important perfumery raw material, exclusively used for imparting *rose like aroma* to wide range of soaps, cosmetic products, triletry goods, tobacco products etc. The principal constituent of the oil is *geraniol* which accounts for its *refreshingly fresh rose-like aroma.*

In India, the grass grows wild in widely scattered forest lands and is particularly rich in parts of open scrub forests of Madhya Pradesh, Maharashtra and Andhra Pradesh. A sizeable quantity of the oil is exported annually, mainly to United Stated, United Kingdom, West Germany, Egypt etc.

C. martini var. *motia* contains around 90 per cent geraniol with traces of terpenes, linalool and citral. *C. martini* var *sofia* contains 39 to 48 per cent geraniol, dipentine (21.1 per cent), carvone (20 per cent), dihydrocarvone (10 per cent), carveol (13.5 per cent), perillyl alcohol (14. per cent).

Palmarosa grows in well-drained sandy loam soils which receive an annual precipitation of 100-150 cm.

Propagation of this plant is done by transplanting the seedlings. The seed rate required is 2.5 kg per hectare and the seedlings are transplanted after 3-4 weeks. An espacement of 60 x 60 cm is adopted.

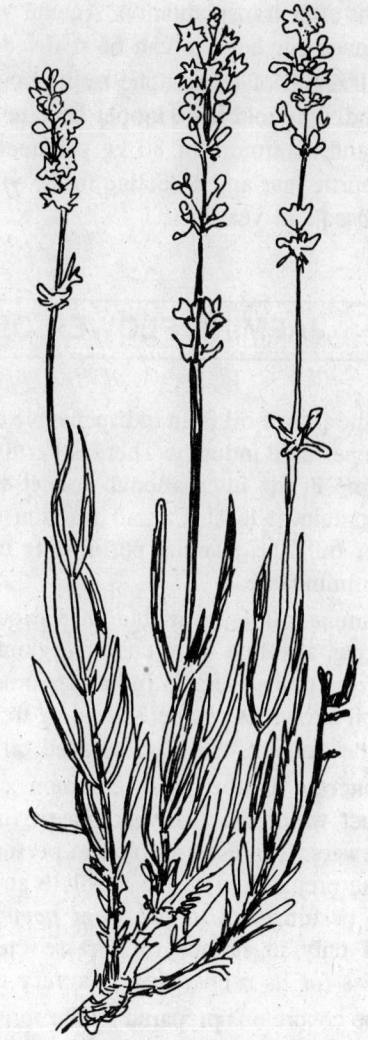

The lavender plant

Fig. 8.5. Lavender—*Lavendula vera.*

Diuron at the rate of 1.5 kg a.i. per hectare is found to be an effective herbicide. 3 tonnes of any organic mulch is between the rows also controls the weeds besides adding organic matter to the soil. Generally 40 : 50 : 40 kg of N, P_2O_5 and K_2O per hectare is applied as basal dressing and 60 kg of N per hectare is applied in three splits as top dressing. Spraying zinc sulphate at the rate of 25 kg per hectare increases the oil content.

Harvesting the upper third of plant is profitable. It is done just before bllming that is, 6 months after transplantation. Annual yield of 15-20 tonnes per hectare can be realised.

Extraction of oil is done by hydro-distillation method. Oil yield is 60 kg per hectare in the first year and maximum of 80 kg per hectare during the fourth year and a decline in the yield is seen in subsequent years.

JASMINE FLOWER OIL

Jasmine flower oil is an indispensable commodity in the perfume industry. There is hardly any good perfume in the international market which does not contain at least a small fraction of jasmine flower oil. The jasmine perfume is undefinable and inimmitable.

Jasmine oil is available in various forms : concrete, absolute from concrete, pomade, absolute from pomad etc. In India, perfumers make a hair oil known as *chameli-ka-tel* by treating fresh jasmine flowers with sesame seed oil.

Concrete is the most common commercial product, which is actually a crude perfume extract of flowers. It is used as such in perfume blends, for the preparation of the absolute and handkerchief perfumes. *Absolute from pomade* is prepared only in France (in Grasse etc.) but the process for its preparation is a very costly one.

The concrete is prepared by the *solvent extraction* of fresh jasmine flowers (*Jasminum grandiflorum* Linn). The jasmine concrete is used as such in various perfume blends and also for the preparation of absolute and handkerchief perfumes. Besides, it is used in various other industrial products, such as, cosmetics and toiletries, soaps, pharmaceuticals, food essences, chewing tobacco, dental preparations, confectionery baked goods, ice creams, room sprays, deodarants etc.

In India, the rate of jasmine concrete varies from Rs. 7,000 to Rs. 10,000 per kg. Though France and Italy are the main producers of jasmine concrete, its production has been decreasing. Egypt, Morocco and other Meditarranean countries have started its production.

Raw materials

Raw materials required for the production of jasmine concrete are fresh jasmine flowers, petroleum ether (B.P. 40°-60°C) or normal hexane (B.P. 65°-70°C), purified to perfumery grade.

Jasminum grandiflorum L. (*Chameli*) is commercially cultivated in Ghazipur, Jaunpur, Ballia and Farrukabad divisions of Uttar Pradesh. Total acreage of land under its cultivation is about 150 hectares. The annual yield of flowers is estimated at about 750-800 kgs per hectare. The flowering season ranges from August to October. In Chitrakoot area of Banda district in UP, however, plants have been found briskly flowering even up to the end of November.

In Tamil Nadu jasmine cultivation occupies about 200 hectares of land. Here, the flowering starts in the second of third week of February and continues up to the end of November. Peak season of the flowering is from May to July end. The yield of flowers per hectare is about 12,000 kg per year.

Jasmine can be grown in any climate, in any part of India. The plant starts giving economic yield of flowers during the 4th year of planting.

Processing unit

The average recovery of concrete from fresh flowers is nearly 0.25 per cent by weight and hence the minimum capacity of the unit should be for processing about 200 kgs of flowers per day. The unit should be able to charge about 500 kgs of the flowers at a time, yielding about 0.125 kg of the concrete. In two extractors, two charges in each can be easily processed in a day, yielding about 0.5 kg of the concrete, i.e., about 15 kgs of concrete per month.

Plant and machinery

1. Baby boiler - 1 (cost Rs. 33,000);
2. Steam-jacketted solvent distillation unit - 2 (cost Rs. 40,000);
3. Extractor (Percolator) - 2 (cost Rs. 10,000);
4. Steam-jacketted solvent evaporation-cum-distillation unit - 1 (cost Rs. 12,000);

5. Vacuum concentrator - 1 (Cost Rs. 2,000);
6. Vacuum pump 1 HP - 1 (Cost Rs. 4,000);
7. Motorised shaking (Horizontal) machine - 1 (Cost Rs. 20,000); 8. Auxiliary equipment (containers for solvent and extracts, weighing equipment, fire extinguishers, exhaust fans, etc. Rs. 5,000. Extraction and solvent distillation equipment may be made of copper (18-20 gauge) or stainless steel.

Purification of the solvent

Sulphur and nitrogenous compounds present in commercial petroleum ether, are removed by repeatedly washing the petroleum ether with concentrated sulphuric acid and then washing it with water to make it acid free.

Washed petroleum ether is fractionally distilled in a distillation column over 5 per cent of its weight of vaseline and the fraction passing in between 40° and 60°C is collected.

The same procedure is adopted with n-hexane and the fraction distilling between 65° and 70°C is collected.

Extraction

Fully opened and freshly collected flowers, preferably collected before sun-rise, are used for extraction which is done in extractors by dipping the flowers in the solvent. The flowers must be fully covered with the solvent.

Each lot of flowers is extracted thrice, by emersing the flowers in the solvent for about 40 minutes for the first time, for 30 minutes for the second time and for 20 minutes for the third time, the solvent being drained out every time. After the last extraction, the flowers in the extractor are gently pressed by any mechanical device or manually, to drain out the solvent adhering to the flowers.

Distillation

The solvent from the extract is removed by distillation which is carried out in the evaporators. When the bulk is reduced to a small amount, it is transferred to the concentrators. Here the solvent is again evaporated. To remove the last traces of the solvent, the concentrator is finally connected with vacuum pump which is run for about 2 hours.

The recovered solvent can again be used for the extraction. Total loss of solvent comes to nearly 15 to 20 per cent. This loss can be minimised by using efficient equipment and taking proper precautions.

The residue left after the removal of the solvent is known as *concrete*, the desired product. This has a semi-solid, waxy consistency, yellow to yellowish brown colour and the odour which is characteristic of fresh jasmine flowers. The melting point of concrete is 50°-51°C; specific rotation (in 10 per cent alcohol) + 10°-12°; acid number 12-15; ester number 72-75; alcohol soluble absolute 50-55 per cent; the concrete should be stored at about 25°C, at a cool place.

9

Spices and Condiments

Important plants yielding condiments are *Zingiber officinale* ginger, adrak, rhizome (Zingiberaceae); *Tamarindus indica* tamarind, imli (Caesalpiniaceae); *Punica granatum* pomegranate, anar seeds (Punicaceae); *Coriandrum sativum* coriander, dhania, cilantro (Umbelliferae); *Foeniculum vulgare* fennel, saunf (Apiaceae); *Cinnamomum zeylanicum* cinnamon, dalchin (Lauraceae); *Curcuma longa* turmeric, haldi (Zingiberaceae); *Syzygium aromaticum* clove, long (Myrtaceae) (Fig. 9.1); *Nigella sativa* kala zira (Ranunculaceae); *Piper nigrum* pepper, kali mirch (Piperaceae); *Amomum aromaticum* cardamom, elaichi (Zingiberaceae); *Mentha viridis* mint, pudina (Labiatae); *Allium sativum* garlic, lasun (Liliaceae); *Cuminum cyminum* cumin, zira (Apiaceae); *Crocus sativus* saffron, kesar (Iridaceae); *Carum carvi* caraway, kala zira (Apiaceae); *Murraya koenigii* mitha neem (Rutaceae); cardamom *Elettaria cardamomum* (Zingiberaceae).

Important *temperate* spices are Allspice (*Pimenta dioica*, Lauraceae); anise (*Pimpinella anisum*, Apiaceae); basil (*Ocimum basilicum*, Lamiaceae); bay (*Laurus nobilis*, Lauraceae); caper (*Capparis spinosa, Capparidaea);* caraway; cassia (*Cinnamomum cassia, Lauraceae);* cayenne (*Capsicum annuum*, Solanaceae); celery (*Apium graveolens*, Apiaceae); chervil (*Anthriscus cerifolium*, Lamiaceae); dill (*Anethum graveolens*, Apiaceae); horse-radish (*Armoracia rusticana*, Brassicaceae); lemon thyme (*Thymus citriodorus*, Lamiaceae); mace nutmeg (*Myristica fragrans*, Myristicaceae); oregano (*Oreganum vulgare*, Lamiaceae); parsley (*Petroselinum crispum*, Apiaceae); peppermint (*Mentha piperita*, Lamiaceae); vanilla (*Vanilla planifolia*, Orchidaceae); thyme (*Thymus vulgaris*, Lamiaceae); spearmint (*Mentha spicata*, Lamiaceae); rosemary (*Rosmarinus officinalis*, Lamiaceae); saffron (*Crocus sativus*, Iridaceae).

Alkaloids : *Piperidine, piperine* present in black pepper, *Piper nigrum* and long pepper *Piper longum*, are the constituents of black, white and long pepper; responsible for peppery taste. *Trigonelline* present in the seeds of fenugreek, *Trigonella foenum-graecum* is used medicinally to relieve stomach upsets. *Trigonelline* is frequently found in human urine. Alkaloids are absent in the members of Apiaceae, (except *Conium maculatum*, helmock). Rosaceae and Labiatae are also alkaloid-free.

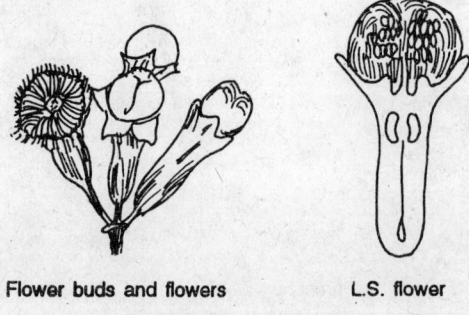

Flower buds and flowers   L.S. flower

Syzygium aromaticum (cloves)

Fig. 9.1. Clove—*Syzygium aromaticum* flower buds and L.S. flower.

BLACK PEPPER

The term *pepper* has been applied to a number of very powerful and 'hot' aromatics. Properly it belongs to the small drupe-like fruit - the so-called berry - of *Piper nigrum* (Piperaceae). Red pepper is *Capsicum*.

The bitter berries of pepper were among the first species of history. Pepper was the chief article of trade between India and Europe.

Pepper is confined to tropical regions of heavy rainfall. The plant is generally grown from cuttings and trained on stakes or trellises or - as shade is desirable in most regions - on living trees. Species with rough bark make the best supports. Pepper vines are severely pruned and trained to bushy form. They are first allowed to fruit at 2-5 years and continue bearing 12 to 25 years (Fig. 9.2).

In terms of quantities traded, black pepper is the most important spice in the world today. The plant is native to jungles of southwestern Asia.

Black pepper is still produced in southwestern Asia, primarily India and Indonesia, but Brazil is now also a major world producer. The United States is the world's largest consumer of black pepper.

Piper nigrum is a woody climber with a stem about 0.5 inch in diameter. It attains a considerable length in the wild; cultivated vines are trained and trimmed for ease in picking. Adventitious roots are produced from the nodes. Leaves are alternate ovate or lenceolate, coriaceous and evergreen. Very small and much reduced flowers grow in catkinlike spikes, which may be pistillate, staminate, or perfect. Fruiting spikes reach a length of about 4 inches, and may have 50 fruits, each about 0.2 inch in diameter and bright red at maturity.

Within the juicy red pericarp, which becomes wrinkled and blackened or drying is a single globular seed.

Ordinary black pepper is gathered when only a few of the fruits of a spike hue turned ripe and red. It is sun-dried; before drying it may be

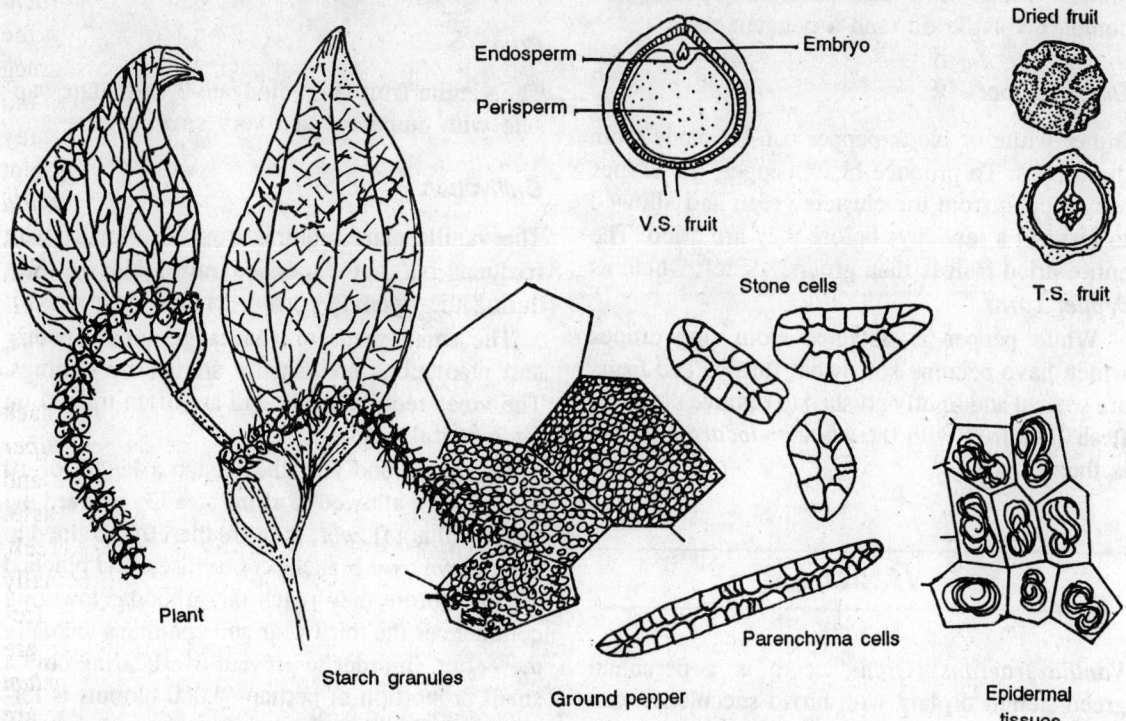

Fig. 9.2. Pepper. *Piper nigrum*—plant and plant parts.

plunged in boiling water. Dried pepper is rubbed by hand to separate the stems, which are winnowed out. In some localities pepper is smoke cured.

White pepper

White pepper consists of the seed only. Two methods are used to remove the pericarp.

1. The berries are allowed to ripen; after collection they are formented in water until the pulp softens and can be trodden off and separated by washing.
2. By milling black pepper in machines which rub off the dried pericarp. The blackish pulverulent portions removed are known as pepper shells. They have been used as adulterants of ground black pepper.

White pepper is less pungent than the black spice to some tastes the flavour is ginger.

Pepper owes its pungency to several substances. Of these the largest in quantity is the alkaloid *piperine*, ranging from 4.5 to 8 per cent. *Piperidine* - which is a derivative of piperine - a complex volatile oil, and a pungent resin.

Black pepper

Either white or black pepper can be made from the drupes. To produce black pepper, the drupes are stripped from the clusters green and allowed to ferment a few days before they are dried. The entire dried fruit is then ground, or left whole as *pepper corns*.

White pepper is obtained from ripe drupes which have become somewhat fleshy. The fruits are soaked and lightly crushed to remove the outer flesh. The fruit with the *white endocarp* exposed is then dried.

VANILLA

Vanilla fragrans (Orchidaceae) is a perennial green-stemmed plant with broad succulent ovate leaves 4-9 inches long. Opposite the leaves occur white roots which twine or tree-trunks or artifi-

cially placed supports. Flowers are in bracted racemes; each flower is about 4 inches long, with linear-oblong pale green perianth. One perianth segment is developed into a trumpet-shaped lip or labellum; this is rolled round a column in which the pollen-mass and the peculiarly modified stigma are separated by a projection, the rostellum. The highly modified structure of orchid flowers usually permits their pollination only by specially adapted insects (Fig. 9.3). In *Vanilla*, this is usually done by certain Mexican bees and humming birds. In other regions of cultivation the requisite species are lacking and artificial pollination is necessary.

Artificial pollination

This is effected by pressing up the rostellum with a small stick, removing the pollen, and thrusting it down on the stigma, on while sticky surface it is held by the pressure of the rostellum, which springs back into place when released. This process is close-pollination, cross-pollination is more difficult and is not commonly done.

Fruit

The vanilla fruit is an elongate-3-carpellate capsule with numerous and very small seeds.

Cultivation

The vanilla plant requires tropical climate, with frequent but not excessive rains. Heavy rains during the ripening period may dislodge pods.

The small seeds of vanilla germinate poorly and plantations are usually started by cuttings. The vines require shade and are often trained on trees or trellises.

By the second year they reach a length of 10 to 14 feet; it allowed to climb steadily upward the vines will not flower; they are therefore trained to hang down over branches or trellises, and pinched off just before they reach the ground. Flowering commences the third year and continues annually thereafter. In order to prevent overbearing only a small proportion of perhaps 4,000 blooms is fertilized. Fruit ripens in 4 to 9 months, depending on the locality. *It is harvested when still unripe*

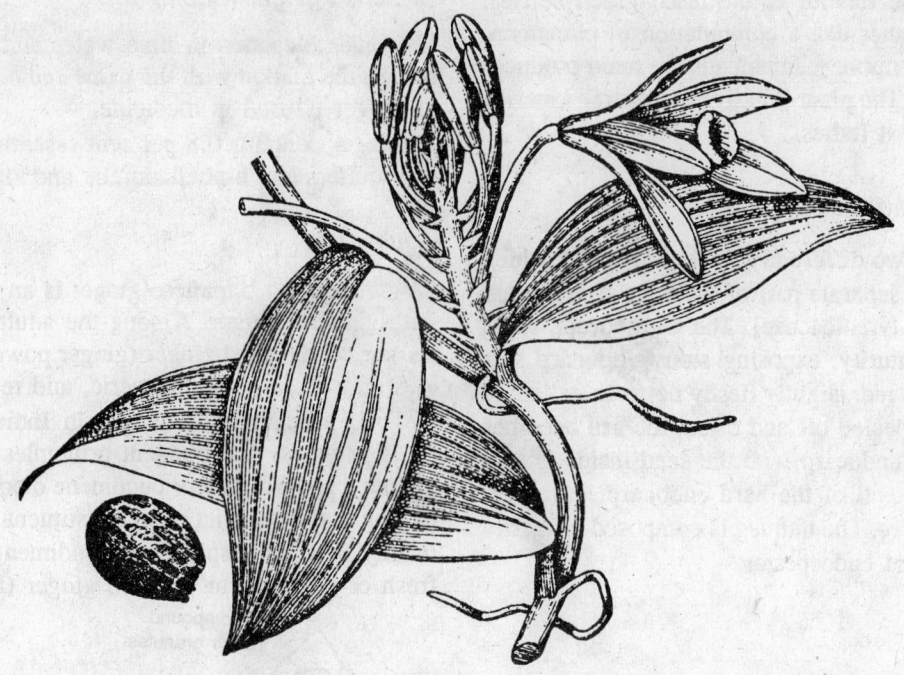

Fig. 9.3. Vanilla (*Vanilla planifolia*) flowering branch with air roots and a seed.

but beginning to turn yellow. Ripe pods are deficient in aroma and are likely to split. After harvest, old stems are cut off, to be replaced by nature before another flowering period.

Curing

The aromatic quality of vanilla develops in the curing process. The pods are piled in heaps in sheds. After partial drying for a few days, they are "sweated". In fair weather this is done by spreading the beans during the morning on woollen blankets in the sun. About mid-day the blankets are folded over the pods, and after sundown the beans are placed overnight in air-tight boxes. The process is continued for about 2 months. The pods become rich coffee-brown in colour and vanillin and other aromatic products develop.

In rainy periods, artificial heat is used in sweating or the pods are periodically dipped in boiling water.

In the cured beans, needlelike or threadlike crystals of **vanillin** appear on the surface.

Vanilla contains 1.5 to 3 per cent of vanillin and very small quantities of other aromatic constituents.

Vanilla extract is obtained by percolating the crushed beans with alcohol.

Vanillin

Vanillin can also be made from guaiacol, a coaltar product; bulbous roots of dahlia; the seeds of cultivated lupine; in the flowers of spiraeas; in the fruit of *Rosa canina*, a European wild rose. Coumarin, found in tonka beans, is similar to vanillin. Seeds of *Coumarauna* (Leguminosae) a tree of Brazil, Guianas contain coumarin.

Uses

Vanilla extract is universally used as a household flavouring, as well as in icecreams, chocolates and other confectionary, and for soda fountain purposes.

Allspice

Pimenta dioica (Lauraceae); the name allspice

refers to the flavour of the dried green berries, which is rather like a combination of cinnamon, cloves, and nutmeg. Jamaica is the main producer of allspice. The plant is native to Central America and the West Indies.

Nutmeg and mace

These are two differently flavoured spices which come from separate part of the fruit of *Myristica fragrans* (Myristicaceae). The fruit (drupe splits open at maturity, exposing stony endocarp surrounded by red, slightly fleshy network called an aril. Once peeled off and dried, the aril becomes *mace*. The endocarp with the seed inside is also dried. Removal of the hard endocarp leaves the round *nutmeg*. The nutmeg is composed primarily of a mass of endosperm.

Ginger

Ginger is obtained from the rhizome of *Zingiber officinale*, Zingiberaceae. The plant is cultivated in India, Jamaica, Sierra Leone, Nigeria, Southern China, Japan, Taiwan and Australia. In India, Andhra Pradesh, Karnataka, Kerala, Madhya Pradesh, Tamil Nadu and West Bengal cultivate ginger. About 70 per cent of the total ginger production in India is confined to Kerala alone. Ginger is ranked third in value among all the spices exported from India.

Ginger is a slender perennial herb, 30-100 cm tall, with robust branched rhizome. Ginger requires a warm and humid climate. Ginger is always propagated vegetatively by portions of the rhizome which have at least one good bud. The cuttings are planted in March and April and the rhizomes are ready to be dug in the following January and February.

Unbleached ginger

The rhizomes are washed in water. Then the outer skin and the underlying parenchyma is scraped away, buds removed, the rhizomes are then dried in the sun.

Bleached ginger (South)

Rhizomes bleached in lime water and dried are sold in the market with the name *sonth*. This type of ginger is used in medicine.

Ginger contains 0.8 per cent essential oil with a sesquiterpene, b-phellandrene and zingiberene.

Adulterants

Zingiber mioga, Japanese ginger is an adulterant with genuine ginger. Among the adulterants are starches, exhausted ginger (ginger powder), cereal products, saw tust, turmeric, and red pepper.

Total production of ginger in India is about 72,000 tonnes. The pungent principles of ginger included non-volatile or cucumene oleoresin containing *gingerol* as its main constituent. Ginger is mainly used as a spice and condiment either in fresh or in the form of dried ginger (Fig. 9.4).

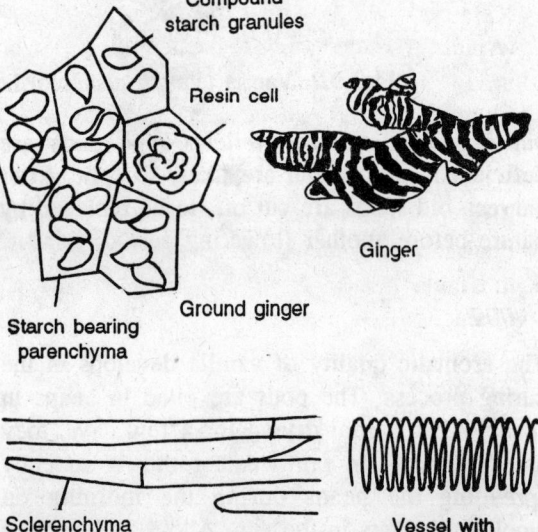

Fig. 9.4. Ginger—*Zingiber officinale* Rhizome and its anatomy.

Ginger is valued in medicine as a carminative and stimulant to the gastro-intestinal tract. The pleasant and aromatic odour of ginger is owing to an essential oil. The rhizome has also been found to be a source of *proteolytic enzymes*. In pulses and vegetables cooking, ginger is used as a spice. Ginger preserve or candy, pickle, ginger-based soft drinks and ginger powder are some of the other outlets from this condiment.

Preservation of ginger

Being a perishable commodity, ginger is available cheaply only for a few months after harvest. For preserving one kg of ginger rhizome about 1 or 1.25 litres of the following solution is needed. Dipping solution : Salt 50 g; glacial acetic acid (concentrated vinegar) 12 ml; potassium metabisulphite 1 g; water 1 litre. After filling the jars (glass or porcelain jars) up to the brim with ginger pieces and dipping solution, seal the jar with tight lid and make the container airtight by putting molten paraffin wax on the lid. Store in a cool, dark place.

Kala zira

Kala zira (*Carum bulbocastanum* Clarke) (Apiaceae)is a perennial crop which is cultivated in Himachal Pradesh (Kinnaur, Lahaul, Spiti and Chamba district; 1,800 m to 3,200 m above sea level). Kala zira is cultivated as a pure crop and also as an intercrop in the apple orchards. The Zira and Saffron Research Station of Himachal Pradesh Krishi Vishva Vidyalaya, located at Kuppa in Kinnaur has been working on the improvement of this crop.

Sandy loam soils are best suited for cultivation. 2-3 ploughings with *desi* plough are necessary. Well rotted farm yard manure at 200-300 q/ha should be applied before the final ploughing. The plant is propagated by bulbs or seeds.

Optimum time for transplanting *bulbs* is mid October to November. Mostly bulbs are used for raising zira crop.

Seed is sown after snow melting, by the end of March/beginning of April in small beds of 3 m x 3 m for raising seedlings. 5-7 kg seed is required for raising seedlings sufficient for transplanting in one hectare.

Kala zira is a crop of about 6 months duration. The crop is harvested in July and the produce is dumped in a room. After 10-15 days, plants are threshed with stick beatings. The seed is passed through various grades in sieves and stored in airtight containers. Average yield 2-2.5 q/ha. After harvest of crop in July, the bulbs embedded in the soil, undergo dormancy for a period of about 6 months.

CUMIN

Cumin is one of the major crops of the spices group. The water requirements of this crop is low. Yield per hectare is good during rabi season in India. The crop is suited for sandy loam areas having less irrigational facilities. In Rajasthan, the seed is sown from 15th November to 30th December. Seed rate 14-15 kg/ha.

The soil is thoroughly ploughed and well pulverised before it is ready for sowing. Seed at the rate of 15 kg/ha treated with 30 g Bavistin is broadcast followed by light planking. Afterwards the field is divided into plots of 2 m x 3 m with necessary irrigation channels. Fertilizers, a dose of 40 kg N and 40 kg P_2O_5, 20 kg K_2O per hectare has been recommended (S.L. Dova, et al., 1984). Half of the nitrogen and full dose of P_2O_5 and K_2O should be incorporated in soil before sowing. The rest 50 per cent nitrogen i.e. 20 kg/ha should be given 40 days after sowing. Application of sulphur dust at 20-25 kg/ha has been found effective for minimising the powdery mildew pathogen present.

First irrigation is to be given just after sowing the seed; second irrigation after 7-12 days after the first; third irrigation after 20-25 per cent seeds have been set.

Weed control by spraying with Nitrogen (tok 6-25) or Basalin. 1 litre/ha is necessary.

Tamarind

Tamarind, imli *Tamarindus indica*, Caesalpiniaceae, is one of the best known Indian trees. Grows to a large size and attains an age of over 200 years. An evergreen tree, thick, dark grey, rough bark with cracks and fissures. Branches spreading. Leaves alternate, compound, 5-12 cm long with slender, channeled axis and 10-20 pairs of nearly stalkless leaflets. Sour to taste stipules fall off early.

Flowers are small, in loose few flowered clusters, yellowish with pink striped petals appearing in May-June. Fruit is slightly curved pod 8-20 cm long, greenish brown, sour pulp and shiny brown, squarish flattened seeds.

The tree is indigenous to tropical Africa but wild plant grow all over the tropical dry savannahs. Plants are usually grown from seed (Fig. 9.5).

The ripe, long brown pods are used primarily for their part, rather sticky pulp. In the United States and Mexico, tamarind pulp is used as a flavouring in sauces. In India, the fruit pulp is an integral part of chutneys and sauces.

The seeds can also be roasted or boiled and eaten after the removal of seed coat.

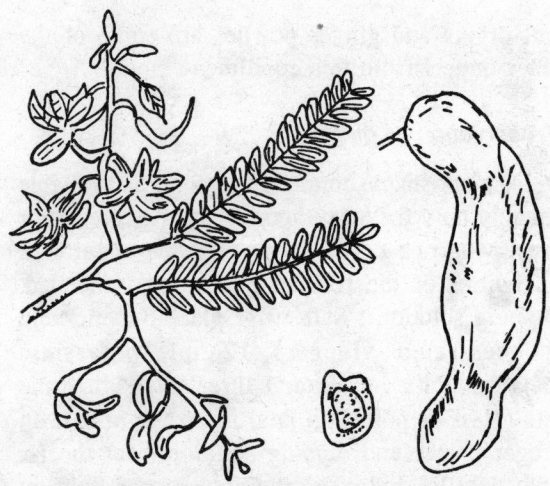

Fig. 9.5. Tamarind—*Tamarindus indica.* Flowering branch, pod and seed.

Carob

Carob, *Ceratonia siliqua* is another Caesalpinoid legume that produces pods used for their pulp. The plant is native to Meditarranean region. The sweet pulp chewed is the mesocarp of the pod. The seeds also been used to make a coffee-like beverage. The plant is propagated by seek. Cyprus is the world's largest producer.

Note : In Figs. 9.6 to 9.15 are given some of the commonly used spices and condiments.

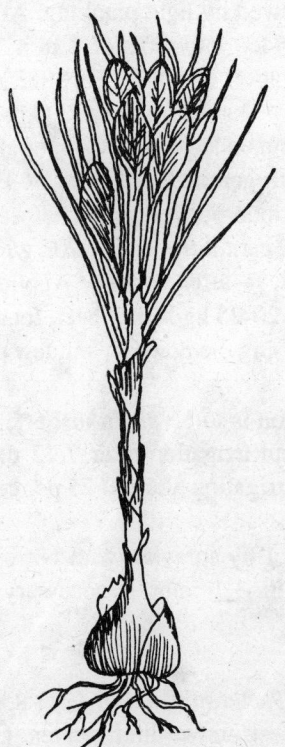

Fig. 9.6. Saffron (*Crocus sativus*) plant in flower.

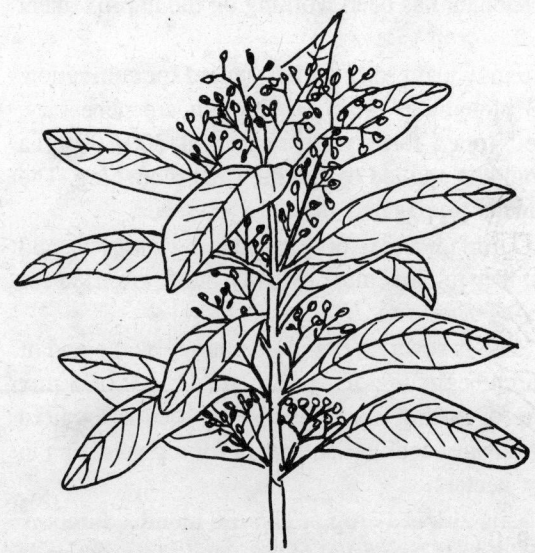

Fig. 9.7. Allspice (*Pimento officinalis*) a flowering branch.

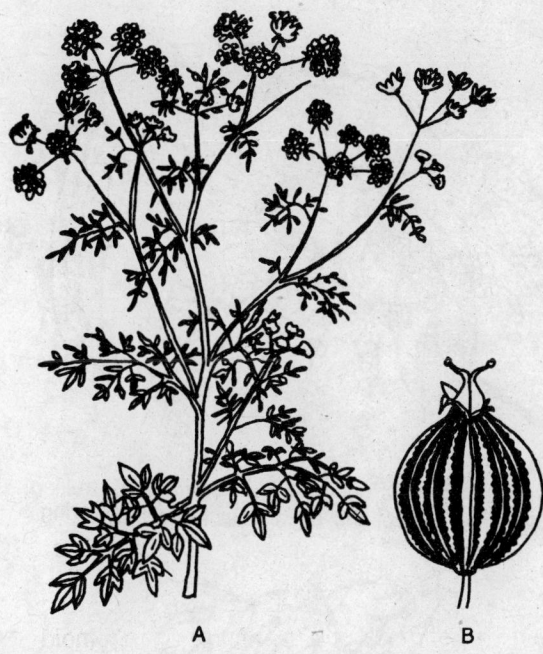

Fig. 9.8. Coriander (*Coriandrum sativum*), A—flowering branch. B—fruit.

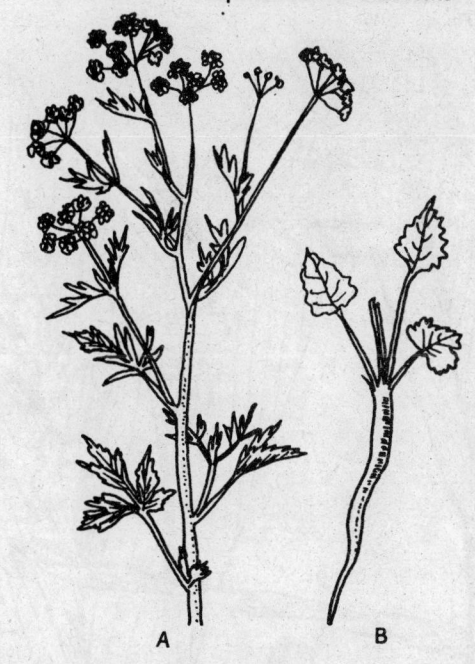

Fig. 9.10. Anise (*Pimpinella anisum*). A—fruiting branch, B—base of the plant.

Fig. 9.9. Cinnamon (*Cinamomum zeylanicum*). Leafy branch with flowers and fruits.

Fig. 9.11. Nutmeg (*Myristica fragrans*) fruiting branch with a split ripe fruit.

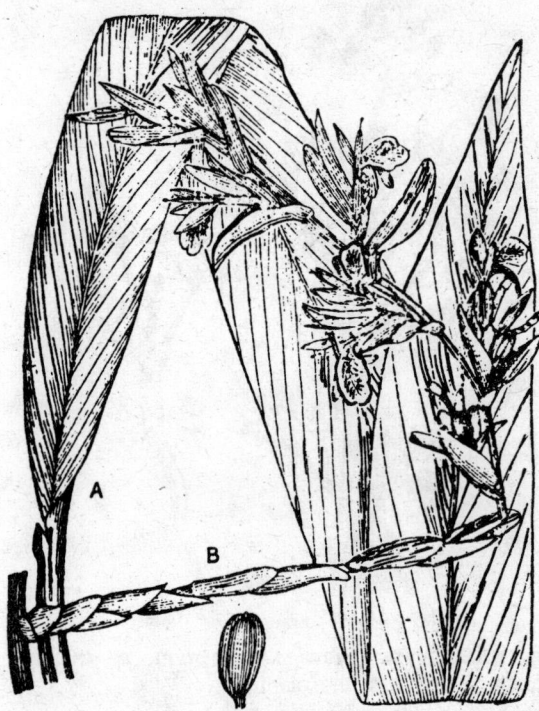

Fig. 9.12. Cardamom (*Elettaria cardamomum*). A—Leaf, B—flowering branch and seed.

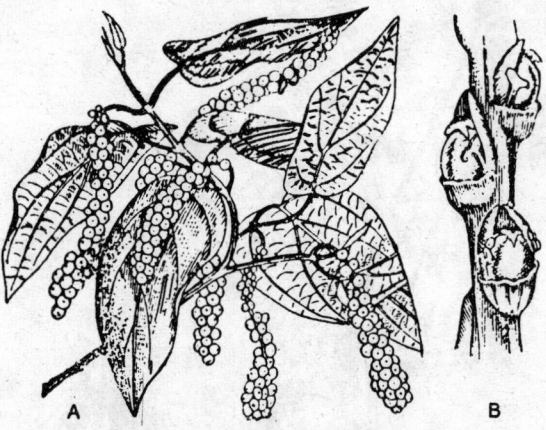

Fig. 9.13. Black Pepper (*Piper nigrum*). A—fruiting branch, B—part of the spike with three flowers.

Fig. 9.14. Red pepper (*Capscium annuum*) a fruiting branch.

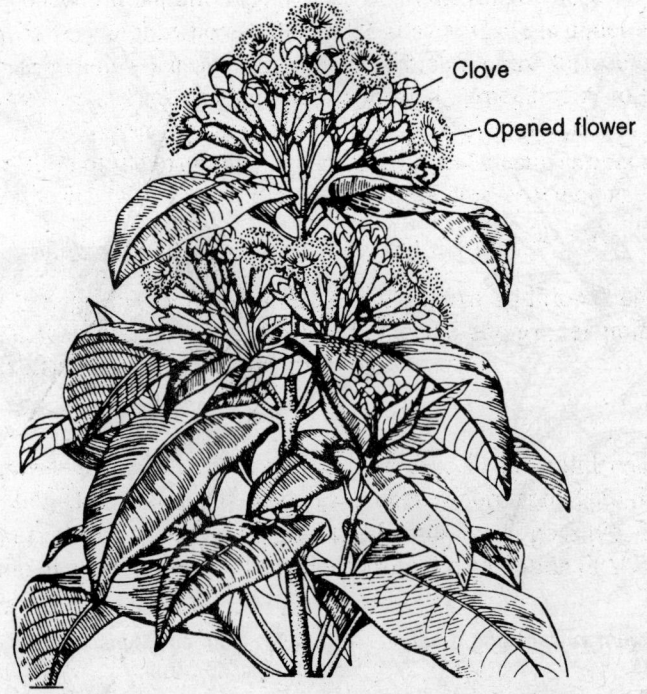

Clove

Opened flower

Fig. 9.15. Clove. *Syzygium aromaticum* twig.

10

Plants that yield fibres occupy second place of importance, next to food plants. Fibres form one of the basic necessities of man. As civilization advanced, the need for vegetable fibres also increasing greatly.

Fibres are the *long sclerenchyma cells* which serve as part of the plant skeleton. They have thick walls and narrow lumen usually with pointed ends. The cell walls contain cellulose as well as lignin. Fibres may occur singly or in groups. According to their origin, fibres are classified into *bast fibres, wood fibres, bundle fibres of leaves* and *surface fibres*. Based on their use, fibres are classified into *textile fibres, brush fibres, rough weaving fibres, filling fibres* and *paper making fibres*.

Some important plant fibres are :

Agavaceae : Agave cantala (maguey); *Agave fourcroyodes* (henequin); *Agave heteracantha* (istli fibre); *Agave sisalana* sisal, hemp; *Furcraea gigantea* var. *willemettiana* (Mauritius hemp); *Phormium tenax* (New Zealand flax); *Sansevieria cylindrica* (bowstring hemp); *Sansevieria hyacinthoides* (African bowstring hemp); *Sans evieria roxburghiana* (Indian bowstring hemp); *Sansevieria zeylanica* (Sri Lanka bowstring hemp).

Apocynaceae : Apocynum cannabium (swamp milkweed, ozone fibre); *Asclepias syriaca* (milk weed).

Bombacaceae : Bombax ceiba (red silk cotton tree, white kapok); *Ochroma pyramidale* (baisa, corkwood).

Bromeliaceae : Aechmea magdalenae (pita); *Ananas comosus* (pineapple); *Neoglaziovia variegata* (caroa); *Tillandsia usneoides* (Spanish moss).

Cyclanthaceae : Carludovica palmata (Panama hat palm).

Poaceae : Arundinaria, Dendrocalamus, Melocanna, Ochlandra, Phyllostachys, Sasa (bamboos). *Legeum spartum* (esparto); *Stipa tenacissima* (esparto).

Fabaceae : Crotalaria juncea (sun hemp); *Sesbania aculeata* (sun hemp); *Sesbania aegyptica* (sun hemp).

Linaceae : Linum usitatissimum flax, linen.

Malvaceae : Abutilon avicennae (flax, linen); *Gossypium* spp (cotton); *Hibiscus cannabinus* (kenaf); *Hibiscus sabdariffa* (roselle); *Pavonia bojeri* (Madagascar hemp); *Pavonia shimperiana* (African) hemp; *Sida cordifolia* (Queensland hemp); *Sida rhombifolia* (Queensland hemp); *Urena lobata* (aramina, Congo jute).

Agave fibres 8000 years old are known from excavations in the Tehuacan Valley, Mexico.

Wood fibres : Xylem (wood) of trees is the world's most prolific source of fibre. Wood fibre can be separated to make *paper* or be dissolved and restructured to make synthetic fibres such as *rayon*.

Moraceae : Broussonetia papyrifera tapa, kapa; *Cannabis sativa* (hemp) (Fig. 10.1).

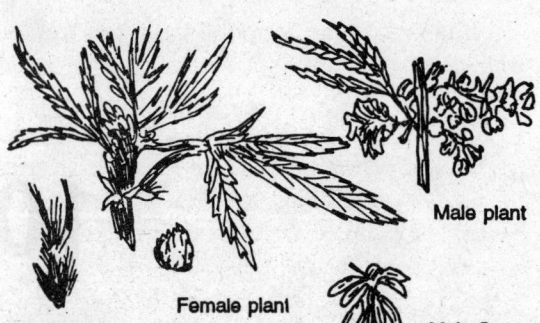

Male plant

Female plant

Female flower

Male flower

Fig. 10.1. *Cannabis sativa*—plant and flowers.

Musaceae : Ensete ventricosa (Abyssinian banana); *Musa textilis* (abaca, Manila hemp).

Arecaceae : Arenga saccharifera (sugar palm); *Attalea* sp (ratha plassava); *Borassus flabellifer* (palmyra palm); *Calamus* sp (rattan cane); *Chamaerops humilis* (dwari fan palm); *Cocos nucifera* (coconut, coir); *Caryota urens* (Kitul fibre); *Daemonorops* spp. (rattan cane); *Metroxylon sagus* (sago palm); *Nipa* sp. (nipa palm); *Raphia* spp. (raffia);*Trachycarpus excelsa* (Chinese windwill palm).

Sterculiaceae : Abroma angustum (devil's cotton).

Tiliaceae : Corchorus capsularis (jute), *Corchorus olitorius* (jute).

Urticaceae : Boehmeria nivea.

Fibres are classified by their use (Table 10.1) or by their structure.

Surface fibres are borne externally on the surface of seeds, fruits, stems, leaves, etc e.g. cotton, kapok. *Soft fibres* or *bast fibres,* from dicotyldonous bark, of which flax, ramie, hemp and jute are most important. *Hard fibres* or *structural fibres,* the fibro-vascular bundles in foliage (leaf) of monocotyledons, such as abaca and sisal. Hard fibres are used mostly as *cordage* (twines and ropes) e.g. abaca. Surface fibres are used chiefly for textiles and as stuffing material. Soft fibres are used mostly for weaving.

Twigs, split stems, sectioned palm leaves are used for brooms, brushes, baskets and for coarse weaving into mats. For brooms, the stems of *Aristida, Sporobolus,* fibres of *Agave,* coconut fibre, the midribs of the leaflets of coconut,

Table 10.1. Classification of vegetable fibres by use

Use	Fibres
Textile fibres	
Bast fibre	Hemp, jute, ramie, linen, plantation fibre
Fruit fibre	Cotton, coir, silk cotton
Leaf fibre	Sisal, henequen, abaca, pineapple
Plaiting and weaving	Pandanus, bamboos
Brush fibres	Sorghum, broomroot
Filling fibres	Silkcotton, milkweed (*Asclepias syriaca*)
Felting fibres	Paper mulberry, lace bark
Papermaking fibres	Cotton flax, hemp, jute, ramie, cereal straws, bagasse, esparto, sisal, Manila hemp, pine, eucalyptus, popular, chestnut

inflorescence of *Arundo donax* are used. Bamboos, cane-palms, willows etc. are used for *wicker work* for baskets, chairs and furniture. *Filling fibres* are used for stuffing matresses, upholstery and life belts. Cotton, the floss of *Calotropis,* the inflorescence of *Aerva tomentosa,* the soft material of coir, silk-cotton, sunhemp fibre and other soft fibres are used as filling material in upholstery. *Plaiting fibres* (mat weaving) are used in hats, matting, thatties, screens or basket making. Leaves of *Pandanus,* stems of cane (*Calamus*), stem of bamboo (*Bambusa* sp), roots of khus-khus grass are used in mats and curtains.

Natural fabrics (tapa cloth) are obtained from the bark of *Broussonetia papyrifera* and *Antiaris toxicaria.*

COTTON

India is one of the major cotton growing countries in the world. In India cotton is grown on about 8 million hectares. Three-fourth acreage of cotton is cultivated under rainfed conditions. Diploid Asiatic species of cotton, *Gossypium arboreum* L. and *G. herbaceum* L. are grown in about 40 per cent of the cotton area. Tetraploid American cotton *G. hirsutum* is grown in India (Fig. 10.2).

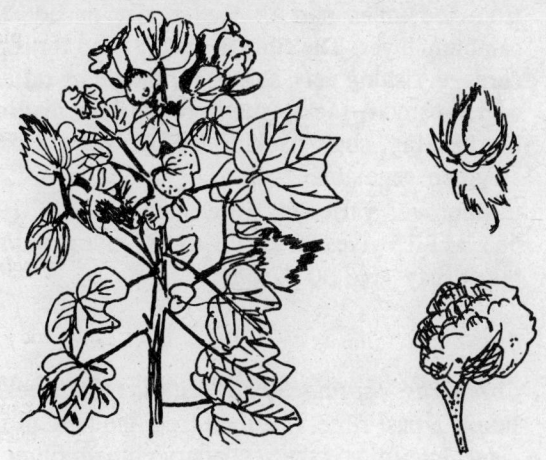

Fig. 10.2. Cotton - *Gossypium herbaceum*. Plant, flower and fruit.

Cotton is indigenous to South East Asia and America. Cotton is a tropical crop. In India cotton is widely cultivated in Karnataka, Andhra Pradesh, Maharashtra, Gujarat. Useful part in cotton for fibre is the *epidermal outgrowths of seed* (*lint*).

The acre yield of cotton in India is the lowest of any major cotton growing country (UAR, USA, USSR). This is due to the cultivation of low yielding varieties. Some work is done to improve cotton in India, incorporation of jassid resistance is *G. hirsutum* from tetraploid wild species *G. tomentosum*; incorporation of resistance and adaptability from Asiatic cultivated species to *G. hirsutum* through interspecific hybridization and back-crossing.

Varieties

Amaravathi, Mahalakshmi, Sangam, Varalakshmi, Sreesailam, Hybrid, Krishna, Deviraj, Deviry Buri 147, B-1007, Budnawar-1, Khandwar-2, SRT-1, MCU-5, Mysore Vijaya, Gujarat-67.

Exploitation of *hybrid vigour* (heterosis) is a land mark in cotton improvement. Balls (1908) reported for the first time hybrid vigour in crosses of upland and Egyptian cotton varieties.

The first cotton hybrid in India was CO-2 (G-*hirsutum*) x SIV-135 (*G. barbadense*), released in 1950, but it was not a commercial success. In 1968, Hybrid-4 an intra-hirsutum cross between G-67 and 'Nectariless', proved to be an instant success. Varalakshmi is an inter-specific hybrid long staple cotton which can yield upto 16 quintals/ha; duration 180-200 days.

But the main bottle neck for heterosis was the complicated technique of seed production requiring huge labour, besides other factors making the hybrid seed much costlier as compared to cereals. Hand emasculation and pollination constituted the only way to produce hybrid seed, as use of cytoplasmic male sterile lines have not been so successful in cotton.

Uses

Cotton fibres form the raw material for the textile industry. Cloth and yarn are manufactured from cotton. *Gun cotton* (nitrocellulose) is obtained by mixing cotton with conc. nitric acid. Cotton seed is used as stock feed because it contains fats, proteins and vitamins; oil is extracted from seed. It is hydrogenated and used as vanaspathi and also in soup making. Now a days it is also used as cooking oil (see under cotton seed oil). Seed cake is also used as cattle feed. Bark is used for the manufacture of low grade paper.

Ramie

Boehmeria nivea (Urticaceae) Ramie or China grass, a bast fibre is regarded as the longest (40-200 mm), toughest and most silky of all vegetable fibres. It has great strength and durability, and is highly resistant to the action of water. The stalks of ramie are not retted to free the fibres, instead, the bark and phloem tissues are peeled from he stems and fibres decorticated by beating and scraping.

Ramie is used for sacks, said cloth, belting, table-cloth, sheeting, nets, threads, cordage, paper etc. Gas mantles of good quality are made from it. Its special use is in the preparation of lustrous, non-creasable fabrics (Fig. 10.3).

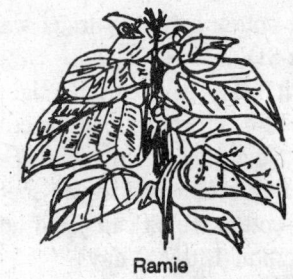

Fig. 10.3. Ramie - *Boehmeria nivea.*

Coir

Cocos nucifera Palmae (Arecaceae) is obtained from coconut fruit. Coir fibre is extracted from the husks of the whole coconut. Mat coir is produced in west coast of India. Curl fibre is short and obtained from unretted husk, used for stuffing upholstery, cushions etc. Bristle fibre is coarse and thick, used in making brushes and brooms.

Coir fibre is used in activated carbon, paper pulp, roofing tiles, writing boards, thermal insulations, high stretch paper, olive oil filters etc. Coir yarn is used for hop wines in United States. Coir bags are used in tea estates. Coir mats are used in ships and boats. Coir is rubberized for making cushion seating for automobiles and railways.

Sunnhemp : *Crotalaria juncea* (Fabaceae)

Sunn hemp (Bombay hemp) is a bast fibre grown on a raw material for sacking and cordage. Sunn hemp is cultivated for fibre and for green manure in Andhra Pradesh, Uttar Pradesh and Madhya Pradesh. The crop is grown on a limited scale in Maharashtra, Punjab, West Bengal and Orissa. The fibre is used for making ropes, canvas, gunny bags etc. A great quantit of the fibre exported is said to be used primarily in the preparation of cigarette papers and high quality tissue papers.

Deccan hemp (Kenaf, Bimli jute, Java jutes, Ambari)

Hibiscus cannabinus (Malvaceae) as a rain fed crop in large areas in Madhya Pradesh, Andhra

Pradesh and Tamil Nadu. It is a common wild plant of tropical and subtropical Africa and is wild or naturalized in Asia. Kenaf fibre is extracted from the inner part of the cortex, outside the cambium layer. The fibre is widely used for rope, cordage, fishing nets, strings for the tieing rafters; coarse canvas, sacks, gunny bags, floor matting, rug backing, chair backing etc. Leaves are used as green vegetable.

Improved varieties : Bimli jute - MT-15, HC-583, MT-15 gives 20 q/ha; HC-583 gives 25 q/ha, but it may give 30-32 q/ha.

Jute

Corchorus capsularis, *C. olitorius* (Tiliaceae). Jute is a bast fibre, obtained from the bark of the plant. Jute fibre is the secondary phloem fibres of the stem. Fibre is extracted by setting the stem. Retting is a process by which the fibres in the bark get loosened and separated from the woody stalk due to the removal of pectins, gums and other mucilaginous substances. This is usually effected by the combined action of water and microorganisms.

Jute is an important textile fibre next in importance to cotton. It is also a cheap fibre used in making certain fabrics, coarse sacking and package material; twines, ropes, paper and as filling material; cloth used for upholstery, linoleum, tapestries and mats.

Jute and Mesta

Jute and Mesta (deccan hemp and roselle) are important cash crops which play a significant role in Indian economy. They are mainly grown in the Eastern States like West Bengal, Bihar, Assam, Tripura, Uttar Pradesh and Meghalaya. Besides, Jute-Mesta is also grown in the States like Andhra Pradesh, Karnataka and Maharashtra. Jute is predominantly a rained crop and covers about 8.5 lakh ha. Mesta is a totally rainfed crop and is grown on about 3.0 lakh ha. The production of Jute and Mesta in 1979-80 was around 80 lakh bales. For 1985-86, the production target of raw Jute and Mesta is 86.5 lakh bales as against the Seventh Plan target of 95 lakh bales.

Flax

Linum usitatissimum, Linaceae is a rabi (winter) crop in India. Fibre is obtained fibrous bundles running the length of the stem and forming a ring in the cortex. The fibre is extracted from the stem by dry-scutching process which enables the fibre being extracted without any pre-treatment of the straw with the help of specially designed machine. The fibre obtained is rough and crude, washing in hot water render it soft. Fibre is used in making strong ropes, turnes and cordages, cheap and rough textiles such as blankets, carpets, hessian-like cloth, galicha, mats, mattresses and newar; linso fabrics, straw boards, gloves, footwear, netting, sports gear, writing paper, parchment paper, cigarette paper, toys etc.

Flax seed oil is used in varnishes, paints and linoleum. Linseed cake is used as cattle feed.

Silk cotton (kapok)

Ceiba pentandra (Bombacaceae) : Silk cotton fibre is obtained from the floss of fruits. The fruits are dried, split open, floss is taken along with seeds, dried in the sun; seeds separated by beating the dried material with sticks.

Kapok finds use in bedding and upholstery industries; manufacture of lifebuoys, bets, waistcoats and other naval life-saving appliances.

Refined kapok seed oil is used for the same purposes as refined cottonseed oil.

Abaca

Abaca or Manila hemp, comes from *Musa textilis* (Musaceae). The fibres are extracted primarily from the *outer peripheries of the leaf bases* that make up the "stem" of the giant herbaceous plant. The fibre is resistant to microbial rotting and to salt water.

The fibre is used for the manufacture of ship's cables and ropes, for strong, sacking, coarse fabrics and tough paper. The fibre is used in tea bags, dollar bills, "Manila" envelopes, Garman of Italian salamis wrapped in clothlike casings and filter-tapped cigarettes.

Hemp

Cannabis sativa (Cannabinaceae) Hemp fibres are extracted from the stems by retting, often followed by scutching and pounding. Well-processed hemp is creamy-white, soft, and has a silky sheen. The fibre is typically used for cordage, rope, canvas, soil cloth and Levi strauss pants. In France, the fibre is used in the manufacture of composition board.

Hemp fibre is a bast fibre obtained from primary and secondary pholem fibres (Fig. 10.1).

Sisal and Henequen

Sisal fibre is obtained from the leaves of *Agave sisalana*, Agavaceae. Henequen fibre is obtained from the leaves of *Agave fourcroydes*. The plants are native to Central America. The outer mature leaves are cut at the base, carted to the factory, and fed between rollers that squeeze out most of the water and turn the soft tissues into an amorphous mush that is scraped away from the fibres. The fibres are then washed and hung in the sun to dry. *A. cantala* is grown in Philippines, Java and Cuba (Fig. 10.4).

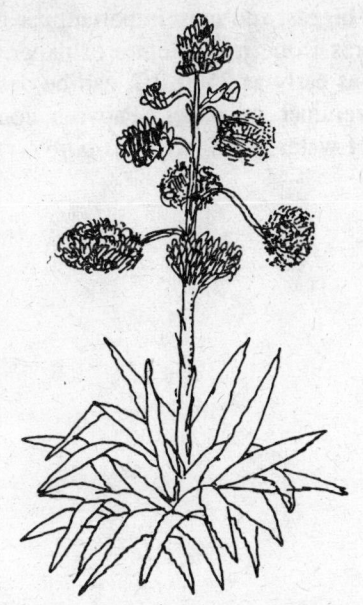

Fig. 10.4. Sisal—*Agave sisalana* plant.

The fibre is mostly used for making ropes, cordage turne coarse fabrics and bushes.

Rosella (mesta)

Hibiscus sabdariffa (Malvaceae) : Roselle fibre is a bast fibre used for sacking, cordage, rope, fishing nets. Bags made of rosette fibre are extensively used in Java for packing sugar. Roselle fibre resembles bimli fibre or jute fibre.

Improved varieties : Roselle NP Sab-5; RT 1, 2, 26; HS 4288, HS 7910, AMU-1; average yields of RT-2 18-18 q/ha; HS 4288 25 q/ha; AMV-1 22 q/ha; HS 7910 25-26 q/ha. HS 4288 may give 30 q/ha.

Congo jute *Urena lobata* (Malvaceae)

Congo jute is a superior jute substitute, used as an admixture in spinning jute. The plant is grown as a fibre crop in Brazil, where it is known as oramine fibre. Congo jute is a bast fibre obtained from the stem. The fibre is used for making ropes, carpets, lessain etc.

PAPER MAKING

The biggest and most important use of vegetable fibres is the manufacture of paper. *Papyrus,* in use as early as 3500 BC, can be described as the forerumer of paper. Papyrus consisted of strips of water seeds (*Cyperus papyrus*) laid in a network, soaked with water and beaten flat. Papyrus was the staple writing surface until is gradually lost its popularity to animal parchment after 200 BC.

A Chinese eunuch named Ts' ai Lun has been accredited with the invention of paper in the year AD 105, but a further 700 years elapsed before the secrets of the technique left China and reached the Arabs. The Moors introduced paper making to Spain in the 12th century and it was a further 300 years before it was in use in the rest of the world.

The basic principles of paper making have not been changed since their inception. Paper is a thin tissue composed of any fibrous vegetable material. The individual fibres are first separated by mechanical or chemical action and then reconstituted in a sheet form, by depositing the fibres on to a wire mesh using water.

The principal raw materials for paper making are *seed hairs* (cotton, once used for high-grade writing and printing paper); *bast fibres* (flax, hemp, jute and ramie which are particularly strong but resistant to bleaching); *grass fibres* (cereal straws and bagasse); *leaf fibres* (esparto, sisal and Manila hemp which have strength and are tear resistant, but hard to bleach); *wood fibres* (*Picea excelsa*, spruce, *Pinus sylvestris*, pine) conifers produce long fibres. Eucalyptus, poplar, chestnut and birch produce shorter fibres, are used in paper making.

11

Woods

Secondary xylem is called wood. Wood is useful as timber and lumber. *Fagus grandiflora* beech (Fagaceae), *Quercus incana*, oak (Fagaceae), *Quercus semecarpifolia*, oak (Fagaceae), *Quercus ilex*, oak (Fagaceae), *Betula lutea*, birch (Betulaceae), *Betula alnoides*, birch (Betulaceae), *Betula utilis*, birch (Betulaceae), *Ulmus racemosa*, elm (Ulmaceae), *Ulmus wallichiana*, elm (Ulmaceae), *Dalbergia latifolia*, rose wood (Fabaceae) (Fig. 11.1), *Dalbergia sissoo*, sissoo, shisham (Fabaceae), *Tectona grandis* teak, *sagwan* (Verbenaceae), *Santalum album* sandal wood, sandal (Santalaceae). Teak wood is used in making furniture. Sal, shisham and rosewoods are also used in furniture. Deodar (*Cedrus deodara*) wood is used in railway sleepers. Pine (*Pinus* sp.) wood is used in furniture and railway sleepers. Sandal wood is used in making decorative boxes, toys, perfumes and toilet soaps. Wood of *Salmalia malabarica* (Bombacaceae) (Fig. 11.2), *Boswellia serrata* (Burseraceae), *Anthocephalus indicus* etc., are used in making matches and match boxes.

India consumes nearly 4.75 lakh cubic metres of match wood. The main timber used (over 60 per cent) is *semul* (*Salmalia malabrica*). Nearly 50 per cent of the matches are produced in large factories located at Ambarnath, Bareilly, Calcutta and Madras and the remaining in the non-mechanical sector concentrated mainly in Sivakasi, Sattur and Kavilpatti areas of Tamil Nadu. About 95 million gross boxes of safety matches are produced per year.

Fig. 11.1. Rose wood, *Dalbergia latifolia* (Fabaceae).

Soft wood is a term applied to any coniferous wood such as pine, larch, spruce and jew. The term *hard wood* is used for timbers which are the wood of broad-leaved i.e., dicotyledonous tree species. Tropical hardwoods supply many of the luxury timbers used for furniture and veneers e.g. mahogany, teak, rosewood and afrormosia. *Afrormosia* is a West African tree, *Pericopsis elata* (= *Afrormosia elata*), whose timber was introduced into international trade about 40 years ago. It is a highly valued hardwood, but due to over-exploitation and poor powers of regeneration the tree is in serious danger of becoming extinct.

Nearly 50 per cent of wood, cut worldwide is useful for fuel; four-fifths of this in the developing world. This firewood is obtained from dry or savannah woodlands, scrubs and brushwood areas.

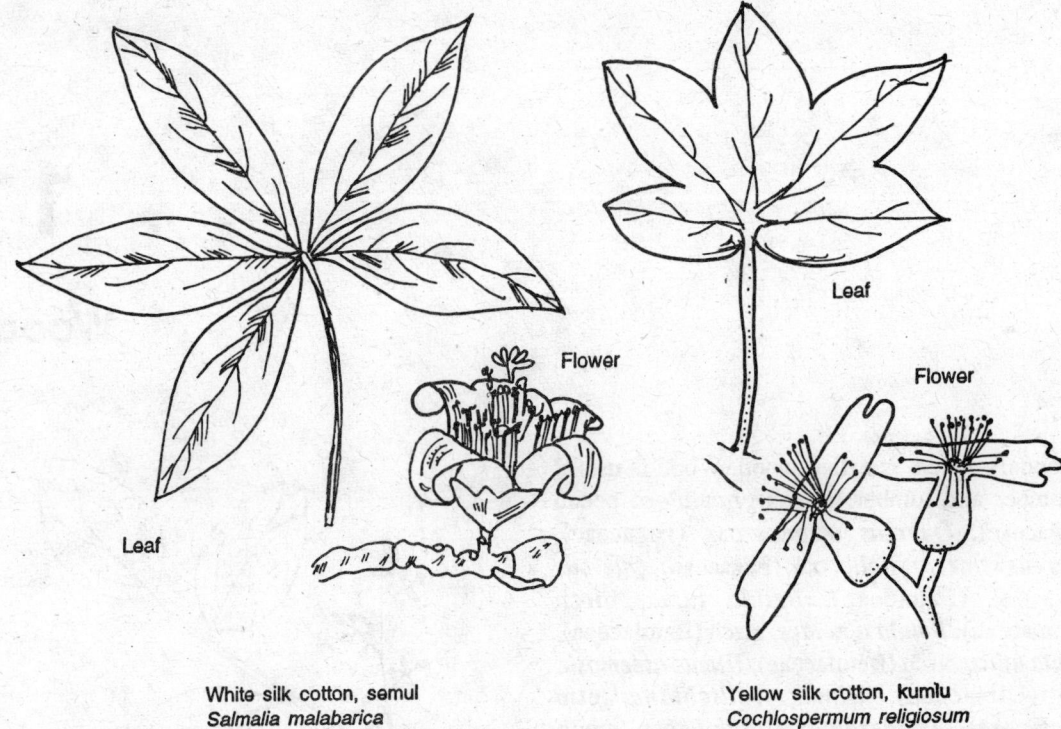

White silk cotton, semul
Salmalia malabarica

Yellow silk cotton, kumĩu
Cochlospermum religiosum

Fig. 11.2. White silk cotton—*Salmalia malabaricum*. Yellow silk cotton—*Cochlospermum religiosum*.

Teak, Tectona grandis a broad-leaved tropical hardwood grow well in monoculture : a southeast Asian species which has been successfully planted within its natural areas and elsewhere in Asia and in Africa (Fig. 11.3).

Mahogany Swietenia mahogany was not only used for furniture and carving, but was irreplaceable material for the construction of parts of machinery before light metal alloys and plastics became available.

Cork is obtained from cork oak *Quercus suber*, Cork has buoyancy and light weight; cellular structure, resilience and compressibility, high resistance to deterioration; resistance to moisture and common liquids. Cork oak is a medium sized evergreen temperate oak, found in areas around Mediterranean. Cork is used for shoe soles, fishing floats, insulating spaceships etc. *Aeschynomene aspera, Ochroma lagopus* are light woods suited for navy and aircrafts.

The wood of *Salix alba* (willow) of Salicaceae

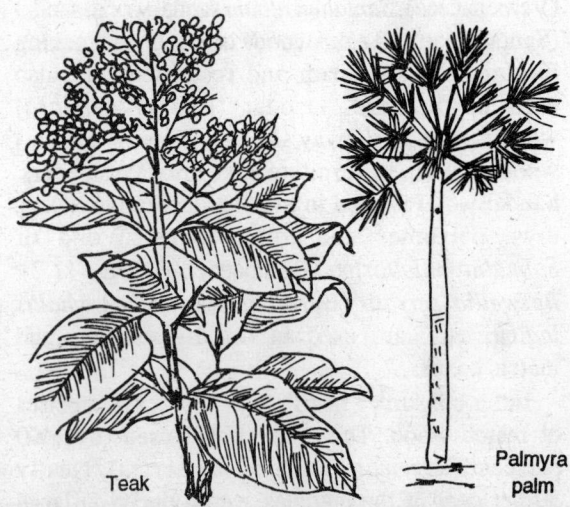

Fig. 11.3. Teak—*Tectona grandis*. Borassus—*Borassus flebellifer*.

is used for making *cricket bats*. The wood of *Morus alba* (mulberry) of Moraceae is used in making *hockey sticks, tennis* and *badminton rack-*

ets, cricket stumps etc. Wood of *Adina cordifolia, Bombax ceiba, Juniperus* sp. etc. are used in making *pencils.* Wood of *Adina, Albizzia, Cedrela, Dalbergia, Morus* etc. are used in making *musical instruments. Rifle parts* are made up of the wood of walnut (*Juglans* sp.). Black wood is obtained from *Acacia* sp. Most durable wood is of *Tectona grandis* (Fig. 11.4).

Hard woods

In the hard woods vessels are present instead of the tracheids. Each individual vessel is called

vessel member and originates from one cell in the cambial tissue. Each vessel is a hollow tube through which conduction of food and water take place. In some timbers the vessels of the early-wood are much larger than those of the late wood with an abrupt change in size. These are called *ring porous woods.* Woods that show the end section of the vessels gradually decreasing in size and with no abrupt change in size of vessels are said to be *diffuse porous woods.* Common hard woods are oak, chestnut, teak, sal, shisham, bass

Fig. 11.4. Teak—Inflorescence and fruits.

wood anjun (*Hardwickia binata*), beech (*Fagus grandiflora*).

Soft woods

In the soft woods, the cells are tiny and shaped like tubes with closed ends which are somewhat pointed. Tracheids are present in the timbers, which are packed closely together, parallel to the length of the tree or log. The spring wood (early word) and summer wood (late wood) are clearly noticeable which defines the growth of rings so well, as seen in many pines. Most soft woods contain resin canals. Bands of medullary rays are seen passing through the wood e.g. cedar, cypress, larch, yew, *Sequoia*, spruce, white silk cotton, red silk cotton (Fig. 11.2).

White teak of Yemane

The *gumhar*, white teak of Yemane *Gmelina arborea,* is a close relative of teak. The plant is indigenous to India, Pakistan, Bangladesh, Burma, Sri Lanka, Thailand, Laos, Vietnam and Southern China.

Gumhar is an unarmed, unbuttressed, deciduous tree, common in moist mixed forests of Assam, West Bengal, Bangladesh and Burma. The tree can attain a 4.5 m girth, 15 m clear bole and a height of 30 m (Fig. 11.5).

Fig. 11.5. White teak—*Gmelina arborea.*

The tree is propagated by sowing fruits. The tree is also propagated by stump planting, stem cutting, grafting and budding.

Uses

The timber is used for construction work, planking, furniture, cabinet work, panelling, carvings, boxes, boat building, agricultural implements, turnery, toys, ornamental carving, picture frames, slate frames, artificial limbs, guns, rifles, musical instruments etc. The wood is suitable for high class plywood. Wood is also used as fuel, as pulp for peper, as fodder for the cattle. Leaves are used as a feed for the eri silkworm.

The fruits, flowers, leaves, root and bark are used in native medicine for treatment of cough troubles, headache, stomachic, laxative, nerve tonic and blood diseases.

Silver oak

Silver oak (*Grevillea robusta*), a member of Proteaceae, is native to sub-tropical coastal areas of New South Wales and Queensland. The plant grows at wide range of attitudes, from sea level to above 2,300 m.

The plant prefers worm, temperate to sub-tropical temperatures. The young plants are frost sensitive. The plant can grow in sandy soils, loams of medium fertility and even in acidic soils. The plant does not tolerate water logging.

Uses

1. Wood is tough, elastic and moderately dense (specific gravity 0.57). Wood is pale pink, heart wood is brown. The wood is excellent for cabinet work. The timber is used in making plywood panelling, air-freight, railroad ties and furniture.
2. The gold coloured flowers have the attraction of honey bees and this tree is designated as honey plant.
3. The leaves can be used as green manure; the dried branches for fuelwood.
4. The bark yields yellow gum; the gum yields on hydrolysis galactose and arabinose.
5. The plant has been used to provide light shade over coffee and tea plantations.

6. Because of the height, good form and attractive flowers, the plant is often planted as a *street tree.*

The Australian blackwood Acacia is a large evergreen glabrons tree mostly of pyramidal form with an erect bole up to 25 m and a dense crown of characteristic shape. The leaves and pods are used as fodder for livestock animals. The tree is useful for windbelts and as an ornamental and shade tree. It is also used as pulp wood. Wood is useful for cabinet work and agricultural instruments.

BAMBOO

Bambusa arundinacea (Poaceae)

41 species of bamboo are found in India. Bamboos are tall 6-16 m, arborescent grasses belonging to the tribe Bambusoideae of the family Poaceae. New growth in bamboo is produced only by apical meristems. Consequently, there is no secondary or radial growth. All growth is vertical and caused by the production of new apical cells and cell elongation. Stems of mature bamboo consist of segmented, hollow tubes which give bamboo a unique combination of flexibility, light weight and strength.

Bamboo is a perennial grass with woody stem. The bamboo cane grows in joints, either straight or unbranched or branching from 10-15 cm upto 40 cm. They arise from an underground rootstock, often forming a large dense mass (Fig. 11.6).

Flowers are formed in small clusters. Flowering takes place after a number of years of vegetative growth. After the flowering is over, the clump usually dies.

Fruit size ranges from wheat grain chaff to the size of first. It is said that the population of jungle mice grows too rapidly when bamboos flower and fruit. The rapid regeneration of bamboo from seed is proverbial.

Bambusa arundinea is found in Orissa, Assam, Eastern Bengal, South and Western India. *B.*

Bamboo

Fig. 11.6. Bamboo—*Bambusa arundinacea.*

vulgaris in Assam, Burma, Java, South America and West Indies. *B. tulda* and *B. balcoa* in Bengal. *Arundinaria aristata* in eastern Himalayas. *A. wightiana* in Nilgiris; *B. polymorpha* in eastern Bengal and Assam. *Pseudostachyum polymorphum* in eastern Himalayas, Assam, Upper Burma and Sikkim. *Dendrocalamus strictus* in the deciduous forests throughout India. Assam bamboo, maggar, *Dendrocalamus hamiltoni* is native to eastern Himalaya and Nepal. Maggar is a wonder bamboo grown in Himachal Pradesh.

Maggar provides an excellent fodder in winter for the milch cattle. The bamboo is used for shuttering, scaffolding, fences, staircases and wooden nails. The culms are easily workable into plaiting fibres, used in wicker-work cottage industry, for making storage bins, baskets, mats, screens, furniture and fan. Bamboo is an ideal plant for checking soil erosion.

Bamboo pulp is an important raw material for paper industry. Bamboo seeds are edible. Tabasheer, a siliceous secretion found in the culms) is a cooling tonic and aphrodisiac, a cure for asthma and cough.

Tendu leaves

These are used as wrappers of tobacco for making *bidis.* In India tendu trees (*Diospyros tomentosa*

L.) (f. Ebenaceae) are found in the forests and the scrub jungle of Madhya Pradesh and Orissa. The new leaves come out in April-May when they are plucked by women and children and made into *gaddies* of 50 each and dried in the sun. There after they are packed in gunny bags and stored for supply to bidi-makers (Fig. 11.7).

Bamboo blossoms

Bamboo blooms and produces seeds only once in its lifetime, just before it dies. Sometimes the event occurs about 120 years after its birth. Bamboo plants usually flower once every 15, 30, 60 or 120 years (a geometric series). Because of its peculiar flowering pattern, plant breeders could not breed improved strains from bamboo seeds. Mascarenhas (1986) worked on shoots of *Bambusa arundinaceae* and *Dendrocalamus brandisii* and fed these in the laboratory with minerals, vitamins, sugar, coconut milk and plant hormones. The process was repeated thrice in a matter of weeks, before the shoots flowered and produced seeds.

In India *Khair* trees are found in Uttar Pradesh, Madhya Pradesh and Rajasthan. There are two factories for making *Kattha* one at Bareilly in Uttar Pradesh and the other at Shivpuri in Madhya Pradesh.

Cutch and *Kattha* are obtained from the heart-wood of *Khair*, mostly in UP and MP. The process is essentially one of extraction. Chips of the wood are boiled in water, more and more of it being added as its quantity decreases due evaporation until a thick decoction containing *Cutch* and *Kattha* is obtained. This is then poured on sand or in bamboo containers lined with gunny, so that *cutch* which is soluble in hot water goes out leaving the suspended *Kattha* as a pink paste. Most of the *Kattha* is made by boiling chips in a battery of earthen pots erected in the forest. This method is wasteful in the consumption of fuelwood and all the *clutch* is lost. There are modern factories in Bareilly and Shivapuri.

In combination with lime, *Kattha* is the main ingredient of *paan* leaf (betel leaf) chewed extensively in India and in the far East. *Kattha* is extracted from the chips of the heartwood of *khair* (*Acacia catechu*) by boiling them in water and then concentrating the extract. When this is colled *kattha* crystallises out and leaves the *cutch* in solution, which can be obtained in solid form by evaporation. *Cutch* was formerly used for dyeing fishing nets but now it is used as a boiler

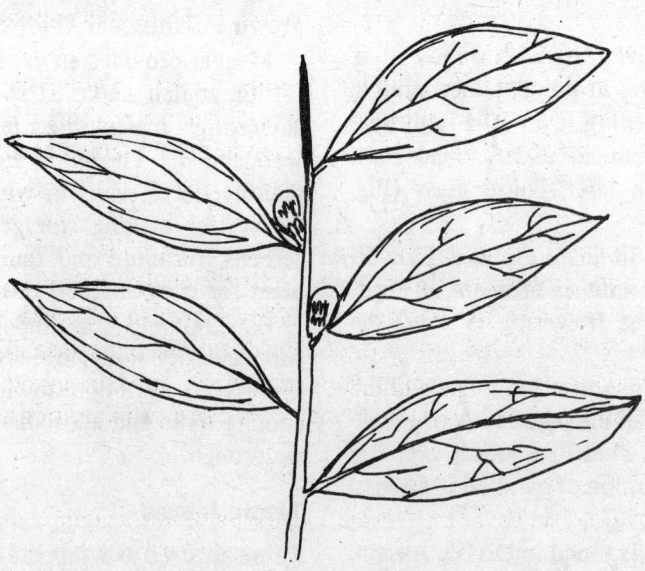

Fig. 11.7. Tendu. Tendu, ebony, *Diospyros melanoxylon* young leaves are used for rolling *bidis*.

compound and also in the machines used for boring.

COMMON TIMBERS

Aglaia (*Aglaia* sp); Alder (*Alnus nepalensis Don*); (*Alnus nitida Endl*); Amari (*Amoora wallichii King*); Amra; (*Spondias pinnata Kurz.*); Ash (*Fraxinus* sp.); Axlewood (*Anogeissus latifolia* Wall). Babul (*Acacia arabica* Willd); Bahera (*Terminalia bellerica* Roxb); Benteak (*Legerstroemia lanceolata* Wall); Bijasal (*Pterocarpus marsupium* Roxb); Birch (*Betula* sp.); Bulletwood (*Manilkara* sp.), (*Mimusops* sp.). Casuarina (*Casuarina equisetifolia* Linn); Celtis (*Celtis australis* Linn); Champ (*Michelia champaca* Linn.), Chickrassy (*Chukrasia tabularis* Adr. Juss); Chialuni (*Schima wallichii* Choisy); Chir (*Pinus roxburghii* Sargent); Cinnamon (*Cinnamonum* sp.). Debdaru (*Polyalthia* sp.); Deodar (*Cedrus deodara* Louden.); Dhaman (*Grewia tiliaefolia* Vahl.); Ebony (*Diospyros melanoxylon* Roxb.); Fig (*Ficus* sp.); Fir (*Abies pindrow* Royle); Gamari (*Gmelina arborea* Linn.); Geon (*Excoecaria agallocha* Linn.); Gurjan (*Dipterocarpus indicus* Bedd.); Hollong (*Dipterocarpus macrocarpus* Vesque.); Hopea (*Hopea parviflora* Bedd.); Horse chestnut (*Aesculus indica* Colebr.); Imli (*Tamarindus indica* Linn.); Indian oak (*Quercus* sp.); Indian olive (*Olea* sp.); Jamcen (*Syzygium cumini* Skeel); Jarul (*Lagerstroemia speciosa* Pers.); Jhingan (*Lannea coromandelica* Merr.' = *Lannea grandis*); Kail, blue pine (*Pinus wallichiana* A.B. jacks); Kanchan (*Bauhinia* sp.); Kardahi (*Anogeissus pendula* Edgew.); Karada (*Cleistanthus collinus* (Roxb.) Benth. and HK; Kasi (*Bridelia retusa* Spreng.); Kathal (*Artocarpus heterophyllus* Lemk.); Khair (*Acacia catechu* Willd.); Khasi-pine (*Pinus insularis* Endl.); Kindal (*Terminalia paniculata* ROH).

Kokko (*Albizzia lebbeck* Benth.); Kusum (*Schleichera trijuga* Oken.); Lakooch (*Artocarpus lakoocha* Roxb.); Lampati (*Duabanga sonneratioides* Ham.); Laurel (*Terminalia tomentosa*); Lendi (*Lagerstroemia parviflora* Roxb.); Mahogany (*Swietenia* sp.); Mahua (*Madhuca* sp.); Mango (*Mangifera indica* Linn.); Maple (*Acer* sp.); Mesua (*Mesua ferrea* Linn.); Mulberry (*Morus* sp.); Mundani (*Artocarpus fraxinifolius* Wight); Myrobalan (*Terminalia chebula* Retz.); Narikel (*Pterygota alata* R. Br.); Padauk (*Pterocarpus dalbergioides* Roxb.); Persian bilac (*Melia azedarach* Linn.); Pitraj (*Aphanamixis polystachya* (Wall.) Parker; Poon (*Calophyllum* sp.); Poplar (*Populus* sp.); Rajbrikh (*Cassia fistula* Linn); Rose wood, black wood (*Dalbergia latifolia* Roxb); Sal (*Shorea robusta* Gaertn. f.); Salai (*Boswellia serrata* Roxb.); Sandan (*Ougeinia dalbergioides* Benth.); Semul (*Salmalia malabarica* Schott. and Endl.; Siris (*Albizzia stipulata* Boivin.; Sissoo (*Dalbergia sissoo* oxb.); Spruce (*Picea smithiana* Boiss.); Teak (*Tectona grandis* Linn. f.); Toon (*Cedrela toona* Roxb.); Walnut (*Juglans* sp.).

12

Medicinal Plants

The study of medicinal plants is termed as *Pharmacognosy*. This branch of science is concerned with the history, collection, identification, selection, preservation, import and export of various plants and plant parts that are drug yielding. There is a closely associated science of *Pharmacology* which includes the study of the action of drugs on human body.

Almost all unani and ayurvedic drugs and medicines are obtained from plants. Several allopathic drugs and medicines and almost all homoeopathic medicines are derived from plants. Biochemic medicines too are derived from plants.

Now a days, well organized farming is done to cultivate medicinal plants, in advanced countries. The drug plants have their medicinal value due to the presence of chemical substances that exert definite physiological influence on human body.

Classification of drugs

I. Drugs derived from lower plants :
   1. From algae
   2. From fungi
II. Drugs derived from higher plants :
   1. Drugs from roots and underground parts
   2. Drugs from stem and aerial branches
   3. Drugs from leaves
   4. Drugs from fruits and seeds
   5. Drugs from bark

Introduction

"With a garden and a well you have enough remedies for a whole town." This may appear oversimplified, but plants have had their place in the healing art and still keep it today in many parts of the world.

The concept of "Signature" depends on the resemblance of the plant or a plant part to the structure or function of an organ of human body on which it is supposed to act. For example, the virginal whiteness of waterlily has been considered a sign of its virtue as an antiphrodisiac, whereas the mandrake which resembles the legs of human body is thought to stimulate virility. Because of this resemblance it has also been regarded as a cure for all bodily ills.

The passion flower, whose flowers bear the symbols of passion, was believed to calm mental and physical anguish. Poppy seeds are enclosed in a capsule resembling the human head and seemed to be indicated for migraine. Pilewort, with roots like tangled clusters of veins, was thought to design it for the treatment of haemorrhoids.

Obviously, the enquiring minds of Greeks could not accept such a simplistic attitude. In the first century of our era, Dioscorides wrote the first known treatise on Materica Medica, in which 600 medicinal plants were listed.

The Chinese also recognized the need for a systematic herbal. One was compiled by Liche-Chen (1518-1593) who spent 30 years writing the Pen-T'sao (pen-origin; t'sao-herbs). He included in it the contents of all the forgotton or incomplete treatises of earlier ages while getting rid, as far as possible, of errors, gaps and flights of fancy. This Chinese pharmacopoeia mentions the use of the chaulmoogra plant against leprosy. Chaulmoogra oil has been used in recent years in conjunction with sulfones.

The Pen-T'sao sings the praises of the camphor tree and the Ma-Huang (*Ephedra sinica*) from which Nagai, a Japanese in 1887 extracted an alkaloid, ephedrine, used in ophthalmology.

In short, among the 250,000 or so plant species, scattered throughout the world, man early drew a distinction between what may be called the 'Flowers of Good' and 'Flowers of Evil'. On this basis, plants have been meticulously catalogued and about 700 of them have been identified as deadly for man as well as beast. The most notorious example is that of Hashish, known as bhang in India and kif in North Africa.

The opium poppy is now classed among the 'Flowers of Evil', although in the past opium was held in high regard for its sleep-giving, pain-killing and other medicinal properties and also as a source of pleasurable sensations. Science has now found synthetic drugs that can be used as a substitute for opium and its derivatives such as morphine, codeine and heroin. The black list of the 'Flowers of Evil' should also include some European plants, such as the corn cockle, most spurges, the foxglove and belladonna, whose cherry-like berries have poisoned many children.

A clear distinction between the 'Flowers of Evil' and 'Flowers of Good' cannot be drawn because it is often the quantity of the drug consumed which determines whether the effects would be beneficial or harmful.

As soon as modern science had taken shape it became possible to produce some of the alkaloids from dangerous plants. The earliest alkaloid was *morphine*, which was extracted from opium in 1806. In 1817 *Emetine*, which was used in the treatment of dysentery and amoebic hepatitis, was obtained from the Brazilian ipecacuanha. The following year saw the advent of *strychnine*, an alkaloid obtained from *Strychnos nux-vomica* and *Veratrine* a hypotensor from the white hellebore.

In 1820, *Colchicine* the active agent of meadow saffron, was isolated. In 1832 *codeine*, a cough sedative was isolated from opium. One year later, belladonna joined the list of essential plants with its product *atropine*.

At the same time, the "good" flowers gave in succession quinine, in 1820 and theobromine, a diuretic alkaloid extracted from *Cacao* in 1842. *Sparteine* was obtained from the common broom in 1851.

Rauwolfia serpentina, which is both useful and poisonous, contains 18 alkaloids. It has been used for over 200 years in the treatment of disorders of every kind from snake bite to psychosis. Research work has shown that this plant to contain substances with sedative properties.

Medicinal plants have been used for various purposes such as cooking, in chemical industry and in medicine. Some of these uses are as follows :

A great number of herbs have been grown both for cooking and for their healing virtues.

The onion has from the earliest times have been valued as a health-giving plant. Onions have been rubbed on the bald scalp in the hope of making hair grow. They have been found to contain antibiotics, but in very small amounts.

The virtues of garlic have been acclaimed since antiquity. It is said that the workers who built the pyramid of Gizeh ate quantities of garlic. The greeks and romans made great use of it and it was introduced into France by the returning crusaders. Now-a-days it is used almost universally in cooking, and in many parts of the world is popularly believed to be a remedy for practically all ills.

Many alkaloids extracted from the plants are used in chemical industry, like *Securine* and *Berberine* from *Securinega suffruticosa* and *Berberis vulgaris* respectively. *Lutenourine* a new vegetable antibiotic from *Nuphar luteum; Diosponine* from *Dioscorea caucasica*; *Tauromisino* from *Artemisia taurica*.

These are many plants which are used in medicine but the major ones are as follows : *Atropa belladonna* is the source of *atropine*, a valuable drug especially used in ophthalmology.

Digitalis purpurea is the source *digitalin* which was used for ulcers of the skin. Now-a-days it is used as a treatment for fevers and even for tuberculosis. *Marrubium vulgare* is used as a remedy for coughs and bronchial affections.

Balm (*Melissa officinalis*) is an aromatic plant that has been used down the centuries in the form of an infusion or cordial for its refreshing and sedative properties. *Rosa canina* has long been used in herb minded families to prepare infusions, syrups and marmalades. A drug used in treating certain heart conditions have been derived from the seeds of jute (*Corchorus olitorius*).

Some drugs from the ocean

Since ancient times products of pharmaceutical application have been derived from the sea. As early as 8th century BC, the Chinese used sea weeds and other algae to treat intestinal disorders.

"Marine Pharmacology" or "Medicinal oceanography" nevertheless is one of the most promising yet unexploited fields in medicine today. It involves the collection of marine organisms that manufacture biologically active substances, their separation and purification and the study of their possible application in the treatment of diseases. The ocean can yield two types of products which can be exploited in medicine :

(i) antibiotics for controlling pathogenic organisms, and

(ii) drugs which act directly on diseases body systems to relieve pain.

Dr. Paul, has isolated a new antibiotic from a marine bacteriophage which has proved valuable in curing certain chest diseases. The antibiotic isolated from green algae can be very useful in treating intestinal disorders. Agar, the well known remedy for stomach complaints, is also produced from algae.

The therapeutic agents in medicinal plants are certain specific groups of substances or individual substances which are the products of plant metabolism. Active principles may be divided into two main groups : *toxic* - those that have a poisonous effect on animal organisms - and *non-toxic* - those that do not cause poisoning but are therapeutically

effective. Needless to say, it is impossible to draw a precise dividing line between the two, for even certain non-toxic plants-aromatic species containing essential oil. for example - may, if used in excessive quantities or over a longer period than that prescribed, impair the function of various bodily organs. The symptoms of mild poisoning, such as sickness, diarrhoea or pains in the stomach or intestines, are not fatal but may be unpleasant or painful. Only those plants that are clearly poisonous when taken even in minute quantities can be definitely branded as toxic; book attention is drawn to the fact that they are poisonous and should not be gathered except by a qualified collector who knows how to take the necessary precautions when gathering them. It is dangerous for any patient to attempt to treat himself with such a drug without the advice and attendant care of a physician. Acutely poisonous plants are used in the treatment of disease in very small doses. They are extremely dangerous in non-qualified hands because the dosage required to achieve the beneficial effect is very close to the dosage which produces symptoms of poisoning. On the other hand, when used by experienced doctors even extremely toxic substances can serve to restore man's health.

The principal groups of active constituents are described below. Besides these the plants also contain subsidiary components which sometimes serve to increase the efficiency of the therapeutically important principles. An example of this action, which is known as synergism, is to be found in laxative drugs, which contain anthraquinone glycosides, such as Senna (*Cassia acutifolia*) or Cascara (*Rhamnus purshiana*).

Birth of a new drug

Plants, animals and minerals form the main sources of our drugs. Scientists all over the country are working hard to test various herbs on modern scientific lines with a view to proving the claims of their medicinal properties. *Rauwolfia serpentina* (Sarpaghandha), the roots of which yield a very useful blood pressure lowering drug, has been mentioned in our Shastras written centuries ago.

A determined and co-ordinated effort by various agencies - ICMR (The Indian Council of Medical Research), CSIR (The Council of Scientific and Industrial Research), the Central Council of Research in Indian Medicine and Homoeopathy may well rediscover many more important drugs from this treasury in the near future.

Herrors of drugs

The use and misuse of drugs has in the past been the subject of strongly held opinions; tea and coffee, indeed, have been suspected of effects that we should not take seriously today.

Sir Aubrey Lewis writes :

"Now the drugs which are mainly causing concern are grouped in six types. Sixty or seventy years ago, the list embraced four of them viz. morphia, alcohol, cannabis and cocaine and in addition chloroform, ether, chloral, sulphonal, phenacetin, tobacco (as a stimulant and intoxicant), tea and coffee. Alcohol and morphine, then as now, were given major consideration.

History

Even as recently as the first half of the nineteenth century apothecaries not only stocked dried medicinal plants in the form of crude drugs for the preparation of various *herbal tea mixtures,* but also used them to make all kinds of *tinctures, extracts* and *juices* which in turn were employed in preparing medicinal drops, *syrups, infusions, ointments* and *liniments.* This period marked not only the peak of the ancient repute of medicinal plants but also the beginning of its decline. The esteem was not unfounded; from the earliest times no other more effective medicinal preparations were available to physicians and drug plants were considered the basis of all treatment. There exist numerous written records and well-authenticated accounts of the collection and cultivation of medicinal plants dating from the beginning of the Christian era and, to a limited extent, from even earlier times. In the Middle Ages knowledge of medicinal plants spread from the monastery garden to the ordinary citizen. Leaving aside the various superstitions and beliefs in supernatural power, popular folk remedies are found in use for the treatment of various illnesses as early as medieval times, and even today in many highly developed countries use is still made of numerous different plants, chiefly in the form of teas.

The second half of the nineteenth century brought with it several important discoveries in the newly developing field of chemistry and saw the rapid progress of this science. Medicinal plants became one of its chief objects of interest and in time chemists succeeded in isolating the pure, active substances, or rather groups of substances which they contained and which, in many instances, have replaced the crude drugs. Then came the first synthetic medicines; they became predominant and gradually pushed the herbal medicines which had formerly been used into the background. This trend still continues, even though the once reigning belief that medicinal plants are of no importance and can all be replaced by man-made drugs is no longer tenable. In the same way that drug plants were rapidly falling into disuse a century ago so today they are regaining their rightful place in the field of medicine, though naturally on a far higher level and with a few better knowledge of their effects on the human organisms. Researches are investigating not only the classical plants but also related species that may contain similar active constituents, as well as hitherto unknown plants which have no previous history of medical use.

Important medicinal plants

Medicinal plants, or rather the parts that are collected and dried - the crude drugs, for example the roots (*radix*), leaves (*folium*), flowers (*flas*), herbage (*herba*) - are the raw material used for the industrial preparation of pure active substances. The synthetic preparation of many of these substances is either unknown at the present time or uneconomical for industrial purposes. These substances are employed either in their pure form or are used for the preparation of new substances, often with a more significant therapeutic action. Examples of active constituents from plants which are now isolated and used by the medical profession in pure tableted form include quinine from *Cinchona*, the anti-malarial drug, digoxin from *Digitalis lanata* used for the treatment of heart failure and vinblastine from *Vinca* species used for certain malignant diseases. The organic chemicals from crude drugs also provide a model which can be copied or modified by the organic chemist to produce a more potent drug or a better drug with less side effects. An example of this latter group are the drugs used as local anaesthetics which are based on the artificially modified chemical structure of cocaine. This is a drug isolated from the leaves of a Peruvian bush. Other examples can be found in the penicillin drugs many of which are semisynthetic but all based on the molecular configuration first isolated from the *Penicillium* fungus.

The drugs used in the treatment of heart disease, are the cardiac glycosides, natural plant products obtained chiefly from *Digitalis lanata*, foxglove (*Digitalis purpurea*), adonis (*Adonis vernalis*) and lily-of-the-valley (*Convallaria majalis*). Ergot alkaloids obtained from ergot fungus (*Claviceps purpurea*), either singly or combined, are the basic drugs of obstetrics, internal medicine and neurology. Opium alkaloids, primarily morphine, obtained from the opium poppy (*Papaver somniferum*) are contained in the countless pharmaceutical preparations serving to relieve pain, alleviate cramps and suppress bouts of coughing. As yet modern medicine has no substitutes for these natural drugs and physicians cannot do without them (Fig. 12.1).

Tinctures

Medicinal plants, or rather the crude drugs, are used in many instances for the preparation of extracts with water, alcohol or ether - thick, thin, fluid or solid may be produced according to the consistency. Alcoholic extracts are termed tinctures. On occasion these are still obtained from the fresh plant by presenting and subsequent thickening of the juice (succus). All these galenicals medicinal preparations made by extracting the desired constituent from the crude drug according to the Galenic method, a technique intro-

Opium poppy - *Papaver somniferum*

Fig. 12.1. Opium poppy—*Papaver somniferum* plant and fruit.

duced by Claudius Galen, a Greek physician of the second century AD - are of minor importance today. However, there are still a number of drugs from which the various pure active principles have not been isolated or there the combination of their active elements has a much more powerful therapeutic action. For instance, an excellent preparation in the treatment of nervous disorders is the tincture or extract made from valerian (*Valeriana officinalis*). The alcoholic extract from a mixture of the leaves of bogbean (*Menyanthes trifoliata*), the top parts of centaury (*Centaurium minus*), the seed vessels of bitter orange (*Citrus aurantium*), and the seed roots of gentian (*Gentiana lutea*) plus a small quantity of cinnamon oil yields the well-known bitter tincture.

Herbal tea mixtures

Drugs from medicinal plants are the basic material for making up herbal tea mixtures, taken either in the form of a decoction or as an infusion. Medicinal teas are of varied composition according to which disease they are intended to treat. They serve as an auxiliary medicine, that is, their mild physiological action promotes that of the primary medicinal preparation. In some instances, especially in chronic ailments, their action is even more effective than that of fast-acting medicines.

Typical drug plants used in herbal tea mixtures are elder berry (*Sambucus nigra*), birch (*Betula verrucosa*), blessed thistle (*Cnicus benedictus*), mullein (*Verbascum thapsiforme*), chamomile (*Matricaria chamomilla*), hawthorn (*Crataegus exyacantha*), common juniper (*Juniperus communis*), restharrow (*Ononis spinosa*), ribwort (*Plantago lanceolata*), small-leaved lime (*Tilia cordata*), peppermint (*Mentha piperita*), bearberry (*Arctostaphylos uva-ursi*), garden sage (*Salvia officinalis*), *Herniaria glabra, Drosera rotundifolia,* agrimony (*Agrimonia eupatoria*), St. John's wort (*Hypericum perforatum*) and centaury (*Centaruium minus*).

The plant kingdom is a limitless source of new species of plants containing active constituents of therapeutic value, and scientists throughout the world are well aware of this. That is why scientific expeditions are organized and extensive test made on the pharmacological activities of the little known plants so collected. An example of the results of such efforts is the discovery of reserpine, an alkaloid which has a hypotensive effect that is, it lowers blood pressure. This was discovered in plan of the genus *Rauwolfia.* Hypotensive alkaloids have also been discovered in the European evergreen Periwinkle (*Vinca minor*) of the genus *Catharanthus,* together with an alkaloid which is, useful in the treatment of certain types of malignant disease.

Moreover, new specific active constituents have been discovered even in well known and much used drug plants, for instance, chamomile oil (the oil of *Matricaria chamomilla*) was found to contain principles with specific anti-inflammatory effects (chamazulene and bisabolol). Members of the carrot family (Umbelliferae), have yielded numerous substances of the furocoumarin group that hold promise for use in internal medicine.

Medicinal plants and the investigation of their chemical composition as well as their therapeutic effects yield knowledge used in the synthetic preparation of new substances or provide substances which in various combinations become powerful medicinal agents.

Thus, for instance, from ergotamine, an alkaloid of Ergot Fungus (*Claviceps purpurea*), it is possible to obtain by chemical means the semi-synthetic alkaloid dihydroergotamine which is very effective in the treatment of migraine (Fig. 12.1). Certain members of the nightshade family (Solanaceae) and of the family Dioscoreaceae contain glyco-alkaloids - solasodine, tomatidine, diosgenine, for example - which serve for the synthetic preparation of steroid hormones, previously isolated only from animal organs or products.

Alkaloids are naturally occurring materials in some plant tissues and are distinguished chemically by the fact that they contain nitrogen. They have marked toxic effects on animal organisms. In the main they are stable, crystalline substances that are both odourless and colourless and also sensitive to high temperatures, at which they disintegrate. Rarely does a plant contain one alkaloid only; generally it contains a group of chemically related components. The responses produced by alkaloids in animals and plants have been studied in some detail and it has been found that alkaloids rank among the most efficient and therapeutically most significant substances known. *They generally occur in all the plant organs but their concentration varies and appears to be greatest just before or at the beginning of the flowering period.* So far no satisfactory answer has been found to the question of their function in the life cycle of the plant. It has been suggested by some scientists that they are waste substances, while others consider them to have hormonal activity or possibly even metabolic function, participating in the biochemical reactions of plant cells.

Those plants that man has not been able to cultivate with success, such as monkshood (*Aconitum napellus*), European white hellebore (*Veratrum album*), autumn crocus (*Colchicum autumnale*), and *Herniaria glabra*, are collected wild, as are those species where single-purpose cultivation would be uneconomic, for example elderberry (*Sambucus nigra*) common oak (*Quercus rubra*), small-leaved lime (*Tilia cordata*) and common juniper (*Juniperus communis*), and last or all those species which require such a specialized environment for growth as is impossible to provide in normal cultivation, such as sundew (*Drosera rotundifolia*), bogbean (*Menyanthes trifoliata*), iceland moss (*Centraria islandica*), and mistletoe (*Viscum album*).

Plants that are grown commercially include those that have been under cultivation for centuries, for example *Anise*.

Medicinal alkaloids

Several hundred alkaloids are known to modern science. The first alkaloid was isolated in 1803 by Friedrich Wilhelm Adam Serturner (1793-1841), a pharmacist's assistant in the German town of Paderborn. It was morphine, the basic alkaloid from opium. This was followed by the isolation of further plant alkaloids, such as strychnine (1818), quinine (1820), conine (1827), nicotine (1828), atropine, hyoscyamine and colchicine (1833).

. The Jimson Weed (*Datura stramonium*) and pepper (*Capsicum annuum*), henbane (*Hyoscyanus niger*), and deadly nightshade (*Atropa belladonna*) are typical alkaloidal plants which contain a group of alkaloids chemically related to the substance tropane. Examples are atropine, hyoscyamine and scopolamine, constituents which significantly affect smooth muscle and thereby bring about the relaxation of muscular spasms. They are also of importance in internal medicine and ophthalmology. Smooth muscle is sometimes known as involuntary muscle because voluntary control of its contractions is not possible as the bladder and somb. Another typical alkaloidal plants is the opium poppy (*Papaver somniferum*), which besides morphine, narcotine, papaverine, codeine and thebaine contains some twenty more alkaloids of lesser medicinal importance. Alkaloids from opium poppy are very effective in relieving pain, relaxing cramps and diminishing or suppressing the desire to cough. Ergot fungus (*Claviceps purpurea*) is another important alkaloidal plant. Ergot alkaloids are widely used by gynaecologists, neurologists and psychiatrists (Fig. 12.2). Mention should also be made of the European, white hellebore (*Veratrum album*), autumn crocus (*Colchicum autumnale*), monkshood (*Aconitum napellus*) and greater celandine (*Chelidonium majus*). Toxic alkaloids such as morphine, codeine or colchicine are taken in the purified form as table or as ingredients of liquid preparations such as mixtures and tinctures.

Other alkaloids such as ergotamine and atropine may also be injected for rapid onset of action.

Nux-vomica

Strychnos nux-vomica, strychnine tree, Loganiaceae beeds contain the alkaloids *strychnine, brucine* alkaloids. They produce a pronounced, usually quite specific action on different areas of the nervous system in man and animal. *Strychnine* is a virulent poison, used as tonic in very small doses. *Brucine* is an extremely toxic alkaloid. *Tubocurarine* blocks neuromuscular activity and thus help relax muscles that retain cataracted under anesthesia. *Strychnine* was formerly used medicinally as a central nervous system stimulant.

Emetine is present in ipecac (*Cephalaeis ipecacuanha*, Rubiaceae). Quinine and quinidine are present in calisaya (*Cinchona calisaya*, Rubiaceae). Quinine is present in *Cinchona micrantha, Cinchona officinalis*. Ergatamine is present in ergot fungus (*Claviceps purpurea*, Hypocreaceae). Vinblastine, vincristine are present in Madagascar periwinkle (*Catharanthus roseus*, Apocynaceae). Reserpine is present in Indian snakeroot (*Rauwolfia serpentina*, Apocynaceae. Strychnine is present in *Strychnos nux-vomica*, Loganiaceae. Heliotrine, lasiocarpine are present in helitrope (*Heliotropium* sp., Boraginaceae). Sparteine is present in *Cytisus scoparius*, Leguminosae. Lupinine is present in *Lupinus* sp., Leguminosae. Cytisine is present in *Laburnum anagyroides*, Leguminosae. Aconite is present in *Aconite sp.* (Ranunculaceae). Ephedrine is present in *Ephedra sinica, E. distachya, E. equisitiva.*

Glycosides

Glycosides occur naturally in plants and are characterized by the fact that chemically they consist of a sugar portion attached by a special chemical bond to a non-sugar-portion. They are readily broken down by the mediation of enzymes into those two parts. Glycosides are substances with a pronounced physiological action on animal tissues and are poisonous to man. They are the product of special metabolic processes in certain plants. Their concentration in the various plant organs varies and is dependent on the age, or ecology, of the plant. These with the most significant therapeutic action are the cardiac glycosides. They exert effects on the muscle tissue of the heart, which consists of specially modified hardworking muscle cells.

During the disease known as *congestive heart failure*, the heart muscle becomes inefficient and the heart enlarged. The *cardiac glycosides* increase the efficiency of the failing heart and return its size to normal. They are often referred to as digitalis glycosides after the plants of the genus *Digitalis* of which they are characteristic constituents.

Fig. 12.2. Ergot—*Claviceps p........a* on rye (*Secale cereale*) plant. Cleistothecia are shown separately.

Alkaloids

Arecoline is present in betel nut palm, (*Areca catehu*, Arecaceae). Coniine is present in hemlock (*Conium maculatum*, Lobeliceae). Atropine is present in belladonna (*Atropa belladonna*, Solanaceae). Hyoscyamine is present in henbane (*Hyoscyamus niger*, Solanaceae). Scopolamine is present in mandrake (*Mandragora officinarum*, Solanaceae). Tubocurarine is present in curape (*Chondrodendocon tomentosum*, Menispermaceae).

Cardiac glycosides are present, for example, in *Digitalis lanata*, the foxglove (*Digitalis purpurea*), and other species of this genus, and also in such plants as adonis (*Adonis vernalis*) and lily-of-the-valley (*Canvallaria majalis*). Cardiac glycosides are natural plant products that are irreplaceable in current medical therapy. They are very potent even in small doses and extremely poisonous. The most commonly used drug in this group is *digitoxin* from *Digitalis lanata* which is commonly prescribed in the form of small tablets. The main problem associated with the use of *Digitalis* is the fact that the useful or beneficial dose is close to the toxic dose. Symptoms of toxicity will be observed by the physician as side effects. This arises because the glycosides are only slowly eliminated from the body; for this reason the lanatosides are sometimes used because they are more easily excreted.

Another important group of glycosides from the medical viewpoint are the *anthraquinone glycosides* which occur in Rhamnaceae, Polygonaceae and Rubiaceae. They are mildly poisonous in larger doses and are generally used for their laxative action in certain disorders of the digestive system. The anthraquinone glycosides have been much abused by the general public as daily laxatives. This regular use of laxatives has been shown to be harmful to the natural rhythm of the body and leads to laziness of the bowels. There are instances, however, when it is necessary to use a laxative, for example, before surgery, during pregnancy or before X-ray of the intestine, and in these cases the anthraquinone glycosides, owing to their mild action, are preferred. Anthraquinone glycosides are present in rhubarb (*Rheum palmatum*) and madder root (*Rubia tinctorum*).

A therapeutic action is also exerted by the 'mustard' or *thiocyanata glycosides*. They are very unstable, being readily broken down by the mediation of an appropriate enzyme which is also present in the plant. Mustard glycosides are distinguished chemically by the presence of sulphur in their structure. They occur chiefly in Brassicaceae, Tropaeolaceae and Resedaceae, and

exert an irritant and disinfectant action besides improving the blood supply to various organs. Applied externally in large doses they cause marked local irritation and even inflammation of te skin. Two examples of such glycosides are sinigrin, found in black mustard (*Brassica nigra*), and glucotropaeolin, present in *Tropaeolum majus*.

Then there are the *phenolic glycosides* which are only slightly poisonous and exert a disinfectant, anti-inflammatory and diuretic action. They are present in many species of Ericaceae. Bearberry (*Arctostaphylos uva-ursi*), contains *arbutin* and *methylarbutin* Cowslip (*Primula veris*), of Primulaceae, contains *primulaverin* and *primaverin*.

Saponins are present in most species of *Primula* are also in many members Solanaceae. The horse chestnut (*Aesculus hippocastanum*), contains the saponin *aescin*.

Anthraquinones are present in cascare (*Rhamnus purshiana*, Rhamnaceae); senna (*Cassia angustifolia*); Alexandrian senna (*Cassia acutifolia*); rhubarb (*Rheum officinale*, *R. palmatum*, Polygonaceae); alder buckthorn (*Rhamnus frangula*, Rhamnaceae); aloe (*Aloe barbadensis*, Liliaceae). Cardiac glycosides are present in foxglove (*Digitalis purpurea*, *Digitalis lanata*, Scropulariaceae); *Strophanthus sarmentosus*, *S. gratus*, *S. hispidus*, *S. kombe*, Apocynaceae, oleander, *Nerium oleander*, Apocynaceae; Queen of the night (*Seleniereus grandiflorus*, Cactaceae); lily of the valley (*Convallaria majalis*, Liliaceae); star of Bethelhem (*Ornithogalum umbellatum*, Liliaceae); squill (*Urginea maritima*, Liliaceae) pheasant's-eye (*Adonis vernalis*, Ranunculaceae; christmas rose (*Helleborus niger*, Ramunculaceae); yam (*Dioscorea elephantipes*, Dioscoreaceae); Saponins are present in licorice (*Glycyrrhiza glabra*, Leguminaceae).

Coumarins are present in wormwood (*Artemisia absinthium*, Asteraceae); mezereon (*Daphne mezereum*, Thymelaeaceae); Wack haw (*Viburnum prunifolium*, Caprifoliaceae).

Saponins

Saponines are natural compounds which have a

chemical nature very similar to that of the glycosides but are distinguished by the fact that they produce a soapy lather when shaken with water. Their most marked physiological activity is that they cause the break up of red blood cells, and are therefore potent blood poisons. This process is known as *haemolysis*. Saponins have a marked irritant effect and some are extremely toxic, such as those occurring in *Paris guadrifolia* and *Agrostemma githago*. However, they also have beneficial therapeutic properties and may act as an expectorant in catarrh of the upper respiratory passages. They are often found together with glycosides. *Digitalis lanata,* for example, in addition to cardiac glycosides also contains the saponin *tigonin,* and foxglove (*Digitalis purpurea*) the saponine *digitonin* and *gitonin. Sapindus emarginatus, S. detergens* yield saponins and are used as soap nut of commerce (Fig. 12.3 and 12.4).

Other groups of active principles

Organic acids, which play an important role in the medicinal properties of many plants. The chief ones are malic acid, citric acid, oxalic acid and tartaric acid. They have a mild laxative effect.

Sugars, which are important auxiliary substances in many plants. The large group of substances to which the sugars belong also includes the mucilages and other polysaccharides, such as starch.

Mention should also be made of plant oils and vitamins. The plant oils are useful dietary components in place of animal fats as it is thought by some experts they have a beneficial effect in the prevention of heart disease and diseases of the body arteries. Vitamins, which are abundant in plant issues, are essential for the maintenance of health.

The importance of medicinal plants today

Medicinal plants are a numerically large group of economically important plants. They include various species or cultivated varieties - cultivars - the active constituents of which are used in the treatment of various diseases. This group also includes plants which not only serve a medicinal purpose but contain aromatic substances used in the cosmetics and food industries. Some species of medicinal plants have a primary use outside medicine, being cultivated for the provision of wood, tannins (used in processing leather). lant fibres for the textile industry, and dyes. This latter group, however, is of little importance today as natural dyes have largely been replaced by manmade substances. Other species provide caoutchouc, oil or fodder for domestic animals, or are used by man as vegetables, fruits or ornamental plants.

Fig. 12.3. *Sapindus emarginatus.* Soapnut tree of South India.

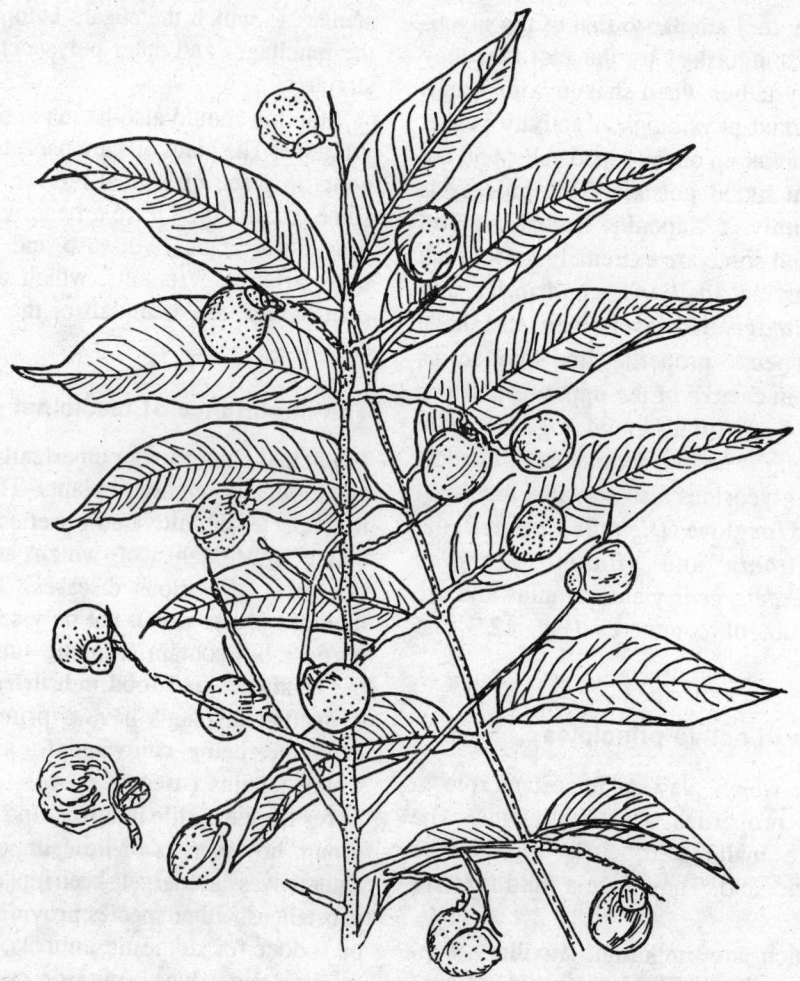

Fig. 12.4. *Sapindus detergens.* Soapnut tree of North India.

Medicinal plants thus include single-purpose species, used only for their medicinal properties and nothing else, and species that can be put to a number of uses. These are in the majority, their chief application being in fields other than medicine and their use in pharmaceutics secondary or even incidental.

Species used purely for medicinal purpose include both poisonous and non-poisonous plants. The first group includes such species as henbane (*Hyoscyamus niger*) and jimson weed (*Datura stramonium*), which contain a group of drugs useful in the treatment of gastric ulcers, as powerful muscle relaxants prior to surgical operations, to control the characteristic tremors of Parkinson's disease and also to prevent travel sickness; European white hellebore (*Veratrum album*), used in the treatment of high blood pressure; *Digitalis lanata,* a patent heart poison used in minute doses as a heart tonic; and Ergot fungus (*Claviceps purpurea*), which contains a group of chemicals useful in the treatment of migranine. Non-poisonous species include such plants as restharrow (*Ononis spinosa*), ribwort (*Plantago lanceolata*), lungwort (*Pulmonaria officinalis*), bearberry (*Arctostaphylosuva-ursi*), and agrimony (*Agrimonia eupatoria*).

Plants with more widespread uses, although chiefly employed for their herbal properties, include elder (*Sambucus nigra*), used also in making refreshing non-alcoholic beverages and home-brewed wines, Blessed thistle (*Cnicus benedictus*), used to make liqueurs, Chamomile (*Matricaria chamomilla*), which is also a source of an essential oil used in the cosmetics industry for bath preparations, and adonis (*Adonis vernalis*), a popular rock garden plant which also contains a group of drugs with a very potent action on the failing heart.

There are some drug plants which have a limited use in medical practice but widespread application in other fields. Examples of these are the onion (*Allium cepa*), garlic (*Allium sativum*), chive (*Allium schoenoprasum*), horse-radish (*Armoracia rusticana*), pepper (*Capsicum annum*), parsnip (*Pastinaca sativa*), and parsley (*Petroselinum crispum*), all known for their use in the kitchen, as are anise (*Anisum vulgare*), caraway (*Carum carvi*), dill (*Anethum graveolens*), coriander (*Coriandrum sativum*), garden thyme (*Thymus vulgaris*) and many others, used mainly as kitchen herbs. Medicinal properties are similarly present in certain purely agricultural plants such as the opium poppy (*Papaver somniferum*), flax (*Linum usitatis-simum*), hops (*Humulus lupulus*) and black mustard (*Brassica nigra*) which, although used primarily for other purposes, are no less important in the treatment of disease.

The opium poppy (*Papaver somniferum*) is a plant cultivated chiefly for its seeds. These are used either for the production of seed oil or else, prepared in various forms, as a cattle feed. Only a small percentage of cultivated poppies is used for the extraction of opium alkaloids, but these are of far greater importance to man besides being more profitable economically since they provide modern medicine with its most powerful analgestics, sedatives and antispasmodics (Fig. 12.1).

Dioscorea

Dioscorea floribunda, D. deltoidea, D. composita are medicinally important for the commercial production of steroidal drugs, which include *corticosteroids* (*cortisone, hydrocortisone, prednisone, prednisolone, beta methasone, dexamethasone,* etc.), *sex hormones* (*testosterene, estrone, pro-*

gesterone etc.), contraceptive steroids for making *oral contraceptive pill* (*derivatives of esterone and 19-nortestosterone and norgestrel* viz., *norethisterone, lynestrenol, estrogen, progesterone,* etc.) and *anabolics* (modifications of testosterone, such as, *methandienone*).

Steroidal drugs are used in rheumatoid arthritis, asthama, inflammatory conditions, rhinitis, ulcers, colitis, hormonal deficiency, muscular dystrophy, pituitary dwarfism, cancer etc. Contraceptive steroids are useful in controlling fertility by suppressing ovulation through antifertility agents in the oral contraceptive pill. Cortisone cures rheumatoid arthritis.

Ox-bile, ovaries and testes contain progesterone, testosterone and cortisone. Ovaries of 50,000 cattle could hardly give 50 g progesterone! *Dioscorea* contains steroidal sapogenin, diosagenin which is very useful source of manufacturing steroidal drugs including contraceptive steroids at a cheaper cost. About 10,000 tonnes of diosgenin were employed per year for manufacturing contraceptives and other steroidal drugs throughout the world.

GINSENG

Ginseng is the mildly aromatic roots of *Panax ginseng, Panax pseudoginseng* (Araliaceae) of Manchuria and Korea and *P. quinoquefolius* an American plant. The drug is highly valued in China. America exports the drug to Asian countries. The Chinese attribute to the drug, various properties and they buy the roots in accordance to the form of individual roots and the perfection of curing. Specimens most prized are those which bear a slight resemblance to the human form; especially good roots used to bring more than their weight in gold.

Nearly all the ginseng now exported in cultivated in the natural shade of the woods or under artificial shade. The plant requires a crop period of several years, rich soil, and proper shade. Losses from disease are often heavy.

American ginseng was cultivated in rich woodlands from southern Minnesota and southern Canada eastward and southward to New England, the mountains of Georgia, and Missouri.

The **Indian ginseng**, *Trichopus zeylanicus* from Western Ghats, is found to have anti-fatigue and immunomodulating properties.

Glycosides (saponins) are present in ginseng.

The root of ginseng which somewhat resembles the legs and even the arms of human body, is considered to be infallible aphrodisiac and a panacea capable of resurrecting the dead. This shade-loving plant grows in remote valleys in eastern Siberia. Ginseng is perhaps the oldest plant still in use and one of the foremost herbal remedies in parts of Asia, where the root is regarded as both a preventive and a cure.

PERIWINKLE

Periwinkle, sadabhar, *Catharanthus roseus* of Apocynaceae has recently come into prominence as a new source of two vital groups of alkaloids vineristine and vinblastine from the leaves and vincamine, raubasin and reserpine from basal stem and roots. Vineristine and vinblastine are used in chemotherapeutic treatment of a wide variety of human neoplasms. Reserpine group of alkaloids are used in controlling conditions of high blood pressure. India exports Rs. 2.5 million worth of drugs per annum.

Periwinkle is a tropical plant, indigenous to caribbean islands and has naturalised in many southeast Asian countries as pot herb grown for its pink or white flowers borne almost throughout the year. In addition to India, several countries in south-east Asia and east Africa new produce and export this raw material to the world market. A few companies in United States and Europe process this raw material.

A mild tropical climate and deep sandy loam to loam soil of medium fertility provide ideal conditions for higher crop yield. But saline -

alkaline lands and waterlogged conditions are unsuitable. Most formers grow it on submarginal lands which could be improved by application of about 10 tonnes of FYM or raising a green manure crop preceding it.

The crop is sown with the onset of monsoon in late June; about 2-3 kg of seed are required per hectare. As the seeds are very small and light, these should be mixed with 8 times of fine sand by volume to ensure even distribution. The seeds should be sown in rows, 45 cm apart and seedlings thinned out at 40 days age maintaining a distance of about 30 cm between the plants. In all, about 74,000 plants should be maintained in a hectare land for optimum yield.

The crop response is significant up to 80 kg N/ha in average sandy loam soils. It detopping is done at 2 cm height at flowering (between 60 to 75 days) stage, its response to application of nitrogen fertilizer increased well up to 120 kg/ha in terms of yield of produce and its alkaloid content. Split application of nitrogen is better.

Periwinkle is a hardy perennial plant and can withstand a long period of drought except its early growth period. Foliage and roots are the commercial products. Foliage yield is maximum in October. When it is picked leaving a few leaves at each plant to continue the growth. AFter this picking, 20 kg N/ha should be applied to hasten growth of new leaves. The roots are harvested after 260 days. Foliage, basal stem (up to 10 cm length), roots (size 15 to 25 cm) from irrigated crop in Peninsular India and north Indian plains is 22 q/ha and 18.75 q/ha respectively.

Indian snake root

Rauvolfia serpentina, sarpagandha, Apocynaceae, is a shrub indigenous to Burma, Sri Lanka, Malaya, Thailand and Java. In India, the plant is found in sub-himalayan tracts especially Dehradun, Siwalik range, Oudh, Gorakhpur, ascending up to 1,220 m; Tamil Nadu, Karnataka, Bihar, Assam and Bengal. The dried root and rhizomatous stalk constitute the drug *Rauvolfia*.

Rauvolfia contains resinous matter with 1.2-1.7 per cent alkaloids. It contains more than 30 alkaloids of which *reserpine, rescinnamine, alstonine, ajmaline, ajmalinine, ajmalicine, serpentine, serpentinine* are important. *Reserpine* exhibits strong hypotensive and sedative activity. A definite lowering of blood pressure in hypertensive states, a slowing of the pulse and a general sense of euphoria follows administration of the extract. In mild anxiety conditions, the drug has a tranquliizing effect. In India *Rauvolfia* is used for the treatment of insomnia and certain forms of insanity. The root decoction is also employed in labours to increase uterine contraction.

Tumba

Tumba (*gudambo ki bel*) *Citrullus colocynthis* Schrad. syn. *Cucumis colocynthis* Linn; *Colocynthis vulgaris* Schrad is native to tropical and subtropical Africa and tropical Asia, is a typical desert creeper or loose sandy soil. The plant is suitable for cultivation on sandy habitats of deserts.

Tumba seeds contain about 21 per cent non-edible oil which is pale-brown to yellow in colour and bitter in taste (iodine value 117.8) soap making industries are largely consuming this oil.

The oil cake is bitter and, is also not relished by animals. But, it can form a good substitute for farmyard manure.

Dried fruits are crushed and given to ailing animals especially to camel for stomach troubles. Fruits are dried by adding a small quantity of *ajwain* to it. Such powdered fruits are mixed with sock salt and given to the person having gas trouble.

The roots are purgative and useful in jaundice and urinary infection.

Cultivation practices

3-5 kg is required for line sowing whereas 0.5-1 kg for spot sowing (1 x 1 m) and 250 gm for 2 x 2 m spacing.

Wax coating - method to germinate dormant seeds.

The seeds have a period of dormancy and contain a wax coating method to germinate dormant seeds :

(i) Wish the onset of monsoons the seeds are cropped in moist gunny bag which is placed 30 cm below the soil for 72-96 hrs. Internal heat of the sand makes the wax coating fragile. Remove the gunny bag from the ground, the seeds are rubbed against the gunny bag. Such seeds can be sun dried. Sowing are made at the appropriate time when needed.

(ii) 70-80 per cent germination can be achieved by treating the seed with concentrated sulphuric acid for 15 min and later wash them thoroughly. If such seeds are allowed to remain in water for 6 hours, they swell up and germinate after 3-4 days of its sowing.

(iii) The seeds maybe soaked in boiling water for about 24 hours, with frequent stirring. This allows the wax coating to wither. Nearly 60 per cent germination could be achieved. Seeds germinate within a week after ground sowing (S.K. Saxena, 1984).

Each creeper bears 8-50 fruits; each fruit yields 10-40 gm dry seeds.

BLACK HENBANE

Back henbane *Hyoscyamus niger* is a drug plant (with black seeds) which is indigenous to Great Britain. In India the plant is distributed along the western Himalayan region extending from Kashmir to Garhwal at altitudes ranging from 480 to 660 m. It is also cultivated in western Uttar Pradesh. *H. muticus* Egyptian herbane is also cultivated in India.

Chemical composition

The dried leaves and flowering tops of the plant constitute the drug which contains the alkaloids hyoscyamine, hyoscine and atropine. Dried roots and seeds are also used.

Medicinal use

A sedative and narcotic in case of maniacal excitement, sleeplessness and nervous depressions. It has frequently been used since long as a relief for enough, specially in an early stage of bronchitis, and in asthma. Apart from its laxative and carminative uses, it is also used for external application in neuralgia, rheumatism, painful glandular enlargements, ulcers and haemorrhoids. Seeds, which contain more alkaloid than the leaves are used in India as a remedy for foothacha.

Cultivation

The plant requires a well drained fertile, light loam or silt loam soil. In the field, double discing followed by planking is done so that field forms a well pulverised seed bed for henbane sowing. At the time of field preparation, formyard manure at 150 to 16o q/ha should be applied. A supplementary basal application of 40 kg/ha each of nitrogen and phosphorus is desirable for better growth and productivity.

In case of transplanted crop, sowing is done in the nursery beds in the first week of October, seed rate being 3 to 4 kg/ha; transplanting is then done after 4 to 5 weeks of sowing when the seedlings are at 4 leaf stage, keeping row to row distance as 60 cm, and plant to plant as 30 cm.

In direct sown crop, seed required is 4 to 5 kg/ha. On account of its small size, seed is mixed with sand before sowing; shallow sowing (5 to 10 mm) below soil surface is usually done in rows 60 cm apart. Germination take place after 5 to 7 days of sowing.

Plants should be uprooted at the first flowering stage. The berbage is then spread in the shade for drying (7 to 10 days in winter). The dried material is then put in gunny bags and stored.

SENNA

Senna (*Cassia angustifolia* and *C. acutifolia*) is an important source of organic laxatives having worldwide demand for use as household drugs. Senna is traditionally grown in Tirunelvelly and Ramnathpuram districts of Tamil Nadu. Senna plant provides high sennosides (glycosides) A, B, C and D containing leaves and pods (pod-shells) for use in medicine. As crops, these plants are maintained as annual herbs rising from 60 to 100 cm in height and are induced to periodical new growth. Leaves are used in herbal tea, bakery products and other home preparations in west and central European countries. UK and United States buy more pods. India exports about Rs. 20 million worth senna per year.

Cultivation

Senna grow well on sandy-loam and lateritic soils of low to moderate fertility with a pH ranging from 7 to 8.5. It is very sensitive to waterlogging conditions, heavy rainfall and low temperature. Once established, the crop withstands moderate saline conditions.

The land is ploughed in summer and exposed to sun for 2 to 3 weeks. Application of 10 per cent BHC to the soil protects young seedlings from attack of white ants and cut-worms. The seed should be treated with Thirum or captan 2.5 gm/kg seed to protect from fungal diseases. The seeds are sown in line 30-40 cm apart at 1.5 to 2 cm depth. A light protective irrigation should be given if there are no monsoon rains.

Seed rate of 5 kg/ha provides 70 to 75 thousand plants/ha. Apply fertilizer at right time - 4 to 5 cartloads of FYM per ha at sowing. 80 kg N and 40 kg P_2O_5 and 20 kg K_2O should be applied. Of this, 40 kg N and the entire dose of P_2O_5 and potash should be given at sowing and is placed at 4-5 cm deeper below the seed. A further quantity of 20 kg N is given at 40 days (just after thinning), 80-85 days, and 110-175 days age i.e. after first and second picking of the leaf. For top dressing urea may preferably used by broadcast in rows and mixed thoroughly in the soil.

Senna could be economically grown under

rainfed conditions (June to October). Senna plant *between 50 and 90 days age produces foliage containing higher concentration of sennosides.* The first picking of leaves between 50 and 70 days age; second picking of leaves between 109 to 110 days; third picking of leaves between 130 to 150 days when the entire plants are removed so that the harvest material included both leaves and pods together.

The harvested crop is spread in a thin layer in an open field to reduce its moisture. Further drying of the produce is done in a well-ventilated drying sheds. The dry produce usually has 8 per cent moisture. The *dry leaves and pods should have light-green to greenish-yellow colour.*

A rapid mechanical drying at 40°C could also be attempted.

A good average crop of senna can give 15 q/ha of dry leaves and 7 q/ha of pods under irrigated and good management conditions. Current price is Rs. 700-800 per quintal of leaves and Rs. 800-1000 per quintal of pods.

Medicinal uses of some common cultivated edible herbs

1. *Fennel : Foeniculum vulgare* ssp. *vulgare* var. *dulce;* used against snake bites and eye diseases.
2. *Hyssop : Hyssopus officinalis;* healing herb, especially for colds.
3. *Sweet bay : Laurus nobilis.*
4. *Lovage : Levisticum officinale* : Root oil good for all manner of complaints (a general cure-all).
5. *Lemon balm;* sweet balm : *Melissa officinalis,* Protective, rejuvenating, sedative.
6. *Mint : Mentha* sp.
7. *Basil : Ocimum basilicum* : Tea can be made from leaves. Ingredient of chartreuse and basil biscuits.
8. *Marjoram : Origanum vulgare* : Used for curing coughs, stomach complaints, tooth ache. Chewing on marjoram is also claimed to keep you in good humour.

9. *Parsley : Petroselinum crispum* spp. crispum highest vitamin A content of any plant and rich in iron. Parsley is purgative, febrifuge and abortifacient.
10. *Rosemary : Rosmarinus officinalis* : The plant was known in Greece as good luck plant and aid against stomach, heart, nerves; essential oil perfume is prepared.
11. *Sorrel : Rumex* sp.; Frenc sorrel *Rumex scutatus,* a tastier form. In the past, the juice of the leaf was used to quench thirst. The plant is used in the treatment of scurvey.
12. *Sage : Salvia officinalis* : Sage makes you sleepy. Sage milk is drunk just before going to bed. The plant is used in making mouth wash tooth pastes.
13. *Summer and winter savory : Satureja hortensis* and *S. montana* : The plant is added to peas, beans etc., for adding aroma.
14. *Thyme : Thymus serpylum,* mother of thyme, is used in kitchen. Thyme syrup is used to cure coughs.
15. *Chives : Allium schoenoprasum* : In parts of Holland, chives are used in a traditional herb stew.
16. *Dill : Anethum graveolens* var *hortorum.*
17. *Chervil : Anthriscus cerefolium* ssp. *cerefolium.* Used to remove gall stones, to purify blood. The leaves must be used fresh.
18. *Celery : Apium graveolens* var. - *secalinum.*
19. *Horse radish Armoracia rusticana* : The plant is used to cure scurvy.
20. *Tarragon : Artemisia dracnculus* : Herb is used in making sour pickles.
21. *Mugwort, wormwood Artemisia vulgaris.*
22. *Borage : Borage officinalis* : Used against infections; give pleasant dreams if kept in a vase beside the bed. Flowers are also edible.
23. *Coriander: Coriandrum sativum* : Seeds were mixed into wine.

Medicinal plants which cure leprosy and skin diseases : Aquilaria agallocha, Cassia fistula, Centratherum anthelininticum, Calotropis gigantea, Psoralea corylifolia, Urginea indica, Ficus bengalensis, Salix caprea, Semecarpus anacardium, Terminalia belerica, Solanum indicum, Cassia tora, Santalum album, Plumbago zeylanica, Baliospermum montanum, Cynodon dactylon. Medicinal plants which cure ulcers, cancer etc : *Withania somnifera, Ficus religiosa, Acacia nilotica* ssp. *indica*; root bark of *Viola odorata* (banafshah) useful in cancer of throat; *Semecarpus anacardium,*

Cassia tora; Plumbago zeylanica; Berberia aristata; bark decoction of *Madhuca indica* is a good remedy for ulcers; *Euphorbia nerifolia* is used to cure abdominal swellings, calculous tumours, bark of *Symplocos racemosa,* leaves of *Lawsonia inermis, Rubia cordifolia* (manjistha), *Caesalpinia cristata, Tecomella undulata* are useful in curing ulcers and tumours.

13

Masticatories and Fumitories (Narcotics)

Masticatories are substances used for chewing purposes. Plants used as masticatories are *Nicotiana tabacum* tobacco (Solanaceae); *Piper betel*, betel leaf, pan *(Piperaceae); Areca catechu* arecanut, supari, drupe fruit (Arecaceae); *Erythroxylon coca* coca, entire leaf chewed (*Erythroxylon cola),* cola, seeds chewed (Sterculiaceae); *Lophophora williamsii* peyote, a cactus, entire stem chewed; masticatory fungi are *Amanita, Psilocybe, Stropharia.*

Fumitories are narcotic substances used for smoking purposes. Plants used as fumitories are *Nicotiana tabacum,* tobacco (Solanaceae); *Papaver somniferum,* poppy latex from fruit is used as fumitory (Papaveraceae); *Cannabis sativa,* hemp, resin from inflorescence used as fumitory (Cannabinaceae).

BETEL VINE

The habit of chewing *pan* and offering betel leaves on auspicious occasions was prevalent in ancient India and is a tradition still in existence. *Piper betle* is a perennial creeper betel vine is grown in rotation with vegetables and banana. The leaves are dark green when mature. The vine reaches a height of 3 m to 4.5 m in a year's time.

Being a perennial crop, betel vine is grown under irrigation. Raised along with banana to provide shade, the first picking is during the tenth month after planting. Leaves are picked at intervals of a fortnight to a couple of months.

In some places in India, after preparing the land, rows of bamboos are planted 4-5 m remaining above ground. Over these are placed *dhaincha* or jute stalks or *ulu* grass. The betel vine garden or *baroj* is fenced all round with the same materials. Gourds and pumpkins are usually planted round the garden to give additional shelter and profit.

Farmers start getting returns from the crop only at the end of the first year. One hectare of land yields about 18 million leaves per annum.

Accumulation of magnesium bicarbonate causes problems in betel vine cultivation. The chemical is easily soluble in water the surface rainoff of irrigation water removes the accumulation of salts.

The wine has a life of 2 to 3 years depending upon climate, soil etc. From the 5th year the yield drops.

Stuff is made from powdered pods of cohoba (*Anadenathera peregrina,* Fabaceae). The snuff is called *yopo,* it is often mixed with lime when taken.

Mescal or red bean, *Sophora secundiflora,* Fabaceae was used in American southwest to induce trances. *Sophora* contains the alkaloid *cystisine.*

In Africa, the alkaloids of kat (*Catha edulis,*

Celastraceae) are ingested by chewing wads of freshly cut leaves usually mixed with small amount of lime. Each lump, or quid, is masticated for about 10 min until all of the juice has been expressed.

In north-central Andes, people dip the leaves of coca (*Erythroxylum coca* or *E. novogranatense*) into lime that they carry in a small bag before chewing them.

South American Amazonian tribes use caapi (*Banisteriopsis*, Malpighiaceae); infusion of mashed bark or stems are chewed.

Arecoline alkaloid from *Areca catechu* (betel nut palm) is used to cure tape worm infections, to reduce liver and as an astringent.

Some narcotic or hallucinogenic plants are sweet flag (*Acorus calamus*, Iridaceae), fly agaric (*Amanita muscaria*, Amanitaceae); yopo (*Anokonanthera peregrina*, Leguminosae); belladonna (*Atropa belladonna*, Solanaceae); caapi (*Banisteriopsis caapi*, *B. inebrians*, Malpighiaceae); boletus (*Boletus manicus*, Boletaceae); angel's trumpet (*Bragsimantia aurea*, Solanaceae); bitter grass (*Calea zacatechichi*, Asteraceae); marijuana, (*Cannabis sativa*); Cannabaceae; ergot, (*Claviceps purpurea*), Hypocreaceae; thorn apple, (*Datura stramonium*), Solanaceae; downy thorn-apple, (*Datura inoxia*, *Solanaceae; horn-a-plenty*), Datura metel, Solanaceae; henbane, *Hyoscyamus niger*, Solanaceae; morning glory (*Ipomoea tricolor*), Convolvulaceae), olololiuqui, (*Turbina corymbosa*), Convolulaceae; peyote button; (*Lophophora williamsii*), Cactaceae; mandrake, (*Mandragora officinarum*), Solanaceae; (*Culebra borrachero*, *Methysticodendron anonaceacum*), Solanaceae; nutmeg, mace, (*Myristica fragrans*), Myristicaceae; she-to, (*Panaeolus sphaect*), Agaricaceae; syrian rue, (*Peganum harmala*), Zygophyllaceae; Indian phytolacca (*Phytolacca acidosa*), Phytolaccaceae; sacred mushroom, (*Psilocybe mexicana*), Agaricaceae; sweet scented marigold, (*Tagetes lucida*), Asteraceae.

Narcotic or hallucinogenic plants have been defined as those which contain *chemicals which in nontoxic doses, produce changes in perception,* *in thought and in mood, but which seldom produce mental confusion, memory loss or disorientation for person, place and time.* Narcotics (*narkoun*, Greek to benumb) are substances which terminate their action with a depressive effect on the central nervous system.

Hoffmann divides psychoactive drugs into 5 groups : Analgesics and euphorics (e.g. opium and coca), sedatives and tranquilizers (e.g. reserpine), hypnotics (e.g. kavakava) and hallucinogens or psychotomimetics (e.g. peyote, marijuana).

Most hallucinogens are derived from plants e.g. morning flory (*Ipomoea tricolor*) and belladonna (*Atropa belladonna*). Ergot derived from the pathogenic fungus *Clavicaps purpurea* which is grown on rye and other grasses contains a highly potent hallucinogen, lysergic acid diethylamide (LSD).

Opium alkaloids : Opium poppy, *Papaver somniferum*, Papaveraceae contains the alkaloids morphine, codeine narcotine and papaverine. Opium is a mixture of some 25 alkaloids including morphine and codeine; from juice of unripe capsules. It is a dangerous habit-forming drug. Codeine is a narcotic closely related to morphine. Morphine is an important narcotic and drug, which acts on the central nervous system. Nacrotine is a powerful narcotic, analgesic; analgesic and drug, which is the chief constituent of opium. Papaverine is a constituent of opium. Theobromine is a minor constituent of opium.

Nicotine alkaloid present in tobacco (*Nicotiana tabacum*, Solanaceae) is contained in all kinds of tobacco. Also used in the manufacture of insecticides and nicotinic acid (stimulates peripheral nerves).

Peyote

Peyote (*Lophophora williamsii*) Cactaceae is a small, globose, gray green cactus native to the United States and Mexico. The plant is harvested by cutting off the top of the stems and leaving the sturdy taproot for regeneration. *The stem tips or* buttons were eaten fresh or dried for later consumption. *Mescaline* is the psychoactive compound which cause hallucinations, after eating this cactus. The *San Pedro cacti Neoraimondia* and *Trichocereus* also produce hallucinations after eating.

Peyote or mescol buttons, a hallucinogen, is obtained from the cactus *Lophophora williamsii*. The cactus resembles a huge carrot with all parts of the plants, except a button-like top, below ground. The mescal buttons are the parts which contain several powerful alkaloids with narcotic properties "*it causes those devouring it to be able to foresee and predict things.*"

14

Beverage Plants

Guava

Guava (Brazilian cocoa) is a dried paste made of crushed seeds of *Paullinia cupana* of the Sapindaceae. Large quantities are employed as a beverage in Brazil. The flavour is astringent and bitter-sweet.

Paraguay tea

Paraguay tea (yerba mate) consists of the leaves of *Ilex paraguayensis*, of the Aquifoliaceae. The tree is a small evergreen somewhat resembling the orange tree in habit. The dried leaves and buds infusion in hot water is drunk. Paraguay tea is drunk on a large scale in Argentina, Brazil and Paraguay. Argentina exports the tea to Uruguay, Chile, Bolivia, Lebanon and Syria.

Cassina is a beverage from leaves and shoots of *Ilex vomitoria*; *yoco* is a beverage prepared from the bark of *Paullinia yoco*. The tap root of *chocory, Cichorium intybus* is a coffee substitute. *Dandelion coffee* is prepared from the tap root of *Taraxacum* sp. Fruits of *barley* are used as a coffee substitute. Fruits of *Ficus carica* are used in preparing *fig coffee*. A herbal tea called *lime* is prepared from the flowers of *Tilia* sp. Leaves of *Cyclopia* sp. are used in the preparation of *cape tea* (*bush tea*).

Chat

Chat, *Catha edulis* of Celastraceae is native to Arabia through eastern Africa to the cape of South Africa. The leaves of chat or flower of paradise is a stimulant of central nervous system. The roots of *Abroma angusta* (**Devil's cotton**) of Sterculiaceae is a menagogue. Leaves of *Miconia willedemowie* of Melastomaceae, **winter berry** (*Ilex verticillata*) are used as beverages.

The kernel of the seed of **cacao** (*Theobroma cacao* of Sterculiaceae) is used in the preparation of Cocoa and chocolate. **Cola** obtained from seeds of *cola nitida* of Sterculiaceae is used as a beverage in Africa. **Cassine** from *Ilex vomitoria* and **yoco** from *Paullinia yoco* are used as beverages in S. America. leaves of **coca** (*Erythroxylum coca* of Erythroxylaceae) widely cultivated in Andean highlands of S. America is used as a beverage and stimulant.

The beverage prepared from the dried leaves of *Camellia sinensis* (Theaceae), which is known in different parts of the world as *tea, chai* or *te* is the most popular non-alcoholic drink in Asia. Tea plant is native to India (Assam) or China. Tea contains 2-5 per cent theine, 13-18 per cent tannin, a small amount of caffeine and volatile oil. When an infusion is made with hot water, the alkaloid and the oil dissolve out resulting in a beverage with characteristic taste and aroma and has a stimulating effect.

Coffee (Fig. 14.1)

Coffee is obtained from the roasted and ground seed of the coffee tree, *Coffea arabica* (Rubiaceae).

Coffee is one of the most popular beverages in the world. The coffee beverage has a bitter taste, characteristic aroma and a stimulating effect. Coffee beans contain 1-1.2 per cent of caffeine which is responsible for the stimulating effect of the coffee drink. Many coffee drinkers prefer coffee mixed with chociry (chicory root obtained from *Cichorium intybus* of Asteraceae). A product called malz coffee is prepared by adding malted barley and wheat to pure coffee. The common adulterants of coffee seeds are - seeds of *Cassia tora, Tamarindus indica*, pea, bean, caramel. Some common substitutes of coffee are seeds of *Cassia tora, Cassia occidentalis, Glycine max,* and *Canavalia ensiformis.*

coca (coca tree); cocaine paralyzes peripheral nerves, dilates pupils of eye; increases body temperature and is a powerful analgesic. *Hygrine* is present in *Erythroxylon coca* (coca tree) is used as masticatory stimulant. *Theobromine* present in *Theobroma cacao* is a heart stimulant and diuretic.

TEA (Fig. 14.2)

Tea is derived from the leaf-tips of *Camellia sinensis* (Camelliaceae) and is consumed by about half of the world's population. It has long been used in the orient as an item of commerce and as a social custom in China dates from the 5th century A.D.; it then spread around the world. Tea was introduced to western Europe at the end of the 16th century after the habit of coffee drinking had become established.

Fig. 14.1. Coffee—*Coffea arabica* plant with fruits.

Alkaloids

Caffeine (Theine) is present in *Coffea arabica, Coffea canephora* (Coffee), *Camellia sinensis* (tea), *Ilex paraguariensis* (yerba de mate), *Theobroma cacao* (cacao), *Cola acuminata* (Kola), *Paullinia cupana* (guarava), *Annona cherimolia* (cherimoya), caffeine is a heart stimulant and diuretic. *Cocaine* is present in *Erythroxylon*

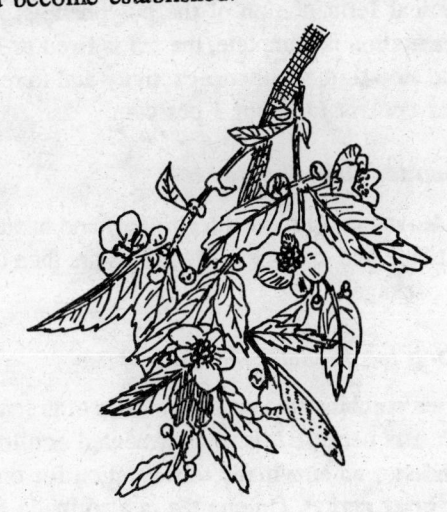

Fig. 14.2. Tea, *Thea sinensis*, a flowering branch (after Baillon).

The characteristic flavour and aroma of tea are provided by the essential oil *theol*. *Tannins (Quinones)*, along with *pectins* and *dextrins*, provide the colour and astringency. The stimulant properties are due to the alkaloid *theine* which occurs in the leaf at a concentration of 2-5 per cent. Theine is identical with caffeine; it is a heart stimulant and diuretic.

Tea plants are small, evergreen trees. Plantations of tea are generally started from seed which has been carefully germinated in enclosures. When sufficiently robust, the young plants are set into field. After about 3 years the trees are pruned and subsequently repruned every 12 years or so. Picking can begin when trees are 4 years old. For fine teas only the 2 or 3 youngest leaves and the terminal bud of each branch are picked. A single shrub can be picked about once a week.

Types of tea

Black tea (Fermented tea)

Tea fermentation refers to the alteration of various chemical constituents. Tea leaves contain caffeine (theine), tannins and polyphenols. The flavour and quality of black tea is determined by the ratios of the products which result from the chemical fermentation of the polyphenols. Once fermentation is complete, the tea is fired or heated, to stop further chemical activity and to reduce water content to about 3 per cent.

Green tea (unfermented tea)

The leaves are shredded (or rolled) and heated to inactivate leaf enzymes. The leaves are then dried and packaged.

Oolong tea (Semifermented tea)

The tea combines the taste properties of green and black teas because it is semifermented or allowed to undergo an enzymatic fermentation for only a very brief period. Onlong tea is also made from *Camellia sinensis* var. *chemisa* and produced primarily in southern China and Taiwan.

Jasmine flowers are used to scent tea.

The leaves are first spread out on a flat surface or placed in drums to wither. The withered, flaccid leaves are then *rolled and twisted*. The purpose of rolling is to crush the cells of the leaves and release the enzymes in the cytoplasm. The rolled leaves are then fermented.

Tea plant accumulates and store aluminium (100-100 mg per gram). Commercial tea yields upto 100 ppm Al^{3+} in a Al^{3+} is considerably reduced by addition of milk with the formation of insoluble phosphate of the metal and thus reducing its bioavailability. However, addition of lemon juice leads to formation of soluble citrate complex of aluminium which is absorbed readily in gastro-intestinal tract.

COFFEE

Coffee is one of the world's international commodities (Fig. 14.1). It is commercially more important than tea although consumed by fewer people. Only two species are of commercial importance : *Coffea arabica,* which produces about three-quarters of the world's good quality crop and *C. canephora* (= *C. robusta*). Congo coffee which is especially valuable for producing beans used in making instant coffee.

The green beans are processed and roasted to develop their characteristic aroma (mainly caffeol and essential oils). The green beans contain upto 3 per cent caffeine by weight.

Indian coffee is *C. arabica*; there are 3 varieties : chicks, coorgs and kents.

C. liberica, Liberian coffee, is native to west tropical Africa and cultivated in Malaya; Mysore and Travancore (India). *C. excelsa,* Excelsa coffee, is native to West Africa and cultivated in Philippines, Java and Vietnam.

In the world, coffee is grown in Angola, Brazil, Columbia, Costa Rica, Dominican Republic, El Salvador, Guatemala, Haiti, India, Sri Lanka, Indonesia, Kenya, Tanganyika, Mexico, Venezuela, Nyasaland, Congo. Annual production of coffee in the world is about 4.5 m mt.

Coffee can be propagated by sowing seed. Seed is sown in nursery beds. Coffee seed takes 35-40 days to germinate. About 180 days old seedlings are plantied. Planting should take place in dull and rainy weather, preferably at the beginning of a wet season. Coffee fruits are picked when they are red ripe. The yield of ripe fruit during the first crop year is about 4 mt/ha. The out turn of clean, dry coffee from ripe fruit is about 16-18 per cent.

Spices such as cardamom are used to give aroma to coffee.

Nestle

A Swiss Company developed the first soluble coffee the fabulously successful *Nescafe* - black in 1938. In India Nestle build a milk products factory as Moga in Punjab. Nestle developed a compact machine *Nespresso* to make espresso coffee at home. The machine uses sealed capsules containing ground coffee that pop open when hot water is forced through them under high pressure. These capsules individual servings, with no need for clean up afterwards. Nespresso has been introduced in Japan, Italy and Switzerland, and may soon appear in the United States.

CHICORY

Perhaps the most common coffee additive is chicory (*Cichorium intybus*, Asteraceae) which can be considered as an adulterant or a flavour enhancer, depending on the season for its addition to ground coffee. Chicory superimposes on the normal coffee flavour a distinctive, pronounced taste that many individuals enjoy. *Dried, powdered roots of chicory plant* have been used alone to produce a beverage, as well as mixed with coffee to stretch the beverage, or substitute for coffee when the beans were scarce. In the United States, coffee with chicory is a regional speciality of Lousiana. Unlike coffee, chicory contains no caffeine.

In India, chicory plant is found wild as a weed in the field and road sides in Punjab, Haryana, Uttar Pradesh, Madhya Pradesh. Chicory is now widely cultivated in Tamil Nadu, Andhra Pradesh,

Gujarat, Karnataka, Kerala, Maharashtra and Punjab.

The chicory crop takes two seasons to complete its life cycle. Application of nitrogen (60 or 100 kg/ha in steckling crop; 80 or 160 kg/ha for seed crop) to the crop in 3 splits (at transplanting, bolting and pre-flowering) resulted in higher seed yields. The crop takes 60 to 120 kg of phosphoric acid/ha potash, 60-70 kg/ha for root crop and 80-100 kg/ha for seed crop per hectare is good.

Seed sown in the first week of June for raising the steckling crop was found to be optimum time and sticklings produced are of medium size suitable for growing seed crop in the next following season. The suitable time for transplanting the stecklings for seed crop was found 15th March to 20th March. The seed crop matures in August-September for harvesting. 70,000 plants and 30,000 plants/ha for steckling and seed crops respectively are near optimum to get maximum steckling and seed yields. In the seed crop, the optimum spacing is 45 cm from plant to plant and 75 cm row to row.

An intensive 2-year rotation of peas with chicory steckling crop-seed crop, can yield 100 quintals of green pods/ha peas, 300 quintals of chicory stecklings used for seed crop. Chicory seed yield is 8-10 quintals/ha.

Many coffee drinkers prefer coffee mixed with chicory. The chicory powder is mixed with pure coffee is different ratios for flavouring and grading.

KOLA

The seeds of kola (*Cola nitida*) are rich in *caffeine* and essential oils. Cola started off the fashion of cola drinks (cola, soda pops) although the flavourings and stimulants are now obtained from other sources. In the original formulation of Coca Cola, Kola, as well as coca, was used to make the beverage. Now, coca leaves from which the cocaine has been removed, artificial flavourings, and caffeine and added. The seeds are used in beverage making. Kola is native to West Africa.

The flowers of the kola plant are borne in axillary inflorescences. The fruits each contain about 8 seeds which are scraped from the harvested pods. The fleshy seed coats are remove before the seeds are allowed to ferment or sweat for a few days to develop their flavour. The seeds (kola nuts) are simply dried and then pulverized into a powder that is mixed with boiling water to make the beverage.

In addition to caffeine, kola seeds contain a glycoside, *kolanin,* which acts as a heart stimulant.

Coca

From very ancient times, the people of South and Central America were in the habit of chewing coca plant leaves for their anti-fatigue and anti-hunger effects. It was not until the end of the 19th century that this plant was brought into focus and it's scientific and medicinal value considered. In 1860 the German chemist, Albert Neimann isolated *cocaine* in the pure alkaloid form, from leaves brought to Europe. It was the first drug to be used as a local anaesthetic. It is too toxic for general use but still is used as a topical anaesthetic in nost and throat surgery and opthalmology.

Cocaline is isolated mostly from the leaves of *Erythroxylon coca* or *E. novgranatense* (Erythroxylaceae). The coca leaves contain 0.5-1.5% cocaine. Cocaine is benzoyl methyl ecgonine, belonging to the coca group of alkaloids e.g., *benzozy ecgonine, tropacocaine, hygrine, cuscohygrine, truxillo cocaine, cocamine, cocaicine* etc. It has the molecular formula $C_{17}H_{21}NO_4$ m.p. 98°C and is an odourless, colourless white crystalline powder.

Cocaine is referred as coke, snow, candy, girl, charlie, big C in slang. Coca paste is the first extraction product during the manufacture of cocaine. It is prepared by adding macerated cocoa leaves to sulfuric acid, kerosene or gasoline. This dried paste contains 40-90% cocaine sulfate, coca alkaloids and other substances. 1 kg of paste usually yields 0.5 kg of pure cocaine. When coca paste is treated with HCl and the product refined, cocaine hydrochloride of more than 98 per cent purity is obtained which is termed as cocaine. Free base (also called cocaine base, cocaine alkaloid, base) contains some of the adulterants found in the illicit cocaine when cocaine hydrochloride is converted to its alkaloid by treatment with an alkali. It is the cocaine alkaloid or benzoyl methylecgonine. It has lower vaporising point than cocaine hydrochloride and thus less is lost when is heated and inhaled. Free base is generally considered to be purer and more potent than coca paste or cocaine sulfate.

Pharmacological action

(i) *Central nervous system* : Stimulates, giving Euphoria accompanied by restlessness and excitement. It can directly act on the brain by easily crossing the blood brain barrier. There is evidence to suggest that Dopamine mediates the major effects of cocaine in brain.

(ii) *Cardiovascular system* : Small doses of cocaine may slow the heart as a result of central vagal stimulation but moderate doses increases heart rate. A large intravenous dose of cocaine may cause immediate death from cardiac failure due to a direct toxic action on the heart muscle.

(iii) *Body temperature* : Cocaine is pyrogenic in nature. The increased muscular activity attending stimulation by cocaine augments heat production.

(iv) *Sympathetic nerve stimulation* : Cocaine blocks the uptake of catecholamines at advenergic nerve endings. Hence, it potentiates the responses of sympathetically innerveted organs to norepinephrine.

(v) *Appetite* : Cocaine can suppress appetite. Very little is known about the nutritional consequences of chronic administration of cocaine nor it is known whether tolerance to appetite suppression develops.

(vi) *Local anaesthetic actions* : It has the ability to block nerve conduction. Cocaine was used in ophthalmological procedures but it causes sloughing of the corneal epithelium and hence was ultimately abandoned.

CACAO

In India, Cocoa was grown in an area of 28,274 ha. with production of 3000 tonnes of dry cocoa beans during 1982-83.

The active ingredient of cocoa or chocolate from the seeds of *Theobroma cacao,* Sterculiaceae is theobromine, a xanthine alkaloid. Upto 3 per cent of theobromine, along with minute quantities of caffeine and large amounts of cacao butter are present in the kernel of the seed, which is ground after fermentation. Theobromine is heart stimulating and diuretic.

There are two types of cacao : *Criollo* - consists of two groups, occurring in Central American countries and Mexico, Venezuela group occurs in South America. The criollo produce the fine cacao of trade. *Forastero* - consists of two groups, Amazanian and Trinitario.

Cacao plant is a woody tree reaching a height of 5-8 m. The tree flourishes within the tropics, mainly between the latitudes 20°N and 20°S. Cacao plant needs warmth and humid atmosphere. The ideal mean shade temerature if 2.6-6°C. Potash rich alluvial absorptive clays with pH 4-5 are good for cultivating cacao trees. Caco is propagated either by seed or vegetatively from buds or cuttings.

The story of chocolate

Many centuries ago in Mexico Montezuma, the famous Aztec emperor used to serve his most favoured guests a delicious frothy drink in gold cups. The drink was nothing but chocolate made from the beans of cocoa trees. In 1521 a Spaniard named Hernan Cortez conquered Mexico. Along with the wealth he carried with him the secret of how to make the chocolate drink. The cocao beans were very expensive and only rich people could afford to drink it. In 1847 the first solid chocolate was made. The cocoa trees were grown in Western Africa. The farmers plant the choicest cocao seeds in a nursery. A few months later the saplings are transferred to a large plantation where they are grown in the shade of big trees. The trees grow to a height of 6 to 8 metres. Now they begin to flower.

Clusters of small pink pale flowers appear on the branches of trees. The flowers produce pods of about 15 cms long. When the pods are ripe yellow they are harvested.

Pods are sliced and opened. Inside each pod lie about forty seeds covered with a white pulp. They are spread over plantain leaves in the sun. Then they are covered with plantain leaves. After somedays the beans turn brown and the white pulp drains away. Then the seeds are taken to the factory and pounded into cocao powder or made into chocolate.

Manufacture of chocolate

Seeds are cleaned and roasted in large roasting ovens. The roasted beans are winnowed to remove the husk and then in a powerful press to squeeze out the fat content (cocao butter). What remains is pressed into cakes. The cakes are crushed and ground into fine cocao powder. It can be used to made drinking chocolate cakes and other delicacies. The remaining mass is placed in a mixing machine called melangeur. More cocao butter and sugar are added to this mass. The mixture is then pushed to and fro for hours to make it smooth in a conching machine and then poured into polished moulds. The moulds travel on a conveyor belt to a cold room to set. The chocolate is ready. The chocolate is wrapped in thin paper foils in order to keep them from spoiling. The wrapper is put last of all.

Alcoholic beverages

1. **Wines** : Wine prepared from the fruit of *Vitis* is utilized in Europe, N. Africa, USA, Australia, S. america, S. Africa. Wines are prepared from the flowers of *Cystisus, Primula, Tussilago, Sambucus* are used in Europe (UK). In N. europe various wines (meads) are prepared from honey (nectar) of flowers.

The bark of *Galipea officinalis* (Rutaceae) a plant of S. America and cultivated in Trinidad is used to flavour drinks. Red wine, white wine, sweet wine, sparkling wine and fortified wine are chief types. Fortified wine has higher content of alcohol due to addition of wine, brandy etc.

2. **Beers : Beer,** ale, lager, stout of Europe, N. America, Australia, New Zealand are prepared from barley grain (*Hordeum*) and hop cones (*Humulus*). **Cider,** cyder of England, France, Spain, USA are prepared from apples (*Malus*). **Ginger beer** of England is prepared from the rhizones of ginger (*Zingiber*). Sake, saki of Japan is prepared from rice grains (*Oryza*). **Pulque** of Mexico is prepared from sap of stems of *Agave*. **Kvass** (Quass) of Russia is prepared from barley grain (*Hordeum*), rye grain (*Secale*) and leaves of peppermint (*Mentha*). **Pombe** (boura) of Africa is prepared from millet seed (*Eleusine*). **Chicha** of S. America is prepared from corn grains (Zea) and Quinoa (*Chenopodium*) seeds. **Sorgo** of Africa, Asia is prepared from *jowar* grain (*Sorghum*). **Kava** of Pacific Islands is prepared from the roots of *Piper methysticum*. **Palm wine** (toddy) C and S. America, Africa, Orient is prepared from sap from decapitated stem apex or maltreated inflorescence of *Borassus, Caryota, Phoenix, Nipa, Raphia, Acroconia, Jubaea, Mauritia* and other palms.

Lager beer is an aged beer produced by bottom fermentation. Bock beer is a strong, dark beer, brewed in spring with new malt and hops. Ale beer is brewed by top fermentation at higher temperature; alcohol content 4-7 per cent. Porter beer is a dark brown beer with burnt taste, made from inferior grades of malt; aged for 40-60 days;

Stout beer is a beer from inferior grades of malt; aged for one year. Root beer is prepared by fermentation of ginger, wintergreen etc. with addition of sugar and yeast.

3 SPIRITS

Brandy, pisco, teogmac, armagnac, marc grappa, bagaceira, Helebranntwein etc. of Europe - France, Spain, Italy, Greece, Germany, S. Africa, S. America, Australia, USA are distilled grape (*Vitis*) wines. **Fruit brandy,** slivovitz, kirsch, calvados, applejack of France, C. and E. Europe, Jugoslavia, USA is distilled fruit-wine of *Malus, Prunus,* fruits, sometimes with some kernels. **Whisky,** whiskey of Scotland, Ireland, USA, **Canada** is prepared from barley (*Hordeum*) grain, corn (*Zea*), rye (*Secale*) grain malted, fermented and distilled. **Rum** of West Indies, Caribbean mainland, USA is made from sugarcane (*Saccharum*), juice fermented and distilled. **Gin of Britain,** Holland, USA is prepared from corn (*Zea*) rye (*Secale*) fermented and distilled **flavoured** with juniper berries (*Juniperus communis*) and botanicals usually angelica root, anise, coriander, caraway seeds, lime, orange peel, licorice, *calamus,* cardamom, Cassia bark, orris root, bitter almonds etc.

Vodka of Russia, Poland, Finland is distilled from potato (*Solanum tuberosum*) starch or wheat (*Triticum*) or rye (*Secale*) flavoured with caraway (*Carum carvi*) seed. **Akvavit,** Aquavit of Norway, Denmark, Sweden is distilled from potato starch, flavoured with caraway (*Carum carvi*) seed. **Raki,** ouzo of Balkan peninsula, Turkey is a distillate of wine, grain (*Triticum*), potato (*Solanum tuberosum*) or molasses (*Saccharum*), sometimes flavoured with anise seed. **Arrack,** a spirit common throughout the Orient is a distillate of fermented rice (*Oryza*) and mollasses (*Saccharum*) or palm juice (*Borassus*) or a mixture of rice (*Oryza*) and *Borassus* (palm toddy). **Mezcal** of Mexico is a distillate of fermented agave (*Agave tequilana*) sap and fibrous pulp. **Tequila** of Mexico is a distillate of fermented agave sap and fibrous pulp.

4. *Various other alcoholic beverages :*
Absinthe of France is a spirit flavoured with oil
of wromweed (*Artemisia absinthium*). **Vermouth**
of Italy, France, Germany, USA is a wine fortified, sweetened and flavoured with herbs like
Artemisia absinthium. **Liqueurs** of Europe, S.
Africa are spirits from grapes (*Vitis vinifera*) or
sugar cane (*Saccharum*) sweetened and flavoured
with herbs.

Distillation liquors are made from many
fermented plant sources, *grape alcohol* being
the most important and providing the basis for
brandy and many locally important spirits. Other
carbohydrate sources include cereals (e.g. *gin,
whisky*), potatoes (e.g. *vodka*), sugar cane
(e.g. *rum*), sap (e.g. *mezcal* from *Agave* sp.) and
fruits (e.g. *calvados* from apples, *slivovitz from
plums*).

Hops in beer, *wormwood* in absenthe, *pine
resin* in retsiña, *juniper berries* in gin and
various herbs in vermouths provide the taste and
flavour.

Distilled liquors are often served diluted
with other drinks of plant origin and fruits
are added to many drinks as a final garnish or
flavouring.

Beers are produced from cereals which first
have to be malted; the starch is converted to
maltose by germination, the action of molds etc.
plants used for beer include wheat, barley, rice,
maize, millet, rye and palms.

Bitter principles

Bitter principles are non-poisonous, non-nitrogenous substances of varied chemical composition
but with one characteristic in common - their
strong bitter, but not unpleasant, taste. They occur
in various plant families, such as Gentianaceae,
Asteraceae and Labiatae.

They are generally used in medicine in the
form of alcoholic extracts, tinctures and medicinal
wines. Taken before meals as prescribed they
stimulate the appetite by promoting the flow of
digestive secretions and thereby aid the digestion

itself. They are also administered to convalescents
for their sedative and tonic properties.

HOPS

Beer contains malt and hops (*Humulus lupulus,*
Cannabaceae). Dutch and Germans used hops
in beer about 746 AD and the English began
to use hops in brewing after 1524. Before the
use of hops, bog myrtle was often added to
beer. Hops became popular because it imparts
a pleasant taste and aroma to beer and add
enzymes that help to coagulate unwanted proteins
in beer.

Hops are dioecious vines (3-6 m) which
produce clusters of flowers (Fig. 14.3). Each

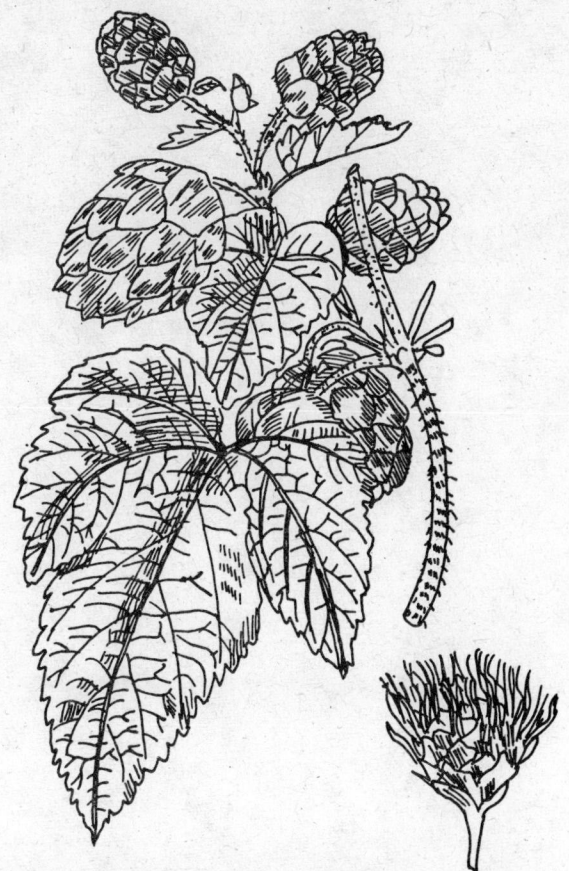

Fig. 14.3. Hops, *Humulus lupulus*. A—Inflorescence.

female flower is subtended by a leafy bract. When mature, the female inflorescences resemble soft fine cones because all of the bracts overlap one another. The bracts of the female flowers possess numerous glands that contain volatile oils. The *hop cone* contains yellow *lupulin* found as glands at the bases of bract and on the perianth of the flower.

Hops is cultivated in Africa, South Africa, north central America, Canada, United States, Asia, Europe and Russia. In India hops is cultivated on a small scale in north west Himalayas. Young shoots of hops have been eaten as vegetable.

15

Dyes and Tannins

Vegetables dyes have been used since earliest times in various parts of the world. They were example, used by the ancient Egyptians, Greeks, Romans and other early civilizations. Wood (*Isatis tinctoria*) was the imperial blue dye of Europe for many centuries and was used by the ancient Britons as a body paint. Indigo (*Indigofera anil. I. tinctoria* and other species) has been grown in India and Africa from time immemorial to provide another blue dye. The antiquity of dye industry in India dates back to 2nd century A.D. which is exemplified from the paintings of Ajanta caves. Madder or Turkey red (*Rubia tinctorum*) was used by the ancient Egyptians and other Middle Eastern peoples to provide a red dye. A yellow dye and food colorant has been obtained from saffron (*Crocus sativus*) since the times of the ancient Greeks. Another yellow dye is obtained from dyer's greenweed (*Reseda lateola*), one of the earliest dye plants known. Many hundreds of plants are still used for dyeing, especially in tribal societies in remoter regions of the world. They include lichens, club-mosses, mosses, ferns, gymnosperms and flowering plants and the parts used range from the bark, twigs, needles, flowers and fruits to the whole plant.

Vegetable dyes are substances which, when dissolved in water are able to colour yarns, textiles, leather, wood and some foodstuffs. They are distinguished from animal dyes and in more recent times, from synthetic dyes. There is currently a resurgence of interest in natural plant dyes as part of home or cottage industries.

Water soluble dyes have to be made insoluble to prevent them running. This is achieved by pretreating the material to be dyed with substances known as *mordants*, which make the dye fast and help attain an intensive coloration. Mordants include alum (used from very early times), common salt, cream of tartar, iron, chromium and tin with which the dyes form an insoluble compound.

Natural dyes are complex organic compounds and are colouring matter obtained from various parts of the plant. Their colour is due to various pigments in plant cells. Red, yellow, blue, green and brown are available from plant pigments. Their chief use is in textile industry. They are also utilized for colouring varnishes, paints and paper.

Wood dyes are obtained from *Haematoxylon campechianum*, haematoxylin (Leguminoseae); haematoxylin is used in histological work and dying of cotton, wool, silk and leather. Some inks are also made.

Acacia catechu cutch, kattha (Mimosaceae) is medicinal; used in dying of cottons and eaten after pasting or pans (betel leaf and betel nut preparation).

Pterocarpus santalinus, red sandal wood lal chandan (Fabaceae) the dye is blood red in colour and is known as *sentalin. Artocarpus integrifolia* and *A. heterophylla* (Moraceae) yields a light yellow dye known as *basanti,* used in colouring of silk robes of Buddhists. *Caesalpinia sappan* (Caesalpiniaceae) known as sappan wood yields a red dye known as *Brazilin.*

Bark dyes are obtained from *Alnus nitida*, black cedar (Betulaceae);*Mimusops elengi* (Sapotaceae) yields a brown dye; *Terminalia tomentosa* (Combretaceae) yields a black dye; *Quercus velutina*, Quercitron (bright yellow dye), Fagaceae; *Rhamnus globosa* and *R. utilis* (Rhamnaceae) yields lokao, chinese green dye; *Casuarina equisetifolia* (Casuarinaceae) yields a brown dye.

Leaf dyes are obtained from *Indigofera tinctoria* and *I. suffruticosa* indigo, neel (Fabaceae); Chlorophyll extracted from leaves of higher plants is used as a colouring material for food, soap, tooth-pastes and powders. *Lawsonia inermis* (Lythraceae) yields an orange dye known as henna, mehndi.

Flower, fruit and **seed dyes** are obtained from *Carthamus tinctorius*, safflower, kaisum (Asteraceae), kusum yields yellow and red dyes. *Crocus sativus* kesar, zefran, saffron; the 3 stigmas of the saffron flower are the *kesar* of commerce, used as a colouring and flavouring material for confectionery, cooked rice and curries. It is of medicinal value.

Butea frondosa dhak, tesu (Fabaceae) flowers yield a brilliant yellow dye. The water coloured with it is used by Hindus on their religious festival of Holi.

Nyctanthes arbortristis, harsingar (Oleaceae) the corolla tube of the flower is orange coloured and yields a yellow dye in water (Fig. 15.1). *Mallotus philippinensis* (Euphorbiaceae), the fruits yield the bright yellow *kamela dye* (Fig. 15.2). *Ochrocarpus longifolius* (Guttiferae), the dry flower buds yield a red dye.

Rhamnus cathartica (Rhamnaceae) yield a green pigment.

Bixa orellana (Bixaceae) yields a valuable bright yellow dye.

Root dyes are obtained from *Rubia cordifolia* (Rubiaceae), the roots yield a bright red dye known as *madar*. Rhizomes of turmeric yield a yellow dye known as turmeric *haldi*. *Berberis aristata* (Berberidaceae) roots yield a yellow dye.

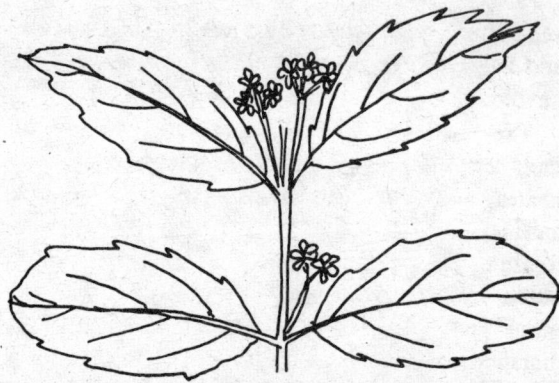

Fig. 15.1. Coral jasmine, *Nyctanthes arbortristis*. The orange-coloured tubes of the corollas was formerly used for dyeing silk.

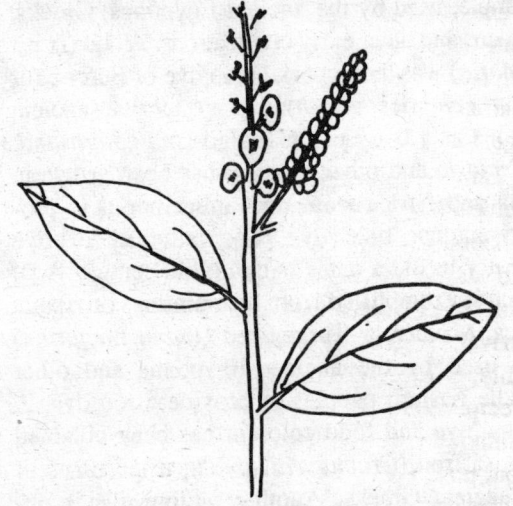

Fig. 15.2. *Mallotus philippensis* (Euphorbiaceae kamala dye) the coloured powder covering the fruits is used as dye.

TANNINS

Tannins are organic, non-nitrogenous plant products which have astringent properties. They are soluble in water and alcohol. Exposed to air they turn dark, their composition changes and they lose their effective properties. It will therefore be obvious that tannin drugs lose their beneficial properties if stored for long periods. Plant families rich in tannins include Rosaceae, Geraniaceae and Fabaceae. Members of the families Brassicaceae and Papaveraceae, on the other hand, are totally devoid of tannins.

Tannins are of medical significance because of their astringent properties; these promote rapid healing and the formation of new tissue on wounds and inflammed mucosa. They are used in the treatment of varicose ulcers, haemorrhoids, minor burns and frostite as well as inflammation of the gums. Internally they are administered in cases of diarrhoea and intestinal catarrh. Tannins occur in crude drugs either as the chief active principle, for example in the bark of oak (*Cortex quercus*) or the leaves of bilberry (*Folium myrtilli*), or as subsidiary components which increase the effectiveness of the main active principles as peppermint (*Mentha piperita*) and garden sage (*Salvia officinalis*). On the other hand, their presence in bearberry (*Arctostaphylos uva-ursi*) is a disadvantage.

Tannins are also widely used in industry for the conversion of hides into leather.

Tannins form dark blue (*pyrogallol tannins*) and dark green (*catechol tannins*) compounds with iron salts; strong solutions of pyrogallol derivatives plus iron have afforded the most durable of *inks* since time immemorial. Tannins precipitate many metallic salts and some organic compounds, such as alkaloids. In alkaline solution, tannins become oxidized in the air to a dark brown colour; pyrogallic acid is used as a reducing agent in photographic development of films and dates.

Tannins are organic compounds that have an astringent nature. On reacting with iron salts, tannins form *inks*. Tannins react with animal skin to form leather. **Bark tannins** are obtained from *Acacia arabica* (Mimosaceae), *Cassia auriculata* (Caesalpiniaceae); *Cassia fistula* (caesalpiniaceae); *Casuarina equisetifolia* (Casuarinaceae); *Ceriops tagal* (Rhizophoraceae); *Acacia decurrens* (Caesalpiniaceae); *Elaeodendron glauceum* (Celastraceae); *Lagerstroemia parviflora* (Lythraceae); *Rhizophora mucronata* (Rhizophoraceae); *Shorea robusta* (Dipterocarpa-ceae); *Soymida febrifuga* (**Meliaceae**); *Terminalia arjuna* (**Combretaceae**); *Zizyphus xylocarpus* (**Rhamna-**ceae); *Quercus* spp. (Fagaceae); *Tsuga canadensis* (Pinaceae); *Bridelia retusa* (Euphor-biaceae).

Leaf tannins are obtained from *Rhus cotinus, R. typhina* (Anacardiaceae); *Uncaria gambier*, gambier (Rubiaceae); *Cinnamomum tamala*, tezpat (Lauraceae); *Anogeissus latifolia* (Combretaceae).

Fruit tannins are obtained from *Acacia scorpioides*, babool (Mimosaceae); *Caesalpinia coriaria*, divi divi (Caesalpiniaceae); *Diospyros embryopteris* (Ebenaceae); *Emblica officinalis* amla (Euphorbiaceae); *Quercus macrolepis*, valonia (Fagaceae); *Terminalia chebula*, herr (Combre-taceae); *Terminalia bellirica*, bahera (Combre-taceae). **Root tannins** are obtained from *Geranium wallichianum* (Geraniaceae); *Rumex hymenosepalus*, canaigre (Polygonaceae); *Sabal palmetto* (Arecaceae); **Wood tannins** are obtained from *Castanea dentata*, chestnut (Fagaceae); *Schinopsis* spp., **Quebracho** (Anacardiaceae).

Tannins and tanning

The term *tannin* describes a group of pale yellow to light brown substances that are widely found in plants and which have been used for centuries for dyeing fabrics, *making ink* and in medical preparations. Their main use, however, lies in *tanning leather* - the process of converting animal skins into leather. This consists of steeping skins in infusions of vegetable materials such as bark, wood, leaves, nuts or galls which are rich in tannins. Putrefaction of untreated animal skins is promoted by enzymes secreted by various micro-organisms. During tanning the tannins inactivate these enzymes and combine with the proteins of the skins, converting them into compounds which the enzymes cannot attack and binding together the protein fibres.

The tannins used in commerce are mainly obtained from the bark and wood of a few species which produce exceptionally high concentrations of tannins. Other materials such as leaves, galls and fruits are used less often. Few of these materials can be produced cheaply enough as a crop and most are other collected from the wild or are a by product of some other activity.

Oak barks have been widely used for tanning for thousands of years. In Europe, *Quercus robur* and *Q. petraea* are mainly used, while in the New World *Q. montana*, *Q. alba*, *Q. prinus*, *Q. borealis*, *Q. dentata* and other species are important. Bark of hemlock spruces (*Tsuga canadensis* and *T. heterophylla*) is an important source of tannin in North America and is preferred for tanning heavy skins. Wattle (*Acacia decurens* and *A. dealbata*) bark has a tannin content of up to 40 per cent these fast-growing trees can be harvested at 5-10 years old. Eucalyptus bark (from *Eucalyptus astringens* and *E. wanleo*) is widely used in Australia. Although mangrove bark is rich in a tannins it is not so popular since the leather it produces is not of good quality. Bircit (*Betula*) and conifer barks are the main tannin sources in Russia.

Quebracho from South America (*Schinopsis balansae* and *S. lorentzil*) produce a wood rich in tannins. Chestnut wood is also used to some extent, notably from *Castanea dentata* in North America and *C. sativa* in Europe. Tannins extracted from autumn leaves of sumacs (*Rhus typhina* and *R. glabra* in America and *R. cordira* in Europe) produce fine leathers in landbinding.

Oaks

The Fagaceae, or Oak family, includes several woody genera which are noted for their richness in tannin. Various members of it have long been utilized in the leather-making industries of both Europe and America. The genus *Quercus*, including the oaks proper, is the largest group of the family, and its representatives with us are both numerous and widespread. Only a single American oak species, however, is of major tanning significance. The closely related *Lithocarpus* Tanbark Oak has recently assumed prominence, and *Castanea*, the Chestnut, is now our major domestic tan-stuff.

The chestnut oak

Early American tanners utilized the barks of *Quercus alba*, the white oak, and *Q. rubra*, the red oak; but for many years the dominant oak tan has been the bark of *Q. prinus*, the chestnut oak. This species is a large and vigorous forest tree, sometimes reaching 100 feet in height, with thick and deeply furrowed greyish or blackish bark. The chestnut oak is a tree of dry soils, rocky lands, and hill and mountainsides.

Chestnut oak bark is peeled between March and June, and air-dried for about six weeks in the woods. The timber is used to some extent for lumber, ties, and fuel. In tanning, both bark and extract are used. Oak bark is used particularly for sole and other heavy leathers. It contains 8 to 14 per cent of tannin in which both *catechol* and *pyrogallol* are probably represented.

Nutgalls

Most species of oak are subject to the attack of small wasplike insects which deposit eggs in the rapidly growing tissue of twigs or leaves. The hatching and growth of the larvae results in a remarkable stimulation or hypertrophy of growth and the production of a more or less globular excrescence known as a gall. *These galls contain remarkably large quantities of tannin.* Galls are common on many species of oak, but are so light in weight and so expensive of collection that they find no industrial use. Commercial nutgalls resulting from the growth within the twigs of the larvae of *Cynips tinctoria* of the Hymenoptera, are from *Quercus infectoria* of Asia Minor. They somewhat resemble small brown or greyish tuberculate marbles. They are extremely heavy for woody growths, sinking readily in water. They contains 50 to 70 per cent of *pyrogallol tannin,* and 2 to 4 per cent of *gallic acid.* Their reacting properties with iron salts have been known for ages and have been used in making ink.

Hemlock bark

The tannin is obtained from the bark of *Tsuga canadensis* and *T. heterophylla*. Tanning extract is made by grinding the bark and leaching it with hot water. It is generally concentrated to a liquid with 25 per cent tannin content, but it may be fully dried. Hemlock has been especially valued for tanning sole and heavy leather and for sheep skins. It contains 8 to 10 per cent of catechol tannin and affords a firm, heavy, dark red leather.

Valonia

A nutgall tan-product from the a corn cups of *Quercus aegilops*, valonia oak. The ripe acorus are shaken from the trees, dried on the ground, and fermented in heaps for several weeks, during which time the across contract and separate from the cup valonia is very rich in a mixture of catechol and pyrogallol tannins. Valonia is used for tanning high grade light coloured leathers alone, it yields a brittle product and is commonly combined with other tans.

Sumac galls

A Nutgall or *Rhus*. *Rhus* galls are collected in China and Japan are used in the manufacture of tannic and gallic acids.

Other nutgalls

Larvae of the insect *Cynips tinctoria* of the thymenoptera form galls on the plant *Quercus infectoria*. The galls somewhat resemble small brown or greyish tuberculata marbles. They are extremely heavy for woody growths, sinking readily in water. They contain 50 to 70 per cent of pyragallol tannin and 2 to 4 per cent of gallic acid. They are employed in the manufacture of inks, tannic acid, pyrogallic acid and allied chemicals.

Tanbark oak

Tannin is obtained from the bark of *Lithocarpus densifolia*, a tree of California.

Quebracho

Tannin extracted from the heavy, dark red heartwood of *Aspidosperma quebracho*-colorado and *A. colorado* of Apocynaceae, trees of Argentina and Paraguay. Quebracho is a relatively cheap, quick-acting tan. It is employed in all classes of leathers, but particularly for some kinds of light leathers. In combination with hemlock or oak it is used for bag, patent and automobile leathers; for sole leathers and other heavy goods.

Mangrove

The thick heavy bark of *Rhizophora mangle* yields 5 to 45 per cent of catechol tannin. In Brazil, the leaves of the mangrove are used in tanning.

Mangrove is a very cheap tan, generally used in combination. Its extract also has limited use in dyeing browns on cotton.

Wattle

The wattles are species of *Acacia* of South Africa, Australia and Tasmania. Wattle bark is generally imported in the form of extract.

Myrobalans

Myrobalans are the dried fruits of *Terminalia chebula* and other species of *Terminalia*. The fruits are about an inch in length and resemble dried plums; their tannin content is chiefly in the pericarp. The fruits is dried on the ground with considerable core, since its value is spoiled by rain-wetting. The content of pyrogallol tannin ranges from 30 to 40 per cent. As a tan, myrobalans are blended with other materials and used for a considerable variety of leathers. They also have limited use in dyeing blacks, as dyeing mordant is fixing colours, and in ink-making (Fig. 15.3).

Divi-divi

Divi-divi is the pods of *Caesalpinia cortaria*. The fruits contain 40 to 14 per cent of pyrogallol tannin and are used in blend with myrobalans, mangrove, quebracho and other materials.

Sumac

The sumac of commerce is chiefly the dried leaves and twigs of *Rhus coriaria* (Anacardiaceae). Scientian sumac contains 20 to 35 per cent of pyrogallol tannin and is used mainly for tanning, though to some extent for dyeing. As a tan, it is employed for the fine light-coloured leathers used in bookbinding, glove-making, etc.

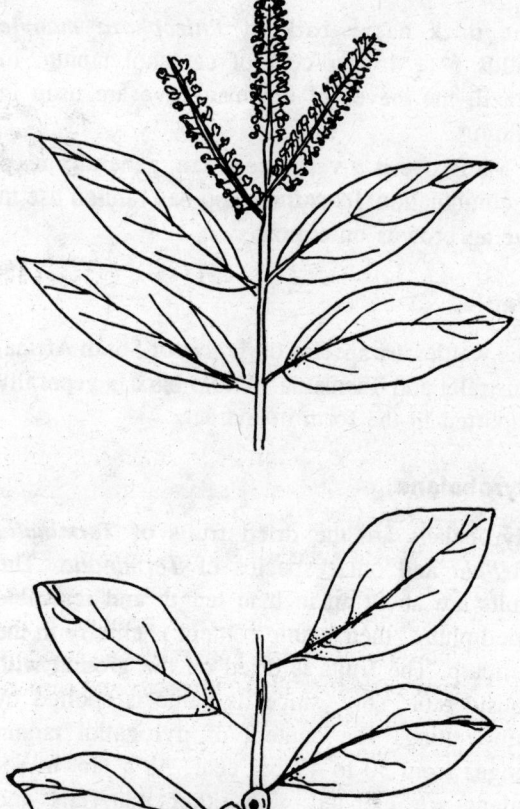

Fig. 15.3. Myrobalans.

Gambier

Gambier is the dried extract of the leaves and twigs of *Ourouparia gambir* (Rubiaceae). This climbing shrub is native to southern Asia. The twigs and leaves are extracted with boiling water and the product evaporated, stirred, hardened, and cut into cubical blocks. It contains 35 to 40 per cent of pyrogallol tannin.

As a tan, gambier is employed in connection with sumac for calfskins and sheepskins, horsehides, rabbit skins, etc. It is rapidly penetrating and is generally blanded with oak or hemlock when used for heavy leather. It is used as a mordant in tanning and as a brown dyestuff. Gambier is an ingredient of boiler compounds for the prevention of scale formation.

Marking nuts

Dhoby nut. Black dye is obtained from the seeds of *Semecarpus anacardium* (Fig. 15.4) which is used to mark the clothes by washerman.

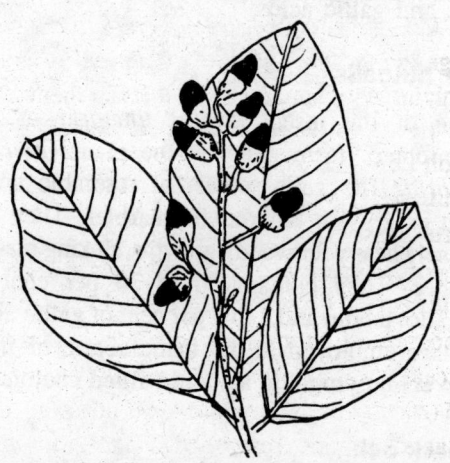

Fig. 15.4. Dhoby nut *Semecarpus anacardium*.

16

Gums and Resins

Gums, which are degradation products of the cell-walls of woody species, are exuded by some trees when they are wounded. These are used as mucilage, for sizing, in confectionary or as a medicine. The most important are those of *babul, dajasal, kulu, salai* and *dhaora*. Of these the gum from kulu trees called *kareya* earns foreign exchange worth about a crore of rupees. It is used for mixing with ice-creams to thicken them. *Kulu* trees occur in large number in Sheopur and Damoh divisions of Madhya Pradesh and large quantities of gum from the latter used to be exported.

Important gum and resin yielding plants are *Acacia senegal* khor, kumta gum arabic from bark (Mimosaceae); *Agathis ausiralis* copal from bark (Pinaceae); *Rhus vernicifolia* sumac, resin from wood (Anacardiaceae); *Pinus longifolia* pine, chil, resin andturpentine from wood (Pinaceae); *Ferula asafoetida* asafoetida, hing, gum resin from root and stem (Apiaceae); *Commiphora myrrha* frankencene and myrrh from wood (Burseraceae); *Boswellia* spp. frankencense and myrrh from wood (Burseraceae); *Burser* spp. elephant tree, copal from wood (Burseraceae); *Abies balsamia* Canada balsam from wood (Pinaceae); *Larix* spp. turpentine from wood (Pinaceae).

True gums are decomposition products of cellulose. This decomposition occurs by a process known as gummosis. They usually occur in plants of dry regions.

Important types of gums

(i) *Gum arabic* : From *Acacia senegal*; gum is used in medicine, in manufacturing ink, in cloth dyeing, in textile, mucilage, paste, polish and confectionery industry.

(ii) *Gum tragacanth* : From *Astragalus gummifer;* used in calico printing, other industrial purposes, gum serves as an adhesive agent for pills and troches and suspension of insoluble powders.

(iii) *Babul gum* : From *Acacia arabica*; used in sizing paper; for adhesive purposes; calcico printing; in sweets (Fig. 16.1).

(iv) *Karaya gum* : A pseudo-gum obtained from bark of *Sterculia urens*; used as a substitute of gum tragacanth; it is superior to tragacanth in viscosity; gum used in the manufacture of glycerine, soaps, creams etc. (Fig. 16.2).

(v) *Kino gum* : Obtained from *Pterocarpus marsupium*; used in medicines.

(vi) *Bengal kino gum* : From *Buta frondosa* (dhak); used as an adulterant for kino gum.

Gum is obtained from *Moringa pterygosperma, Aegle marmelos (Bengal quince), Albizia lebbeck (siris), Albizzia odoratissima (gum khota), Albizzia procera (white siris), Anogeissus latifolia (gatty gum), Mangifera indica* etc. (Fig. 16.3).

Oleo-resins

(i) *Peru balsam* : Obtained from *Myroxylon pereirae,* contains benzoic acid. used as a

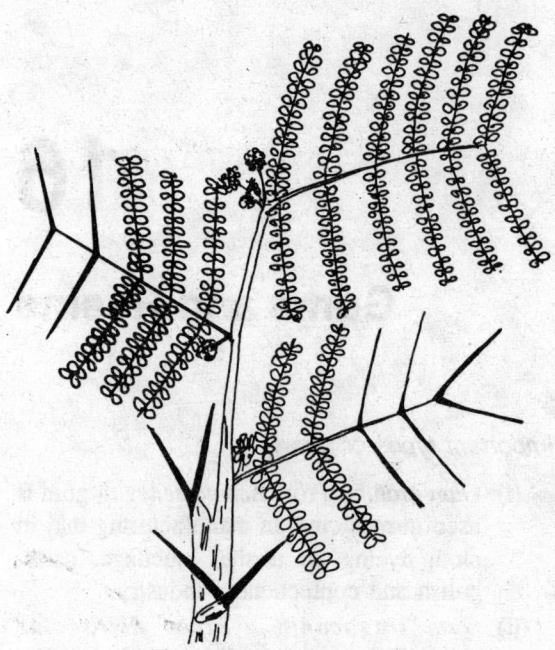

Fig. 16.1. Babul, *Acacia nilotica* (= A. arabica).

fixative with soothing qualities; used in lipstick perfumes.

(ii) *Toru balsam* : Obtained from *Myroxylon balsamum;* used as a fixative and in lipstick manufacture.

(iii) *Canada balsam* : It is a turpentine obtained from the pine *Abies balsamina*. The secreted liquid is viscous, yellowish or greenish in colour. It is mainly used as a mounting medium and for making permanent microscopic preparations, and as a cement for optical lenses.

(iv) *Benzoin* : Obtained from *Styrax tonkinense* (Siam benzoin) and *S. benzoin* (Sumatra benzoin); used as a fixative.

(v) *Frankincense* (myrrh) : An oleo-gum resin (gum olibanum) obtained from *Boswellia carteri* and *B. frereana*; used mainly in the manufacture of an incense.

M-resins

Asafoetida (oleo-resin) is present in *Ferula foetida, Ferula rubricaulis* (Apiaceae). Benzoin is the primary component in the resin of storax (*Styrax*

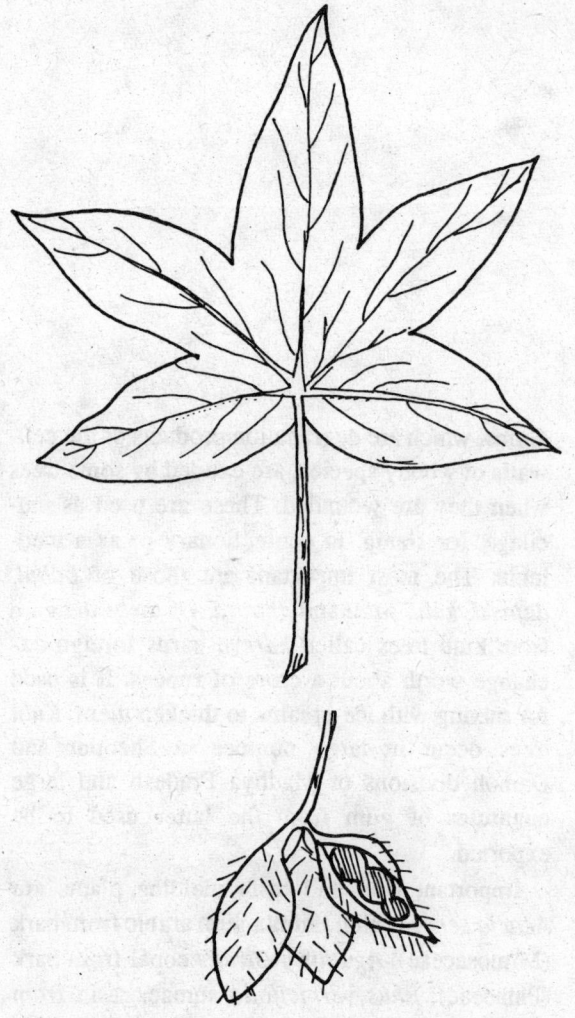

Fig. 16.2. *Sterculia urens* (Sterculiaceae) Kulu, karayer gum tree.

benzoin, Styracaceae). Cinnamein is the primary component in the balsam of Peru *Myroxylon balsamum*, Leguminosae. Storesin is the primary component in the oriental sweet gum *Liquidamber orientalis*, and *Liquidamber styraciflua*, Hememelidaceae. Urushiol is the primary component in the pepper tree, *Schinus molle*, Anacardiaceae.

(i) *Myrrh* : An aromatic gum-resin from *Commiphora* Myrrh was esteemed for its use in incense and perfumes.

(ii) *Asafoetida* : A gum-resin obtained from the cortex of fleshy roots of *Ferula asafoetida*.

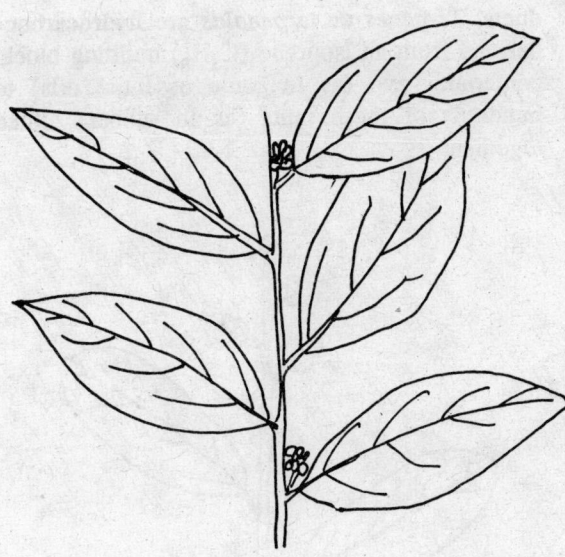

Fig. 16.3. Dhaura—*Anogeissus latifolia* dhavda gum is used in textile printing.

It is used in sauces, curries for flavouring purposes. It is used medicinally in the treatment of coughs, asthma and aid in digestion.

GUAR GUM

Guar (*Cyamopsis tetragonoloba*, Fabaceae) is grown in the arid and semi-arid tracts, particularly in north-western parts of India. At present, guar is commercially grown in India, Pakistan and United States and to a limited extent in Australia, Brazil and South Africa.

Guar gum

The gums is useful in industries loke tobacco, petroleum, mining, textile, cosmetic, oil and pharmaceuticals, explosives, purification of potash, sizing paper and food industries. Maximum guar gum is being used in paper industry. The petroleum industry used to consume a large amount of guar gum. The gum was added to the water in drilling operations to fracture rocks and to brines used to maintain pressure in pumping operations.

Guar gum can be used in edible products like ketchup etc. It is being used as a stabilizer in ice creams, ice crystals and for poor heat thick resistance in the finished icecreams; also in stabilization of ice-pops and sherbets, frozen foods, cheeses, prefillings, icings, dog foods and puddings.

The gum meal can be used as poultry feed.

Gum processing : There are 4 stages in the gum processing. Cleaning of seeds, splitting of seeds, Dehusking and Pulverising.

The seeds are first cleaned and sieved, the smaller seeds are separated out and are then splitted in 'chakis'. The splitted seeds after removing the proteinaceous meal is passed through 3 mills, where the husk is removed from the gum. The *gum splits* then go through the flaking mills, which are then pulverized. After the pulverization, seiving is done to get the required mesh, fine, coarse etc. Then the *sieved gum* is passed through the blenders for blending. Blending is done to make it homogenous, then the gum is packed for marketing.

For marketing, gum is packed in one PUC bag and double gummy bag with 80 kg gum/bag. For pharmaceuticals - Dedca P225 is used.

Gum derivatives

Hydroxyproxypyl guar gum is a good flocculent and used in many mining applications.

In waste water treatments cationic guar gum can replace some of the inorganics used such as alum.

High gum yielding types

Guar varieties IC547/P3, 642/P2-1, 3267, 9004, 9065 (Sona), 10352/P3, 11521, PLG605, RGC144, N.C.11, P-109-1, EC36954/P3, 11521, PLG605, RGC144, NC11/P109-1, EC36954/P29 which contain on an average 29.6 to 31.7 per cent gum content on whole seed basis. Out of these, IC9065 (Sona), and IC11521 are high seed yielders as well and have wider adaptability.

RESINS

Resins form a group of heterogeneous collection of substances with complex and variable chemical composition and with some common physical properties. The main constituents of many resins are esters known as resin esters, complex acids known as *resin acids* and substances of unknown constitution known as resenes. Some are found to contain phenolic substances with strongly irritant properties, while others contain bitter substances and have a strong purgative action. The chemistry of most of the resins is still obscure and recent investigations of some of these have resulted in the isolation of active constituents as distinct chemical substances, bringing them under other classes of compounds. For example, the purgative action of the resin from *Ecballum elaterium* has been traced to one of its components, *elaterin,* and the insecticidal properties of the resin from different species of *Derris* have been traced to the components, such as *rotenone, deguelin, toxicarol,* etc., which are pure chemical entities and can no longer be retained under the category of resins.

Resins having purgative properties are found to be present in *Podophyllum hexandrum, Garcinia morella, Citrullus colocyanthis, Ecballium elaterium, Convolvulus scammonia, Ipomoea hederacea, Ipomoea purga* etc. Resins having blistering or irritant properties are found in *Anacardium occidentale* Linn., *Heligarna arnottiana, Rhus* sp., *Semeca pus anacardium* (Fig. 16.4), *Euphoribia* sp., etc. And resins having insecticidal or toxic properties are known to be present in *Derris* sp., *Tepl rosia* sp., *Calotropis* sp., *Cannabis sativa* etc.

Resins are also exudations of certain trees. They differ from gum by being soluble in alcohol. The most important resin of India is that of chair. The tree is tapped in Punjab, Himachal Pradesh and Uttar Pradesh.

Resin is a plant exudate that is insoluble in water out soluble in organic solvents, and hardens on exposure; chemically resin is a mixture of either *terpenoids* or *flavonoids*. In plants, resin is secreted from epithelial cells lining the resin ducts. *Terpenes* or *terpenoids* are hydrocarbons derived from an isoprene (C_5H_8) building block, containing two (as in some essential oils) to hundreds of these units (as in rubber) joined together.

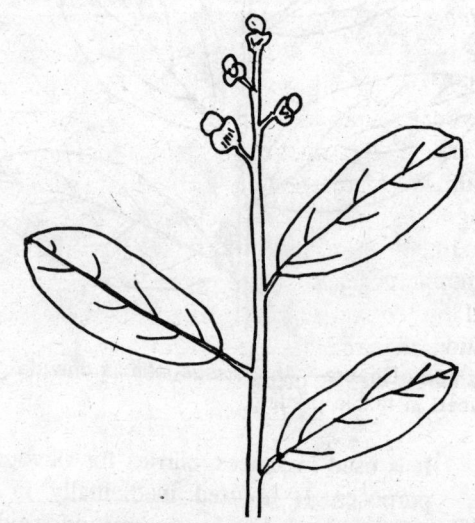

Fig. 16.4. *Semecarpus anacardium,* marking nut tree; the juice of the fruit is used for marking linen; also used in varnish, paint, plastic industry.

Resins are oxidation products of different essential oils and are complex and varied so far as their chemical composition is concerned. Resins are secreted in certain cavities called resin canals. Resins are highly antiseptic and they prevent plants from decay and transpiration too is reduced by them.

True resins

Sal damar is obtained from *Shorea robusta*; white damar is obtained from *Vateria indica;* black damar is obtained from *Canarium strictum*. These resins are used in varnishes, for water proof coating, in medicine, for sizing paper, in mats, in sealing wax preparation, in incense and perfumes.

Butea monosperma, Ficus religiosa, Cajanus cajan, Zizyphus jijuba; used in making sealing wax; in paints and polishes. Lac consists of a resinous substance produced by the female *Coccus lacca,* an insect which is found on the twigs and branches of Shellac is prepared from stic-lac - a resinous substance secreted on the twigs of many trees by an insect called *Tachordia lacca.*

Asafoetida is a *gum-resin.* The fragrant *gigal* (*Bdelellium*) is a gum-resin. Balsams, turpentine are *oleo-resins.*

Lac is used to produce magenta coloured dye, obtained by soaking lac bearing twigs in water. The dye is used as a medicine, for dying silk, and as an article of cosmetics for decorating the feet and palms by maidens. Some lac was also used for making bangles. Western countries became interested in the lac resin chiefly for making grammophone records soon after they were invented by Edison and as an excellent electric insulation material. In India Bihar and Madhya Pradesh are the main producers of lac. India meets nearly 70 per cent of the world demand.

Kauri

A softer varnish resin from *Agathis* (*Dammara*) *australia,* of New Zealand. The tree is a huge one, broad leaved. The best quality and the principal supplies of resin are from fossil deposits. The resin is found in lumps a few feet below the ground surface. The colour of Kauri ranges from clear white to dark brown; the lightest and best is found in dry soil and the darkest in swamps.

Dammar

The term applies to a most miscellaneous group of resins, mostly originating in India and the East Indies. The principal dammars of commerce are from living trees of *Shorea* and other genera of Dipterocarpaceae of Malaya and the South Sea Islands.

Dammars are readily soluble in alcohol, wood alcohol, and turpentine and are used in spirit varnishes.

Copal

The true copals are from *Trachylobium* of the Leguminoceae. The hardest and best grade of varnish resin is Zanzibar or East African copal, from fossil remains in East Africa. The elemi copal comes from *Canarium* of Burseraceae.

Copal is a hard, lustrous resin yielded by various tropical trees; used for varnishes and other materials where a double surface is desired. Dammar is an oleoresin from a variety of tropical trees, primarily from Southeast Asia.

Copals are heat-treated to drive off water and volatile oil and then employed with linseed or China-wood oil in making oil varnishes.

Dragon's blood

The term is applied to a red, resinous substance from the fruits of *Calamus draco* and other palms of Malaya and the Moluccas. The fruits of *Calamus draco* are spheroidal, an inch or less in diamter, with a sealy surface upon which the resin exudes as they ripen. It is separated by heating the fruits in sacks. It is then boiled in water, kneaded, and molded into sticks or lumps for market. Dragon's blood is used chiefly for coloring varnishes and stains.

Amber

This fossil resin is attributed to the extinct conifer, *Pinites succinifer.* The principal source is fossil beds along the shores of the Baltic Sea in East Prussia, where it is mined. Lumps of it are occasionally cast up by the waves. Amber is in high regard and charms and amulets are made from it. It is still used in the rites of the muslim religion. Amber was once used in varnishes.

Uses of turpentine and rosin

Turpentine is chiefly employed as a solvent in paints and varnishes. It mixes with and thins oils, resins, and oil-pigment mixtures and evaporates quickly when the film is spread. It also has some oxidizing, or drying, effect. Turpentine is likewise used in shoe polishes, in oils and greases, sealing waxes, and various other substances. It is the chemical basis of synthetic camphor (Chapter 14).

Small quantities of specially purified turpentine are used in medicine.

Large quantities of rosin are used in the paper industry for sizing-imparting luster and weight and hindering absorption of ink or moisture. Rosin also enters paints and varnishes. A major use is in combination with alkalis in soap-making. Smaller quantities of rosin are distilled into rosin oils which are employed in lubrication. Sealing wax and linoleum also demand resin.

Mucilages are products of metabolic activity in plants. They consist of chains of chemically linked sugars known as polysaccharides, partially soluble in water in which they will swell and form a gel. It is for this reason that drugs which exhibit this property are called mucilaginous. Mucilages are employed medicinally for their local action, that is their beneficial effect on deseased tissue at the point of contact. For example, mucilaginous drugs exert a favourable action in catarrh of the upper respiratory passages in that they soothe abraded mucous membrane and diminish irritation. Their effect in the intes-tines is also excellent. Not only do they check undesirable or fermentation processes but they also promote a mild laxative action due to the initiation of regular, rhythmic peristalsis. These properties are employed in pediatrics, for example, in the treatment of intestinal disorders in infants.

Some of the more important mucilaginous drugs include the flowers of common mallow (*Flos malvae*), flax seed (*Semen lini*), iceland Moss (*Cetraria islandica*), and plantain seed (*Semen psyllium*) of the species *Plantago psyllium* and *Plantago indica*. In some medicinal plants, such as mullein (*Verbascum thapsiforme*), mucilages are not the chief active constituent but are of therapeutic importance as auwiliary components promoting the action of the main active principle.

Medicines from mucilaginous drugs are prepared by extraction in cold water since boiling causes them to lose their efficacy.

17

Rubber and Latex

Rubber is obtained from the milky white latex which is produced by several species of woody plants of Apocynaceae, Asclepiadaceae, Asteraceae, Euphorbiaceae, Moraceae, Papaveraceae and Sapotaceae. Latex is a mixture of hydrocarbons, oils, proteins, resins, acids, sugars, enzymes, salts and caoutchouc. Latex is usually white in colour and is an emulsion.

Important rubber yielding plants are hevea rubber or para rubber, *Hevea brasiliensis* and *Hevea benthamiana* (Euphorbiaceae); Panama rubber or castilla rubber, *Castilla elastica* (Moraceae); ceara rubber, *Manihot glaziovii* (Euphorbiaceae); India rubber or Assam rubber, *Ficus elastica* (Moraceae); Lagos silk rubber, *Funtuma elastica* (Apocynaceae); landolphia rubber, *Landolphia kirkii* (Apocynaceae); guayule rubber *Parthenium argentatum* (Asteraceae); dandelion rubber, *Taraxacum koksaghyz* (Asteraceae).

Other minor rubber yielding plants are *Ficus vogelli* kobo rubber (Moraceae); *Manihot dichotoma, Manihot heptophylla, M. piauhyensis*, jeovie rubber (Euphorbiaceae); *Funtuma latifolia* (Apocynaceae); *Taraxacum megalorhizon* (Asteraceae); *Mascarenhasia elastica, M. arborescens, M. lisianthifolia, M. anceps, M. longifolia, M. velutina, M. geayi, M. kidroa* (Apocynaceae); *Landolphia awariensis, L parvifolia* (Apocynaceae); *Citandra elastica, C. orientalis* (Apocynaceae); *Carpodinus chylorrhiza, C. hirtusa* (Apocynaceae); *Cryptostegia grandiflora* (Apocynaceae); *Marsdenia verucosa (Asclepiadaceae); Euphorbia tirucalli, E. resinifera*,

E. balsamifera, E. intisy (Euphorbiaceae); *Raphionacme utilis* (Asclepiadaceae); *Bleekrodea tonkinensis* (Moraceae); *Urceola elastica* (Apocynaceae); *Alstonia scholaris* (Apocynaceae); *Dyera costulata* (Apocynaceae); *Chondrilla ambigua* (Asteraceae); *Micrandra minor* (Euphorbiaceae); *Sapium jeumani* (Euphorbiaceae); *Jatropha aconitifolia* (Euphorbiaceae); *Palaquium gutta*, gutta percha (Euphorbiaceae); *Mimusops batata* chickle (Sapotaceae); *Achras zapota*, chickle (Sapotaceae).

GUAYULE

Guayule, pronounced wy-oo-le, *Parthenium argentatum*, Asteraceae, is a rubber plant for arid and marginal lands. It is an excellent renewable source for petroleum derived polyisoperene rubbers.

Guayule is a shrub, native to upland plateaus in Mexico and Texas which have sub-tropical climate with low and erratic rainfall. The plant can live for 30-40 years or even more under desert conditions where rainfall may be less than 250 mm.

In 1910 about 50 per cent of the US rubber was extracted from wild guayule shrubs. During world war II nearly 12,000 hectares were cultivated with this crop under the Emergency Rubber Project. From 1910-46, US imported more than 150 million pounds (68 million kg) of Mexican guayule rubber.

Guayule is a shrub and comes to commercial production in only 3 years. No tapping of latex is required. Two-thirds of the rubber is in the stem and branches, and the remainder in the roots. The whole plant can, therefore, be processed for rubber. Leaves do not contain rubber and these may be eliminated by drying.

Until recently, 95 per cent of the rubber used was polybutadeine (synthetic rubber). Radial tyres, which have swept the market during the last decade, need more resilience than that afforded by butadiene polymers. Consequently, manufacturers incorporate large amounts of natural rubber into radial tyres.

Quality of rubber

Guayule rubber and Hevea rubber are similar. Both are hydrocarbons, polymers of isoperene and both are approximately of the same molecular length and weight. Guayule rubber volcanises like Hevea rubber. It has the properties that allow it to flow properly in molds and extruders and like the Hevea rubber it has the nature and tack crucial for tyre manufacture.

In India experimental planting of guayule could possibly be established in the arid regions of Rajasthan, Gujarat, Maharashtra and other parts with arid lands.

This plant reproduces apomictically. The varietal characteristics can be perpetuated through seeds. The rubber is contained within cells in the entire plant but the roots and stems are particularly rich. Guayule shrub may live as long as 50 years and thus represent a living stockpile of rubber. Yields up to 12 per cent (dry weight) have been obtained from mild plants and over 20 per cent from improved varieties can be obtained.

Rubber extraction

The shrubs are macerated in the presence of water in a pebble mill. The crushed shrub is floated in the floatation tanks, where the rubber particles float and the water-logged plant debris sinks. The rubber is next carried over a vibrating screen to free it from water. It then goes through a drier where entire moisture is removed and, lastly, to a press which compresses it into 45 kg blocks for shipment to the market.

Vulcanization

Rubber is heated with various quantities of sulphur (only about 2 per cent of soft rubber, as much as 40 per cent of hard rubber) retained its firmness at normal temperatures of heat and did not become tacky.

Cold vulcanization

Thin films of rubber could be changed from the plastic to the elastic state by the use of sulphur monochloride. This change could be accomplished without heat.

18

Sericulture

Silk is a natural fibrous substance, obtained in the form of a long, continuous filament from cocoons (dupal nests) spun by a large variety of moth-caterpillars, known as *silk worms* (*Bombyx mori*). Fine-texture silk is the product of the mulberry-feeding silkworms (domesticated type) belonging to the family Bombycidae.

In India sericulture dates back to antiquity. Mulberry culture spread to India by about 140 BC from China through Khotan. Silk industry gained importance during the 17th century and flourished during the 18th and 19th centuries in Bengal, Mysore and Kashmir states. India is now the fourth largest producer of raw silk in the world. According to the Internation Silk Association, production of raw silk in the world in 1968 was estimated at 37,622 tonnes of which Japan produced 20,755, China 9,000, USSR 3,000 and India 1,745 tonnes. Other countries producing raw silk and Brazil, Bulgaria, Republic of Korea, Iran and Italy.

Sericulture is one of the most integrated rural industries, ancillary to agriculture, serving India's economy by lending a broad base for rural employment The non-mulberry raw silk industry- *eri*, *muga*, and *tasar*, affects directly or indirectly the life of nearly a million persons in Assam, Bihar, Orissa and Madhya Pradesh. *Erisilk* : The moth *Philosamia ricini* whose larvae feed on castor plant (*Ricinus communis*) produce eri silk. Worms of this species feed also on leaves of *Keseru* (*Heteropanax fragrans*), cassava (*Manihot esculenta*), Papaya (*Carica papaya*), sankru

(*Jatropha multifida*). Worms of *eri* moth are also found to feed on leaves of potato, cabbage, sissoo (*Dalbergia sissoo*). Cocoons of *Phylosamia cynthia* are brick-red coloured whereas those of *P. ricini* are white in colour. *Eri* silk is mainly produced in Assam; other *eri* silk producing states are Bihar, Manipur, West Bengal and Orissa.

Muga silk

The wild silkworm *Antheraea assama* produces muga silk; caterpillars of this species feed on leaves of *som* (*Machilus bombycina*) and produce white or amber-coloured coccoons, from which is obtained the *muga* or *moonga* silk. The much esteemed creamy white *muga* silk and *mozankuri* silk are obtained from worms fed on leaves of *champa* (*Michelia oblonga*) and *mozankuri* (*Litsea cubeba*) respectively.

Tasar silk

The wild silkworm *Antheraea mylitta* produce *tasar* silk; the cater pillers of this species feed on leaves of *sal* (*Shorea robusta*), *asan* (*Terminalia tomentosa*), arjun (*T. arjuna*), *ber* or *kul* (*Zizyphus jujuba*), semul (*Salmalia malabarica*), peepal (*Ficus religiosa*, fig (*Ficus glomerata*) etc. Colour of the cocoons varies from brown to different shades of tan; cocoons of this species yield fine copperish coloured *tasar* or *tussah* silk.

Chinese and Japanese silks

Wild silk moths of different varieties are also

domesticated in other countries of the world. Of these, important ones are, the Chinese *tasar* silk moth, *Antheraea pernyi* which yields *shunting* silk occurring in China, Siam and Indo-China. The Japanese *muga* silk moth is *A. yamamai*. The moth *Attacus atlas* is widely distributed in Sri Lanka, North India to Assam, Malaya, Java, China, Philippines; the caterpillars of this species cocoons which yield various silken filaments such as Fagora, Tagore and *Ailanthus* raw silk.

Race, which is popular among the sericulturists yielding around 35 kg per 100 layings with a renditta of 12, compared to 18 to 20 kg yield and 16 to 18 renditta with the conventional multivoltine race, *Pure Mysore*.

The four new bivoltine races viz. *KA, NB402,* and *NB18* are very popular and yield 45 to 50 kg per unit of 100 dfls reared with the renditta around 7.5.

Rearing improvement

In the past sericulturists experienced 2 or 3 crop failures out of 5 or 6 each year. Silkworm diseases used to take a heavy toll. With the new technology it has become possible for the farmers to harvest cocoon production of the order of 40 kg/100 dfls as compared to the 15-20 kg a decade ago.

Control of uzi fly

In 1980, the sericultural states of South India were threatened by the uzi-fly menace. By providing a nylon net enclosure around the rearing stands and the working space the files can be kept out and the crop loss which was 40 to 50 per cent at the initial stages has been reduced to 3-5 per cent. Use of levigated China clay on the spinning worms and mountages by dusting is another simple technique to prevent the uzi fly attack.

Inter-specific hybrids of tasar

The indigenous oak feeding tasar silkworm (*Antherea roylei*) remained unexploited. During 1960 attempts were made to evolve a productive strain through inter-specific hybridization with its popular Chinese counterpart.

Silk production in India

India enjoys the privilege of having all the four varieties of natural silk viz. mulberry, *tasar, eri* and *muga* and ranks third among the mulberry silk producing countries of the world. It occupies the coveted second place in *tasar* silk production and has the monopoly in bringing out *eri* and *muga* silks. The current annual production of mulberry silk, *tasar, muga* and *eri* are 5488, 279, 173 and 37 metric tonnes respectively. If the potential available in the country is appropriately exploited, India could emerge as the leading silk producing country in the world.

Karnataka accounts for 60 per cent of the total mulberry raw silk production of the country followed by Andhra Pradesh (15 per cent), West Bengal (12 per cent), Tamil Nadu (10 per cent), Jammu and Kashmir (1 per cent), the contribution from other states being about 1 per cent.

Upgrading silkworm breeds

Sericulture in India is practiced since time immemorial. In the beginning only the local silkworm race, *pure Mysore C. Nichi* and *HS6* were available. Hybrids of these races were reared for industrial cocoon production which were low yielders with low filament length and poor reelability.

With the evolution of improved techniques of silkworm rearing under tropical conditions and as a result of breeding and hybridization experiments conducted at the Central Sericultural Research and Training Institute, Mysore and Berhampore, new multivoltine and bivoltine races have been evolved during the last 10 years.

Small quantity of *silk worm pupae oil* - the waste product obtained after extraction of raw silk from cocoons, is being produced.

MULBERRY

Mulberries, a few of the *Morus* species are valued for their foliage which constitute the chief feed for mulberry silkworm (*Bombyx mori*). The most important type of mulberry grown in India for rearing silkworms is *M. alba* var. *multicaulis* which is native to China or Philippines. It is a fast-growing type adapted for cultivation as a field crop and giving high yields of large, tender and thick leaves. The variety kanva-2 gives 30 per cent higher yield; varieties 530, 536, 541 and 554 have double potential of leaf yield compared to local varieties.

Mulberry grows well in sandy or heavy loam and black cotton soil. The average rainfall required is 60-100 in/annum, spread uniformly throughout the year. The plant does not stand water-logging or shade.

The yield of mulberry leaves varies from 10,000-14,000 lb. Per acre annually under irrigated conditions in the Mysore region.

Fruit is the ripened ovary enclosing seeds. Fruits are edible, also used as vegetables and for making pickles. The study of fruits is called *pomology*. Ripened fruits contain high percentage of pectin sugars. Most tropical fruits come from woody perennials which cannot germinate and produce fruits in a few months. Mango, banana, apple, grapes, oranges, guava, papaya, pomegranate, custard apple, grapes, oranges, guava, papaya, pomegranate, custard apple are the common tropical fruits.

Table 19.1. Common tropical fruits

Common name	Botanical name	Family
Apple	*Pyrus malus*	Rosaceae
Pear	*Pyrus communis*	Rosaceae
Cherry	*Prunus avium*	Rosaceae
Peach	*Pruncus persica*	Rosaceae
Plum	*Prunus domestica*	Rosaceae
Watermelon	*Citrullus vulgaris*	Cucurbitaceae
Grape	*Vitis vinifera*	Vitaceae
Date palm	*Phoenix dactylifer*	Arecaceae
Sweet orange	*Citrus sinensis*	Rutaceae
Orange	*Citrus reticulata*	Rutaceae
Custard apple	*Annona squamosa*	Annonaceae
Guava	*Psidium guajava*	Myrtaceae
Papaya	*Carica papaya*	Caricaceae
Pomegranate	*Punica granatum*	Punicaceae
Pineapple	*Ananas sativas*	Bromeliaceae
Mango	*Mangifera indica*	Anacardiaceae

(Contd.)

Common name	Botanical name	Family
Banana	*Musa paradisiaca*	Musaceae
Bread fruit	*Artocarpus altilis*	Moraceae
Carambola	*Averrhoa carambola*	Euphorbiaceae
Citron	*Citrus medica*	Rutaceae
Fig	*Ficus carica*	Moraceae
Jack fruit	*Artocarpus heterophyllus*	Moraceae
Lemon	*Citrus lemon*	Rutaceae
Litchi	*Litchi chinensis*	Sapindaceae
Lime	*Citrus aurantifolia*	Rutaceae
Loquat	*Eriobotrya japonica*	Rosaceae
Malay apple	*Syzygium malaccense*	Myrtaceae
Mammey sapota	*Pouteria sapota*	Sapotaceae
Sapodilla	*Manilkara zapota*	Sapotaceae
Melon	*Cucumis melo*	Cucurbitaceae
Bitter orange	*Citrus aurantium*	Rutaceae
Passion fruit	*Passiflora edulis*	Passifloraceae
Pummelo	*Citrus maxima*	Rutaceae
Rose apple	*Syzygium jamkos*	Myrtaceae
Squash	*Cucurbita maxima, Cucurbita moschata, Cucurbita pepo*	Cucurbitaceae
Ber	*Zizyphus jujuba*	Rhamnaceae
Phalse	*Grewia asiatica*	Tiliaceae
Amla (Indian gooseberry)	*Phyllanthus emblica*	Euphorbiaceae
Persimmon	*Diospyros virginiana*	Ebenaceae

(Contd.)

Common name	Botanical name	Family
Bael	*Aegle marunelos*	Rutaceae
Mulberry	*Morus alba, Morus nigra*	Moraceae
Cashew fruit	*Anacardium occidentale*	Anacardiaceae
Cherimoyer	*Annona cherimolia*	Annonaceae
Soursop	*Annona muricata*	Annonaceae
Bullock's heart	*Annona reticulata*	Annonaceae
Tree sorrel	*Averrhoa bilimbi*	Oxalidaceae
Palmyra fruit	*Borassus flabellifer*	Arecaceae
Christ's thorn	*Carissa carandas*	Apocynaceae
Natal plum	*Carissa grandiflora*	Apocynaceae
White sapote	*Casimiroa edulis*	Rutaceae
Star apple	*Chrysophyllum cainito*	Sapotaceae
Tree tomato	*Cyphomandra betacea*	Solanaceae
Halwa tendu	*Diospyros kaki*	Ebenaceae
Gaub persimmon	*Diospyros peregrina*	Ebenaceae
Wood apple	*Feronia limonia*	Rutaceae
Timla	*Ficus auriculata*	Moraceae
Gular	*Ficus glomerata*	Moraceae
Daburi	*Ficus hispida*	Moraceae
Anjini	*Ficus palmata*	Moraceae
Gagjaria	*Ficus rumphit*	Moraceae
Governer's plum	*Flacourtia ramantchi*	Flacourtiaceae
Strawberry	*Fragaria vesca*	Rosaceae
Avocado	*Persia americana*	Lauraceae
Cape gooseberry	*Physalis peruviana*	Solanaceae
Manila tamarind	*Pithecellobium dulce*	Mimosaceae
Mysore raspberry	*Rubus niveus*	Rosaceae

Common **temperate** fruits are avocado (*Persia americana*, Lauraceae); apple (*Malus pumila*, Rosaceae); apricots (*Prunus armeniaca*, Rosaceae); black-berries (*Rubus* spp. Rosaceae); blue berries (*Vaccinium corymbosum, Vaccinium angustifolium*, Ericaceae); sour cherry (*Prunus cerasus*, Rosaceae); sweet cherry (*Prunus avium*, Rosaceae); cranberries (*Vaccinium oxycoccos, Vaccinium macrocarpon*, Ericaceae); European black currant (*Ribes nigrum*, Grossulariaceae); red currant (*Ribes sativum*, Grossulariaceae); gooseberry (*Ribes uva-crispa*, Grossulariaceae); fox grapes (*Vitis lambrusca*, Vitaceae); wine grapes (*Vitis vinifera*, Vitaceae); loganberry (*Rubus vitifolius*, Rosaceae); olive (*Olea europpa*, Oleaceae); peach (*Prunus persica*, Rosaceae); pear (*Pyrus communis*, Rosaceae); plums (*Prunus domestica*, Rosaceae); Quince (*Cydonia oblonga*, Rosaceae); black raspberry (*Rubus occidentalis*, Rosaceae); red raspberry (*Rubus idaeus*, Rosaceae); Chilean strawberry (*Fragaria chiloensis*, Rosaceae); Virginia strawberry (*Fragaria virginiana*).

BANANA

The Malayan peninsula or south-east Asia are thought to be the original home of the cultivated banana as the fruits have been used by man in these regions. It was one of the first fruits grown by man. References to bananas in India occur during the 16th century BC, and one is of particular interest since it suggests the existence of a mutation 2,000 years ago. From 600-300 BC Indian writings often refer to bananas.

The Islands of Honduras and Jamaica are among the chief banana exporting countries today. It is cultivated extensively, in Jamaica, India, Costa Rica, Cuba, Honduras, Northern shores of Columbia, Central America, Canary Islands and the West Indies. The chief regions in India which grow banana extensively are Tamil Nadu (Travancore-Cochin), West Bengal and Gujarat (Fig. 19.2).

Banana belongs to the family *Scitamineae*, genus *Musa*. According to Cheesman, Linnaeus used originally two specificnames for the edible bananas, viz., *paradisiaca* and *sapiantam*, but later three species were generally accepted as follows :

1. *M. paradisiaca* for those which yield fruits of a starchy consistency, not usually consumed, except after cooking, and popularly known as plantains.
2. *M. sapientum* for all desert bananas except of dwarf habit.

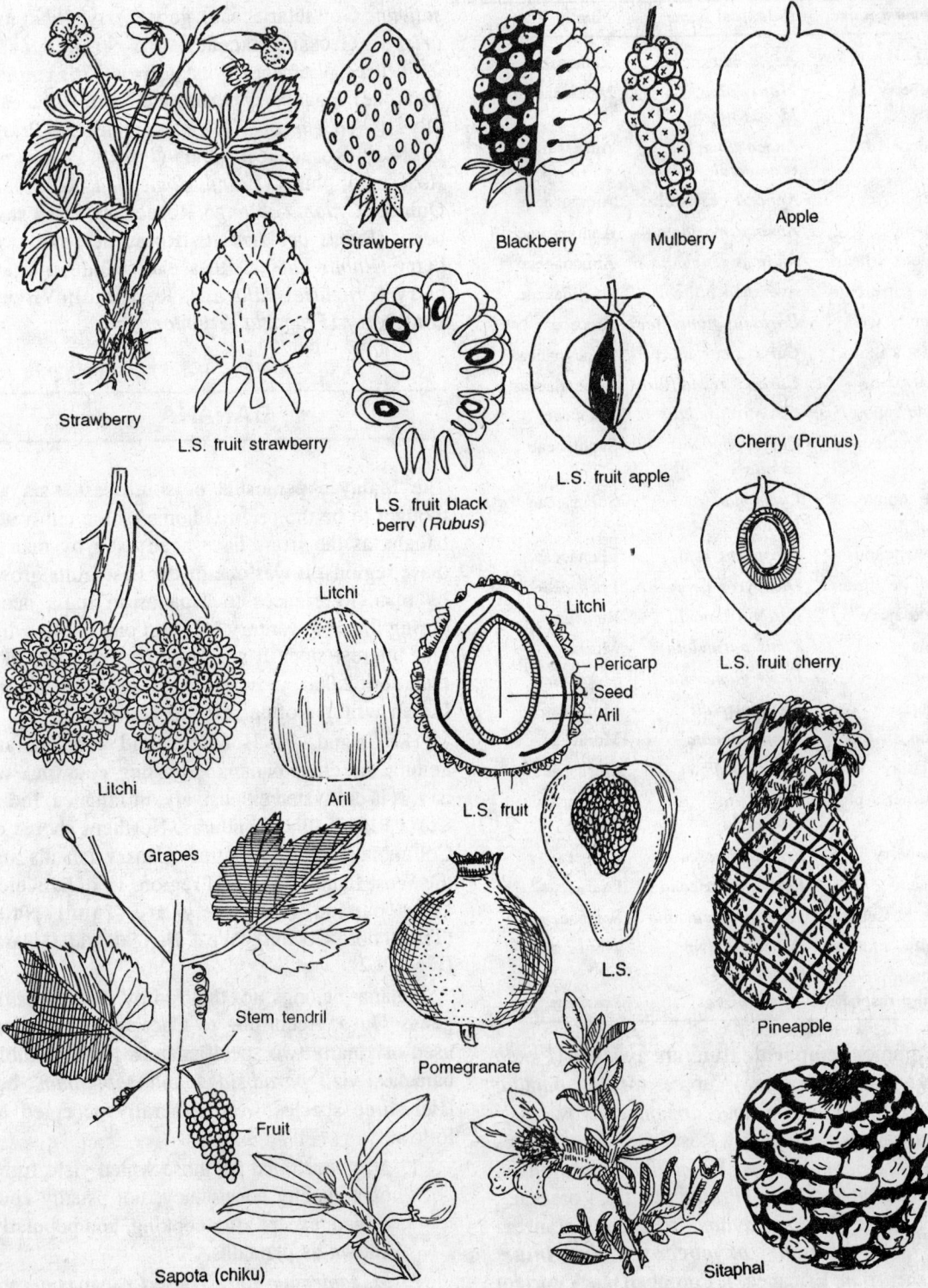

Strawberry

Strawberry

Blackberry

Mulberry

Apple

L.S. fruit strawberry

L.S. fruit black
berry (*Rubus*)

L.S. fruit apple

Cherry (Prunus)

L.S. fruit cherry

Litchi

Litchi

Pericarp
Seed
Aril

Litchi

Aril

L.S. fruit

Grapes

Stem tendril

Fruit

Pomegranate

L.S.

Pineapple

Sapota (chiku)

Sitaphal

Fig. 19.1. Different fruits.

Banana tree and flowers

Fig. 19.2. Banana—*Musa paradisiaca* plant, flowers and fruits.

3. *M. cavendishie* (Lamb) for dwarf bananas, known as Canary or Chinese bananas.

Varieties

Mindoli, Rajeli, Sonkel, Velchi, Lal kel, Bassrai, Walha, Multheli, Poovan (Karpura Chakrakeli), Nendran, Sirumalai, Monthan, Maurituius, Rasthali, Pacha Nadan, Bathisa, Chakrakeli, Bontha are important cultivated varieties of banana in our country.

Description

The **banana** is a herbaceous plant reaching a height of more than 30 feet in some cases, although its only true stem is the underground **rhizome**. The *pseudostem* is formed of the bases **of the leaves**, and from its centre emerges the **infolrescence**, which is an elongated *spike*. In most varieties the spike bends over, so that the bunch of fruit hangs down. One set of roots extends *horizontally* in the top 2 feet of the soil, while another grows *vertically* to a depth of about 6 feet. The main roots are nearly uniform in thickness, and possess numerous small rootlets. The flowers are arranged in clusters of two spiral rows each and are of three types. Those at the base of the spike open first and are *pistillate*. Towards the end of the spike are the *neutral flowers,* with neither pistils nor stamens well developed, and the *staminate flowers*. In most cultivated bananas the fruits are seedless, and pollination is not required.

Ecology

Banana is essentially a tropical plant and grows in warm humid climates better than in cold dry climates. Tracts where there is a chance of frost occurring during the winter, are not safe for its cultivation. Low temperature damages or kills the plant tissues and hinders the development of the fruit, affecting its quality. Soils which are best suited for banana have a fair quantity of *lime and humus.* A soil which is about 2 feet deep with a *porous subsoil* constitutes the minimum requirements for successful banana cultivation. Rainfall about 70"-80" per annum is good for its cultivation. Banana is usually propagated by *suckers,* a large number of which spring up from the base of the mother plants.

Cultivation

Before planting suckers of banana, the land should necessarily be well tilled. The size of pits taken for planting suckers and the spacing distance from plant to plant depends on the variety and nature of the soil. In heavy black soils, pits are generally larger and deeper being sometimes as big as 1½ feet by 1½ feet each. In lighter soils, it is not necessary to have such big pites. The bananas are planted 5-6 feet apart in the case of dwarf varieties, and 9-10 feet apart in the case of tall varieties. 800-1, 250 suckers are planted to an acre depending upon the variety. Banana is often planted as an amidst areca-nut and coconut plantations in Malabar and along with mango trees in E. Punjab and Gujarat.

Sword sucker

Which has a vigorous shoot, thick at the base and sharply converging towards the end with few linear leaves from which the name, *sword sucker* has arisen. Such suckers are preferred to ordinary suckers. Suckers less than 3 feet in height and selected from healthy and mature plants. These are found superior as planting material. Large suckers might bean fruit earlier than smaller ones but their bunches are generally small.

In Brazil and Queensland *corms carrying 2-3 eyes* only are used for planting.

Planting is done in February-March or in July-August in some places. Banana is a heavy surface feeder. Caster cake, sulphate of ammonia, sulphate of potash, super-phosphate are given as top dressings at regular interval of one month for a period of 3-4 months.

Harvesting

The first crop of fruits is ordinarily borne in somewhat more than a year after planting. Successive crops are secured at intervals of 5-10 months, depending on the climate and the management. Bananas are harvested while still green, but after they have reached almost their full size and become plump.

The Banana furnishes food in a variety of forms, fodder for the cattle, thatching for the house, and fibre for a number of purposes.

Artificial ripening of bananas

In transporting the fruit to longer distances, it is advisable to select only well developed green bunches, because ripe fruits decay and a great loss might occur subsequently. For long distance, fruit is packed raw and ripened artificially.

At some places in the Deccan, banana bench bunches are heaped one over the other covered with straw and mud and then smoke is blown into the heap. The green fruits change their colour due to high temperature thus generated.

Banana is ripened in some countries by the aid of *gas heaters,* the bunches being staked in specially constructed rooms, *coal gas* or *ethylene gas* is used for their purpose.

Emanations given out from oranges effect premature repending of bananas and, therefore, it is necessary to have separate storage for citrus fruits and bananas.

MANGO (Fig. 19.3)

After the month of August, mango virtually disappears from the Indian Markets. Mango starts ripening in Cochin i.e. Kerala as early as February followed by Tamil Nadu. Delhi market is supplied with mangoes first from South, then from Bihar and UP, the latter produce reaches in June and

July. Punjab mangoes are last to arrive in the market even though they happen to cultivars most popular in other North Indian plains, viz. Langra and *Dusheri*. There are three ways to extend the harvesting season :

   (i) To adopt some cultural treatments including the use of growth regulators which celay the ripening;

   (ii) To develop late-ripening varieties;

   (iii) To improve the storage possibilities of the crop;

   (iv) To explore area which are suitable for the cultivation of mango and have agro-climatic feature which induce late-ripening.

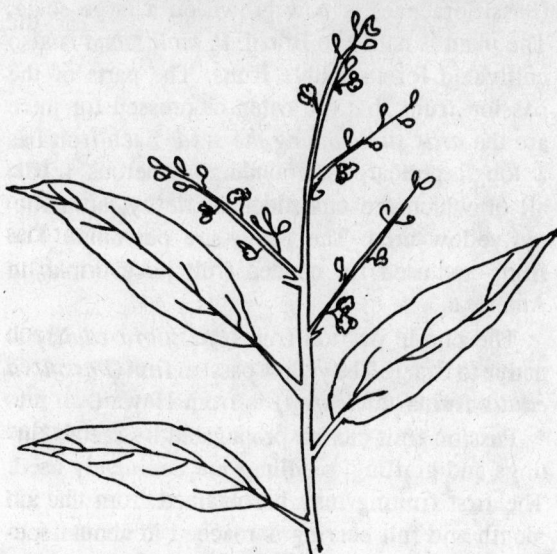

Fig. 19.3. Mango—*Mangifera indica.*

High latitude mangiculture

Mango can grow up to 1,700 m above sea level in the Himalayas. So low hills of Sikkim and Bengal may also be suited for the production of late ripening mangoes. In the Punjab, the following mango-growing areas come under the high latitude : *Dasuya, Mukerian, Gardhiwala, Dhar, Pathankot* and *Dunera.* The tract of high as latitude and low-altitude district of Jammu, Punjab and Himachal Pradesh mangoes ripen precisely aftèr August and hold great promise both for domestic and export markets.

MANGOSTEEN

A globular iridiscent fruit, just about the size of a tennis-ball is seen in the fruit markets of Coimbatore, Ooty and Tirunelveli districts in Tamil Nadu during July-September. The fruit of beau-tiful deep reddish purple hue, is mangosteen (*Garcinia mangostana,* Guttiferae) and is popu-larly referred to as the *Queen of tropical fruits.*

Often grouped among the world's best flavoured fruits, mangosteen originating from the Malayan Peninsula, is grown in Indonesia, Malaysia, Sri Lanka, Philippines and South India. Preferring an ideal elevation of 400-1000 m and an annual precitation of 125-150 cm, this tropical fruit is at present restricted to the lower range of Nilgiri hills and Courtallam range in Tamil Nadu.

Mangosteen comes up well in dry fertile soils with adequate drainage and prefers mild acidic soils rich in humus. The yield and the season of bearing is largely determined by the weather conditions and soil moisture. Watering the seed-lings during the bearing season will help increase the yield and improve the size and characteristics of the fruit.

Seedlings of true tó type nature emerge from parthenocarpic fruits. These are grafted on root stocks of *Garcinia tinctoria* or *Garcinia speciosa,* to make the plants early-bearing. Conventional seedlings take about 8-10 years for yielding fruits.

Seedlings are slow growing; pre-treatment of seeds with gibberdlic acid will help in improved germination and growth of seedlings. The plants are grown at spacings of 8-10 m on either side, and upto 150 trees can accommodated in a hect-are.

The first harvest of the crop will begin in about 10 years from planting and yields will continue for more than 100 years. Each tree will produce on an average 600 fruits in January-March season and July-October season. A maximum of 3,800 fruits per tree in a year is also possible.

Gambog is the only serious disease of the crop and prophylactic spraying with copper fungicide can present infection.

STRAWBERRY

Strawberry is mainly cultivated in temperate and subtropical regions, such as Simla and Srinagar in India, but recently it has been grown successfully in the plains of India as well. Strawberry fruits are attractive bright and delicious.

Cultural practises

Select a bed with a slope to ensure drainage, use sandy loam though other soils are good. Avoid water-logging. For manuring a 34 sq m area, 6 baskets of well-rotted farmyard manure are required.

In the plains planting is done after the monsoon is over and by mid-September the suckers are transplanted to the beds. These are set firmly in the ground, 60 cm apart in rows, the rows a metre apart. Light irrigation is provided every 15 days till the fruits appear.

A top-dressing 10 days after setting, with a mixture of 1 kg ammonium sulphate, 0.75 kg super phosphate and 0.25 kg potassium sulphate to 33 m of rows gives good result. The fertilizer is applied again after a month or so, if plants do not show vigorous growth.

To keep the berries free from mud, wheat or oat or rice straw is spread around the plants before they come into flowering, so that the berries can rest on them. In the plains, fruits ripen in March to May. They are picked individually daily, and stored in a cool place until used, fresh or frozen.

After fruiting is over, the beds are raised, flattened and the slope is maintained to drain off rain-water. The plants develop *runners* and these are planted 10 cm apart in raised beds and then covered with *sirkhi* to avoid run-off during rains. They are kept there till the next sowing season.

A tropical fruit that is gaining international importance is the passion fruit-*Passiflora edulis* (purple variety), *Passiflora edulis* of flavicarpa (yellow variety). The fruit is mainly used for its juice qualities since it imparts a very attractive aroma.

Kiwi fruit

Actinidia chinensis, Actinidiaceae is native to Asia. It has been grown commercially in New Zealand. The plant is a dioecious vine. The fruits are avoid, fuzzy green, delicate in flavour and blends well with other fruits. The slices of the fruit with their translucent, pale green flesh surrounding a narrow ring of tiny black seeds produce attractive deserts. The fruit is popular in America.

Passion fruit

Of the 55 or so edible species, *Passiflora edulis* (Passifloraceae) is now grown on a large scale. The plant is native to Brazil. *P. mollissima* is also cultivated for its edible fruits. The parts of the passion fruits that are eaten or pressed for juice are the *arils surrounding the seed*. Each fruit has a tough pericarp surrounding numerous seeds, all of which are embedded in fleshy, aromatic, red-yellow arils. The plants are perennial. The fruits are used for canned fruit juice drinks in America.

The purple passion fruit (*Passiflora edulis*) is native to Brazil. The yellow passion fruit (*Passiflora edulis* forma *flavicarpa*) is from Hawaii.

Passion fruit can be propagated by seeds, cuttings and grafting, seedlings are commonly used. The first fruiting may be obtained from the 9th month and full bearing is reached in about 16 to 18 months from planting. A fully grown vine can yield about 10-15 kg during season. The juice content varies from 25 to 40 per cent in both the varieties.

Uses

The fruit juice provides an excellent flavour for pies, cake, frosting, puddings, sauces and salads, desserts, punch base etc. It is considered an excellent mix for alcoholic beverages such as vodka, gin and rum. It is also used as a flavourant in jams, butter, conserves, jellies, marmalades and hot spiced beverages.

Passion fruit is an attractive ornamental plant in residential colonies.

Extraction of juice

Wash fully ripe fruits and cut into halves by stainless steel knife. Scoop out the pulp (seed and mucilaginous material) by spoon and mix an equal quantity of water (1 cup pulp + 1 cup water).

Macerate the pulp and water with hand, remove seeds and squeeze out the juice after tying the whole mass in a muslin cloth. Add more water to the pulp helps in separation of seeds from pulp and more recovery of colour and water soluble constituents from mucilaginous material. A kilogram of fruits yields more than 700 ml dilated juice.

Passion fruit squash

Juice 300 g; Sugar 450 g; water 250 g; Citric acid 8 g or 2 teaspoons (levelled); Preservative (potassium metabisulphite) 500 mg or 1/8th tea spoon (levelled). Mix the sugar and water and heat it till it comes to boiling. At this stage, add citric acid after dissolving it in a small quantity of water. Filter the syrup through muslin cloth and cool it. Mix the juice and syrup and add the preservative which has been dissolved in a small quantity of water. Fill the product in 600 to 750 ml sterilized squash bottles and close them with screw type caps. Met the paraffin wax or bees wax and apply it all sides of cap to make the bottle airtight. This product could be store for more than 6 months and could be consumed after dilution with two-folds of water (one cup squash + 2 cups of water).

OPUNTIA

The native Americans knew many uses of cacti; cacti was a major source of food before the arrival of Columbus. Many people even now live for months on *Opuntia* (prickly pear) fruits. They also make marmalade, honey, and cheese out of the fruit. The famous Mexican liquor, 'pulaque', is made from the sap of agave and other succulents flavoured by the fruits of *Opuntia*.

Besides yielding food and drink, the native Americans used to obtain fibre, wooly hair, spines knitting needles and medicinal compounds from *Opuntia*.

The scale insects, used in the manufacture of red dyes, feed or *Opuntia*.

Some fruits of Indian deserts

Ker (*Capparis decidua*) *pilu* (*Salvadora oleoides*), *gondi* (*Cordia gharaf*), *jharber* (*Zizyphus nummularia*), *Carissa carandas* etc. (Fig. 19.4).

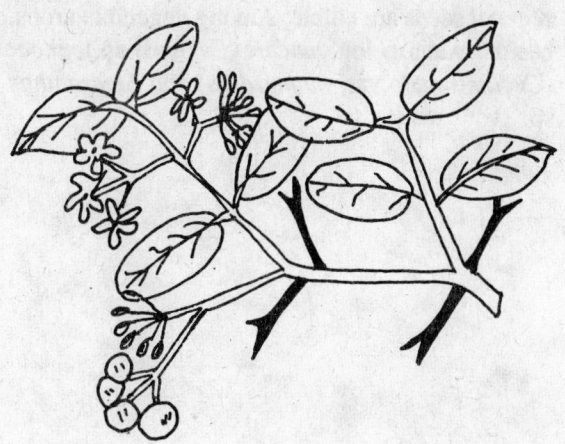

Fig. 19.4. *Carissa carandus.*

Capparis decidua

Much branched staggling shrub. Fruit is a berry, ovoid to subglobose, 0.5-1.5 cm in diam., full of seeds, glabrons, green when urine and scarlet-red on full maturing. Fully mature fruits have slimy and sweet pulp. Fruits are used for making pickle, dehydration and vegetable purpose. The fruits are believed to provide relief in *cardiac trouble*.

Salvadora oleoides

A shrub or small tree, grows in highly saline tracts in the desert. The fruit is a drupe, globose, usually turn yellow when ripe and is sweet.

Cordia gharaf

A shrub or a small tree. Its fruit is a drupe, ovoid in shape, yellow or reddish brown in colour (Fig. 19.5).

Zizyphus nummularia

A shrub or small tree. Its fruit is small, round and turn red on ripening. Pulp is sweet but very little. There is another type called *boradi* (*Zizyphus* sp.) growing along the river banks in Jodhpur. The fruits are sold fresh as well as after sun drying.

Prosopis cineraria gives edible pods; *Acacia senegal* seeds are edible. Among vegetable crops, besides watermelon, *machri* (*Cucumis* sp.), *phoot* (*Cucumis melo* var. *momordica*) and *Smaranthus* sp. grow wild.

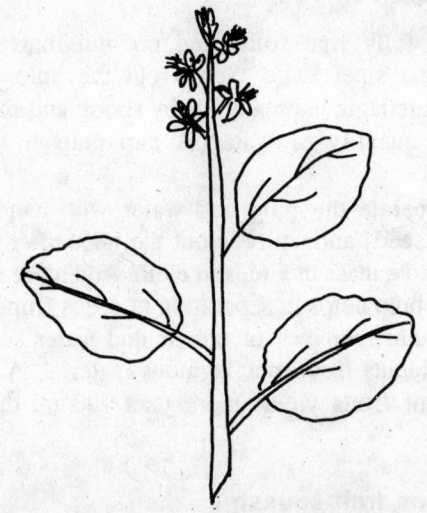

Fig. 19.5. *Cordia dichotoma.* Boraginaceae. The fruit is sweet to taste and edible, and can be picked.

Nut is a one-celled, one seeded dry fruit with a hard pericarp. One of the most expensive tropical nut is *macadamia* (*Macadamia integrifolia*; Proteaceae) native to Australia. *Cashew,* another popular nut is native to the forests to South America. The cashew nut (*Anacardium occidentale,* Anacardiaceae) is the embryo borne inside a hard seed coat. *Brazil nuts* (*Bertholletia excelsa* and other species of Lecythidaceae) are native to South American tropics.

Walnuts, pecans, almonds, chestnuts, filberts are native to warm temperate regions and most of the old world. *Pecans* are the only important temperate nut to come from North America. Pecans (corya, pecan) hickory nuts, butternuts and walnuts belong to the Juglandaceae. *American chestnuts* (*Castanea dentata,* Fagaceae), *European chestnut* (*Castanea sativa*) and *Japanese chestnut* (*Castanea crenata*) are delicious nuts used in candies and pastries, and pureed as a vegetable, and eaten as a plain or roasted nut. *Hazelnuts* (filberts, cob nuts) *Corylus avellana,* Coryleaceae is extensively cultivated as a crop.

Almonds, the stone fruits of *Prunus amygalus,* Rosaceae are grown in mediterranean regions var. *amygdalus* includes wild types found in West Asia, Greece and North Africa, var. *amara* includes cultivated types of bitter almond; var. *sativa* includes cultivates types of sweet almond, almond seeds contain amygdatin, a cyanogenic compound which gives the characteristic almond flavour and odour.

Pistachio nut (*Pistacia vera,* Anacardiaceae) has edible yellow-green kernels salted or incorporated in ice cream cakes or rouget candies. Pistachis is native to Mediterranean and central Asia where the nuts have been cultivated for over 3,000 years.

Other important edible nuts are *Acrocomia aculeata,* gru-gru palm (Arecaceae); *Afzelia africana* (Leguminosae); *Aleurites moluccana* **candle nut** (Euphorbiaceae); *Anacolosa luzoniensis,* galo (Olacaceae); *Amphicarpaea monoica* **hog peanut** (Leguminosae); *Areca catechu* **betel nut** (Arecaceae); *Arachis hypogea* **peanut** (Leguminosae); *Arenga pinnata* sugar palm (Arecaceae); *Attalea funifera* (Arecaceae); *Bactris gasipaes* peach palm (Arecaceae); *Balanites aegyptiaca* soapberry (Balanitaceae); *Brabejum stellatifolium* **wild chestnut** (Proteaceae), *Buchanania lanzan* (Anacardiaceae); *Butyrospermum parkii,* **shea butter tree** (Sapotaceae); *Calatola laevigata* (Icacinaceae); *Calodendrum capense,* **cape chestnut** (Rutaceae); *Caryota mitis* (Arecaceae); *Castanea mollissima* **Chinese chestnut** (Fagaceae); *Castanosperum australe,* **Australian chestnut** (Leguminosae); *Chrysobalanus icaco,* cocoplum (Rosaceae); *Copernicia cerifera,* Brazilian wax palm (Arecaceae); *Corypha umbraculifera* (Arecaceae); *Durio zibethinus* (Bombacaceae); *Hydnocarpus kurzii,* **chalmoogra nut** (Flacourtiaceae); *Lithocarpus corneus* (Arecaceae); *Madhuca indica,* **illupe nut** (Sapotaceae); *Moringa oleifera* ben nut (Moringaceae), *Phytelephas macrocarpa* ivory-nut (Arecaceae); *Shorea macrophylla,* false illipe nut (Dipterocarpaceae); *Semecarpus anacardium* marking nut (Anacardiaceae) (Fig. 15.4); *Trapa bispinosa* (Trapaceae).

Cashew nut

Cashew (*Anacardium occidentale*, Anacardiaceae) was introduced into India by the Portuguese in the early 16th century. Today, it is grown in about 4.20 lakh hectares mostly in peninsular India. India produces about 1.80 lakh tonnes raw nuts annually. India exports about 65,000 tonnes processed nuts annually earning over Rs. 148 crores every year. The precious, delicious cashew kernel is most loved and ardently admised in many countries (Fig. 20.1).

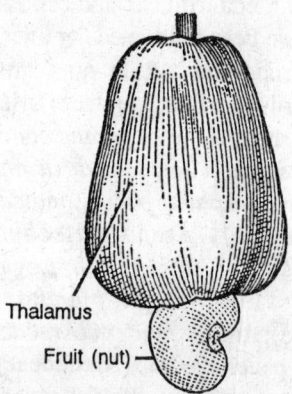

Thalamus

Fruit (nut)

Fig. 20.1. Cashewnut—*Anacardium occidentale*.

Cashews are produced on curious fruits. The *nut* is the embryo borne inside a hard seed coat. This structures sits atop a large, fleshy, inverted pear-shaped organ. The fleshy portion of the fruit called cashew apple is crushed and fermented into a wine called *kaju*.

Vegetative propagation of cashew by air layering, inarching, side grafting, veneer grafting, patch budding, bud grafting. Cashew is usually propagates bu nuts.

Hybrids

Hybrids 5 (Ansur 1 x Venture 56), Hybrids 11 and 19 (Midnapore Red x Venture 56) have given high yields at a very early age.

Cashew plantations in forest wastelands can generate additional employment in the rural sector. Cashew helps to fix sand dumes in Orissa.

Average yields is different states are estimated to vary from 140 to 1126 kg/ha or 0.80 to 6.49 kg/tree.

Cashewnut is a major foreign exchange earner in India. Despite this, cashew is generally considered to be a waste land crop and very little attention has been given to its cultivation. In order to pay better attention to the development of cashew, Government of India had set up a Directorate of Cashewnut Development at Cochin.

A World Bank aided Multi State Cashew Project with an outlay of Rs. 38.36 crore is in operation in Kerala, Karnataka, Andhra Pradesh and Orissa. The main aim of the project is to increase production of cashewnut through area expansion and improvement of existing plantations.

PECAN-NUT

The pecan (*Carya illinoensis*, Juglandaceae) is a big tree and bears edible nuts somewhat resembling the walnut. USA, Mexico, Israel, Australia, South Africa, Brazil, Taiwan, China, Hongkong and Iran grow pecans. The major pecan importing countries are Canada, Mexico, Netherlands, Luxemberg, Italy, Germany, France, England, Ireland, Denmark and Belgium. In India pecans are cultivated in Almore, Kumaon, Simla, Mukteswar and Chaubattia.

Uses

Pecans are marketed primarily as shelled kernels and consumed like walnuts. Pecans are used in bakery, confectionery and icecreams.

Production

USA is the biggest producer of pecan nuts, with a production of 1,14,940 metric tonnes during 1977. This is followed by Mexico, which produces about 19,950 metric tonnes.

21

Vegetables

The group of plants known as "vegetables" is virtually impossible to define in a satisfactory way. It refers to a very diverse collection of species which provide roots, stems, leaves, flowers, flower buds, fruits, or seeds which are eaten raw or cooked in some way. Most vegetables belong to the flowering plants, but some are found in the algae and fungi. Examples of algae used as vegetables are *Laver* and *Nori* derived from species of *Porphyra, Sea Iettuce (Ulva lactuca)*, and species of *Oarweed (Laminaria)*, known as *Kombu* or *Kobu,* used in Japan. Numerous macrofungi are used as vegetables such as *mushrooms, ceps, chanterelles, morels, trufflles* etc.

Nutritionally, vegetables vary widely. Some, such as legumes and pulses, provide a rich source of protein, some, such as potatoes and sweet potatoes, contain large quantities of carbohydrate, while the avocado is rich in fat. Apart from these examples, vegetables are normally eaten as a source of vitamins and minerals and roughage. The vitamin content varies widely; raw carrot contains, for example, 11,000 international units of **Vitamin A** per 100 g. closely followed by *kale* with 10,000 units, compared with 30 units in *parsnip* and only traces in potato. The soybean is the richest vegetables in *Vitamin B₁* (thiamine), while the fresh fruits of the green or bell pepper (*Caspicum* species) are a very rich source of **Vitamin C,** with per 100 g. Vegetables rich in vitamins are the spinach leaves, turnip tops and parsley leaves. Algae have high iodine content.

Vegetables are rich in minerals. They contain over 50 elements. The mineral requirement for a human organism is only to a small extent met by the consumption of vegetables. A wider variety of crops will help to meet these requirements more fully.

It should be noted that the mineral complexes of meat, fish, and bread products create an excess of acid in the organism which disrupts the protective mechanisms and metabolic processes. Vegetables neutralize these substances, this providing for normal metabolism and an alkaline blood reaction. The daily human requirement for iodine is between 100 and 200 millionth of a gram. *Watercress* (0.448 mg per 1 kg of dry matter), *tomato,* and *sweet pepper* contain the largest amount of *iodine.*

Classification of vegetable plants

There are many classification systems for vegetable crops, but the most common are the following :

Botanical classification

Vegetable plants belong to the Angiospermae, which is subdivided into two classes; monocotyledonous and dicotyledonous. These classes are further divided into families, genera, species, sub-species, and botanical varieties. All basic vegetable plants belong to mainly 18 families.

Classification of vegetables based on the part of plant used as food

This classification is widely accepted in commercial vegetable growing. Vegetables are divided into groups depending on the part of the plant used as foodstuffs.

According to this classification, vegetables are divided into leafstem, fruit-bearing, flowering (broccoli, cauliflower), root and tuber plants.

Mortenson and Bullard (1968) suggest one more classification system for tropical vegetable crops based on their importance and marketability.

1. **Crops of high commercial value** : Onion, peanut, cabbage, sweet pepper, hot pepper.

Watermelon, melon, cucumber, yam, sweet potato, tomato, manioc, beans, and brinjal.

Crops of limited commercial value

Garlic, celery, cauliflower, broccoli, taro, lettuce, luffa, potatoes and certain species of beans.

Crops grown for a local market

Shallot, beet, turnip, Chinese cabbage, green mustard, cabbage, Brussels sprouts, kohlrabi, endive, chicory, jute, pumpkin, artichoke, carrot, fennel, okra, gumbo, parsley, radish, watercress, chayote, spinach, corn, etc.

Crops grown mainly for home consumption

Chives, dill, chervel, horseradish, asparagus, parsnip, portulaca, rhubarb, and many spices and aromatic herbs.

A classification system for vegetables based on the season of their preferable cultivation is also used.

Classification of vegetables by the growing season in the tropics (Blumenfeld, 1968)

1. **Rainy-season vegetables** : Okra, tomato, cucurbita, radish, beans, hot pepper, brinjal, sweet potato.

2. **Winter vegetables (Cold-season vegeta-**

Table 21.1. Botanical classification of vegetable crops

No.	Name of the family	Main vegetables in the given family
Monocotylendons		
1.	Alliaceae	Various species of onion
2.	Asperagaceae	Asparagus
3.	Araceae	Taro (*Colocasia esculenta*)
4.	Dioscoreaceae	Yam
Dicotylendons		
5.	Aizoaceae	Newzealand Spinach
6.	Chenopodiaceae	Beet, Swiss, chard, spinach, goosefoot
7.	Asteraceae	Topinambour, chicory, endive, lettuce, artichoke, scorzonera
8.	Brassicaceae	Various species of cabbage, broccoli, kohlrabi, rutabaga, turnip, mustard, horseradish, radish, crambe, cress
9.	Convolvulaceae	Sweet potato
10.	Cucubitaceae	Various species of pumpkins, watermelon, melon, chayote, marrow, scallop squash, cucumber, sponge gourd, bittergourd, pointed gourd
11.	Euphorbiaceae	Manioc
12.	Fabaceae	Peas, beans, China bean, cow peas, soya bean
13.	Malvaceae	Okra
14.	Polygonaceae	Rhubarb, sorrel
15.	Solanaceae	Early potato, brinjal, tomato, pepper, ground cherry, pepino
16.	Apiaceae	Carrot, parsley, celery, parsnip, dill, anise, chervel, coriander, fennel, caraway
17.	Lamiaceae	Basil, hyssopus, marjoram, balm, mint, rosemary, savory, thyme, stachys
18.	Poaceae	Maize of corn sugar

bles) : Cabbage, cauliflower, carrot, tomato, beans, peas, onion, hot pepper, radish, turnip, spinach, beet root, lettuce and kohlrabi.

3. **Summer vegetables (Warm-season vegetables)** : Cucurbits, brinjal, okra, gumbo and many others under irrigation.

Cold-season vegetables include those whose plant parts, i.e. roots, stems, leaves and buds or unripe flowers (except beans, peas, and hot pepper) are used as food products. *Warm-season vegetables* are used in their ripe and unripe forms.

Vegetables are also classified into the following groups based on ecology

1. **Frost-and winter-resistant :** Perennial vegetable plants (perennial species of onion, garlic, asparagus, rhubarb, horseradish and sorrel) which in winter and autumn can withstand frosts upto - 10°C.

2. **Cold-resistant :** Biennial cabbage plants, root crops (carrot, beet-root parsley, celery, etc.), onion and greens (lettuce, spinach, and watercress). These plants can withstand temperatures 1-2°C for a long time and frosts of -5°C and lower, for several days.

3. **Intermediate between cold-resistant and heat-loving plants :** Potato and dill.

4. **Heat loving :** Tomato, pepper, brinjal, pepino, cucumber.

According to recent FAO data, the annual consumption of vegetables in the world is approximately 250 million tonnes. In different countries, the annual vegetable consumption per capita varies greatly; from 2 kg (in Mauritania) to 195 kg (in Portugal). Most of the developing countries fall under the category of countries with insufficient vegetable consumption (under 25 kg) per capita per annum. Scientists recommend that a person consume approximately 100-120 kg of vegetables per year. To provide the population with the necessary varieties of vegetables, many developing countries need to significantly increase the production of vegetables.

Common tropical vegetables

Apium graveobus celery, ajmud (Apiaceae) leaves used as vegetable salad. *Asparagus officinalis* (Liliaceae), bulb and young shoots are used as vegetable, *Allium capa*, onion (Liliaceae) tunicated bulbs, leaves are eaten, *Brassica campestris* mustard, sarson, (Brassicaceae), leaves used as

pot herb, *sag, Brassica compestris* var. *rapa* turnip, shalgam (Brassicaceae), roots used as vegetable; *Brossica oleracea* var. *acephala* borecole, kale (Brassicaceae) leaves eaten, *Brassica oleracea* wild cabbage (Brassicaceae) leaves eaten; *Brassica oleracea* var. *caulorapa* kholrabi, knolkhol, gand gobi (Brassicaceae), stem eaten; *Brassica oleracea* var. *capitata* cabbage band gole eaten; *Brassica oleracea* var. *botrytis* cauliflower, phool gobi, inflorescence eaten; *Brassica oleracea* var. *gemmifera* Brussel sprout, button gobi (Brassicaceae) buds used as vegetable; *Bauhinia variegata* kachnar (Caesalpiniaceae) flower buds are eaten.

Colocasia antiquorum taro, *kachalu, arvi* (Araceae), corm and leaves are eaten, *Cicer arietinum* gram, chana (Fabaceae) fresh green seeds used as vegetable; *Cajanus cajan* pigeon pea, *arhar* (Fabaceae) fresh unripe pods and seeds used as vegetable in Karnataka; *Capsicum annuum* and *Capsicum frutescens* red pepper, lal mirch (Solanaceae) green fruits (berries) are used as vegetable; *Cucumis sativus* cucumber, khira (Cucurbitaceae) young fruits (berries) are eaten as vegetable; *Cucumis melo* var. *utilissima* kakri, far (Cucurbitaceae) young fruits eaten; *Cucurbita maxima* great pumpkin, walaiti kaddoo (Cucurbitaceae) fruits (berries) eaten; *Cucurbita pepo* vegetable marrow ghia kaddoo (Cucurbitaceae) fruit (berry) is eaten, *Cucurbita moschata* musk, melon pumpkin, *halwa kaddoo* (Cucurbitaceae).

Citrullus vulgaris var. *fistulosus dil pasand, tinda* (Cucurbitaceae), green fruits (berries) eaten as vegetable; *Chenopodium album* white goosefoot, *bathu* (Chenopodiaceae) leaves used as potherb; *Chenopodium murale* goose foot, *chulai* (Chenopodiaceae) leaves used as potherb; *Dolichos lablab* beem, *sem* (Fabaceae); pods and seeds eaten; *Daucus carota* carrot, *gajar* (Apiaceae) roots eaten; *Glycine max* soya bean, *bhat* (Fabaceae) seeds eaten; *Hibiscus esculentus* lady's finger, *bhindi* (Malvaceae) green fruits (capsule) are eaten; *Helianthus tuberosus* jerusalem artichoke, *hathichoke* (Asteraceae) tubers eaten; *Ipomoea batatas* sweet-potato, sakar kandi (Convolvulaceae) roots eaten; *Lepidium sativum* cress, halion

(Brassicaceae) leaves used as pot-herb; *Lactuca sativa* lettuce, *salad* (Apiaceae) leaves used as vegetable salad; *Lycopersicum esculentum* tomato, *tamatar* (Solanaceae) green and rice berries are eaten, *Lagenaria vulgaris* syn. *L. ciceraria* bottle gourd, lanki (Cucurbitaceae) fruit (berry) eaten; *Luffa aegyptiaca* syn. *L. cylindrica* loofah, bath-sponge, *ghia tori* (Cucurbitaceae) fruit (berry) is eaten; *Luffa acutangula* loofah, *kali tori, moongi tori* (Cucurbitaceae) fruit (berry) eaten; *Musa paradisiaca* banana, *kela* (Musaceae) unripe fruit, rhicome, pseudo-stem pith, flowers are eaten, *Momordica charantia karela*, bitter gourd (Cucurbitaceae) fruit (berry) eaten; *Momordica cochinchinensis* syn. *M. muricata* kakora (Cucurbitaceae) fruit eaten; *Medicago falcata methi* (Fabaceae) leaves used as pot-herb (sag); *Murraya koenigii* curry leaf, mitanim (Rutaceae) leaves used as pot-herb; *Nelumbium speciosum* kanwal, lotus (Nymphaeaceae) rhizome (*bhea*) used as vegetable; *Pisum sativum* pea, *mattar* (Fabaceae) seeds eaten; *Phaseolus lanatus, lobia,* lima bean (Fabaceae) pods, seeds eaten; *Phaseolus vulgaris,* French bean, kidney bean, *rajmah* (Fabaceae) fresh bean used as vegetable; *Phaseolus mungo,* green gram, *mung* (Fabaceae) fresh seeds eaten); *Phaseolus radiatus* mash, urd, black-gram (Fabaceae) fresh seeds eaten; *Phaseolus aconitifolium moth,* moth bean (Fabaceae) fresh seeds, pods eaten; *Salmalia malabaricum* syn. *Bombax malabaricum* silk-colon, *simbal* (Bombacaceae) flower buds are used as vegetable; *Solanum tuberosum,* potato, *alu* (Solanaceae) stem tubers eaten; *Solanum melongena* bangan, brinjal (Solanaceae) fruit (berry) eaten; *Spinacea oleracea palek,* spinach (Chenopodiaceae) leaves used as pot-herb; *Trigonella foenum-graecum* methi, fenugreek (Fabaceae) leaves used as pot-herb; *Trichosanthes anguina* snake gourd, *parel, chichinda* (Cucurbitaceae) fruit (berry) is eaten; *Trichosanthes dioica parwal, palwar* (Cucurbitaceae) fruit used as vegetable.

BULBOUS VEGETABLES

Leek

Leek, *Allium porrum* L. *Kirath, valaiti piaz* is native to the Meditarranean region. The bulbs are edible, having a milder and more delicate flavour than the onions. Bulbs are eaten raw when tender and cooked with vegetables or in soup. Leek is a *cool season crop.* The plants are first raised in the nursery by sowing seeds during August-October and planted out when 15 cm tall, at a spacing of 15 cm. Seed rate is 5 kg per hectare.

Blanching

This is covering the plant to a certain height with earth so as to bleach them, which improves its quality. For this purpose, plants are put in up to their centre leaves in trenches or pits which are heavily manured and to earth up soil as they grow.

Rhubarb

Rhubarb, *Rheum rehaponticum* L. is native to the Mediterranean region and extensively cultivated in Europe and United States. Rhubarb is a perennial crop. The large, thick leafy stalks are edible. The plant thrives well in the hills and *cool localities.* It is propagated by division of roots and also by seeds.

A deep, rich, well drained soil is selected and the divided roots are set out at a spacing of 75 x 75 cm. Water accumulation near the roots is harmful. Flowering spikes are removed as and when they appear. Harvesting of leaf stalks begins in the second year. Not more than three-fourths of the shoots from a plant should be removed.

The plant yields for about 5 years.

Globe artichoke

Cynara scolymus L. *hathichak* is a popular crop in Europe. The globular flower heads in bud stage are edible. These are eaten raw or boiled and served with butter or pickle. The plant is propagated from seed or suckers or by division of roots.

Seeds are sown from August to October in nursery beds. Seedlings are transplanted at a spacing of 120 x 130 cm in the field. The plant yields for 2 or 3 years.

Asparagus

Asparagus (*Asparagus officinalis* L. var. *altilis root mooli, shatawar*).

Asparagus is native to Europe; used by the Romans from 2000 BC.

Ecology

The crop is well adapted to *tropics* and *sub-tropics*. A rich, friable, well drained soil is required. Muck soils in which the water table is not too high, produce good crops.

Cultural practises

Sowing in the hills is usual during early spring and transplanted in next spring. In the plains, July-November is common period for sowing. 2.5 to 4 kg of seed is required for one hectare land. Seeds are sown in seed beds 15 to 22 cm apart and 1 to 2 cm away from each other and at a depth of 2 cm. The seedlings are left in the nursery for 2-4 months. The plants are transplanted in trenches which are 30 cm wide and 45 cm or more deep and 1 metre apart. The plants are set at the centre of the trenches at a spacing of 30 to 45 cm. Annual application of 5 tonnes of green leaf or other bulky manure may have to be given. Every winter, top dressing with a mixed fertiliser 500 kg/ha may be given. Irrigation should be frequent and at regular intervals of 7 to 10 days.

The shoot, the growing point of the stem before branching occurs, is called *spear*. These are edible. Tender shoots are cut in the second year. Yield : 2,500 to 4,000 kg/ha.

Asparagus is a perennial crop and the crop can be harvested annually, for about 10 years.

CODE CROPS

Sprouting Broccoli

Sprouting broccoli (*Brassica oleracea* var. *italica*) is considered to have long been grown and extensively used in Italy. It is a vegetable of the highest quality; ranks very high from the nutritional point of view. This is an annual, grown in the same manner as cauliflower. Each plant produces one large, central, *green head* which is edible.

Ecology

Soil : It requires a cool and moist growing season. Rich, heavy loam soils with good drainage and liberal supply of moisture are ideal for growing. Light soils with plenty of organic matter are also suited.

Cultural practices

The seed is first sown in seed beds and seedlings are transplanted afterward. The seed required for sowing in one hectare land is 600 to 750 gm. The seedlings, when six weeks old, are transplanted at a distance of 45 cm to 60 cm between plants in the rows and about 60 to 75 cm between rows. The seedlings should be transplanted on ridges, and not in furrows, as these will be badly affected in rainy season.

The correct stage of cutting occurs shortly before the buds begin to open, while the heads remain compact. The total stem cut is about 20 to 25 cm. The auxiliary shoots which are also edible, are cut at the same stage as the central heads. Green sprouting broccoli is one of the most perishable vegetables so it should be marketed immediately after harvest.

Brussels sprouts

Brussel's sprouts (*Brassica oleracea* var *gemmifera* Zenker) has originated in the Northern Europe. The edible part is the *large bud* or *miniature head* borne in axil of each leaf of the plant. The solid little *heads* or *sprouts* ranging from 2.5 cm to 5 cm are tender and delicious (Fig. 21.1).

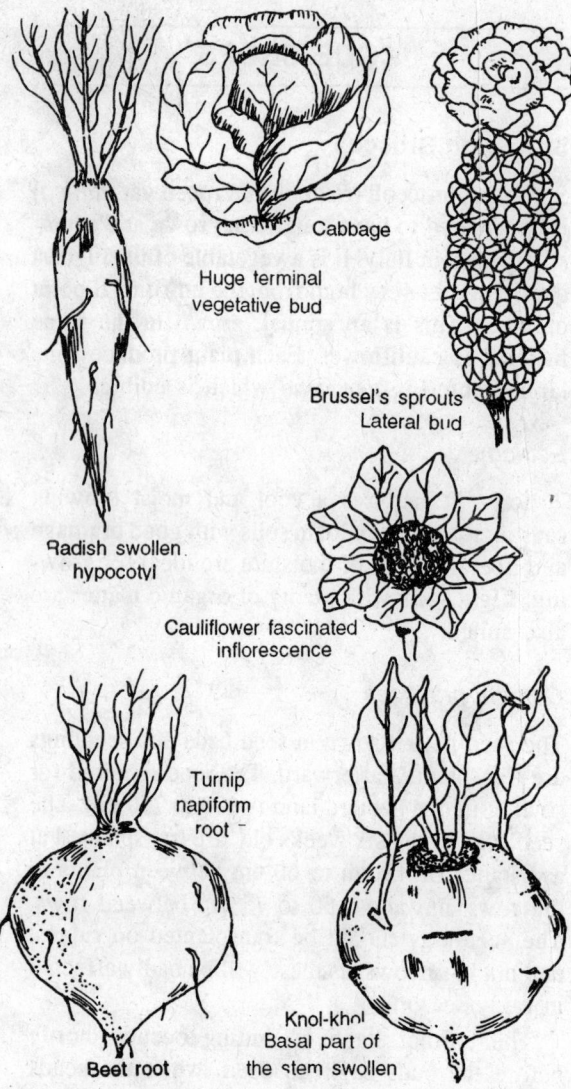

Cabbage

Huge terminal
vegetative bud

Brussel's sprouts
Lateral bud

Radish swollen
hypocotyl

Cauliflower fascinated
inflorescence

Turnip
napiform
root

Beet root

Knol-khol
Basal part of
the stem swollen

Fig. 21.1. Different vegetables.

Chinese cabbage

Chinese cabbage (*Brassica pekinensis* and *Brassica chinensis*) is used as a pot herb and as salad. It was grown in China for over 1500 years. *Brassica pekinensis* is a *heading type; forms an erect, moderately compact, somewhat cylindrical head. Brassica chinensis* is an *open leaf type*; and develops a clustor of leaves without forming a distinct head.

Cultural requirements are similar to it quickly forms flowering stalks as soon as the weather changes after which the heads are not fit for consumption. The crops is harvested when the heads are fairly firm. The crop can be kept for few weeks in cold storage.

LEGUMINOUS VEGETABLES

Cluster beans

Cluster bean (*Cyamopsis tetragonoloba* L. Taub. *guar, gawarphali*) is a popular (summer) crop in India. It is a good forage, green manure and vegetable crop.

Ecology

Well drained sandy loam soil is suitable. It is warm season crop and does not stand cold winters.

Cultural practices

12 kg seed is required to sow in one hectare land. Sowing is done during the rainy season. Sowing is done in rows 60 cm apart. Pods are formed from the base and are produced as the plant grows. Each leaf axil may have a cluster of 3 to 6 pods and these should be plucked carefully when they are still tender. The yield of green pods varies from 5,000 to 6,000 kg per hectare.

Cultural practices

Ecological and cultural requirements are the same as for. The quality is improved by slight freezing in winter. Sprouts are ready for picking in about 4 months from the time of sowing. The picking is first started from the lower portion of the stem by first breaking of the leaf below the sprout and then breaking away the sprout. By this method the stem is encouraged to grow and push out new leaves in the axil of which sprouts are formed.

CUCURBITS AND VINE CROPS

Snake gourd

Snake gourd (*Trichosanthes anguina* L. *padwal, chichinda*) is an annual plant of exceedingly rapid, growth and of climbing habit. The plant is native to India or the Indian Archepelago. *Tender fruits* are used as vegetable. The fruit is a large, greenish-white gourd often reaching up to 150 cm length (usually 60 to 90 cm long) and 8 cm thickness.

Ecology

It is a warm season crop; not grown alone an altitude of 1,500 metre. Any good garden soil, which has been well manured, will suit the crop.

Cultural practices

The crop is sown from April to July, and during October-November in places where winter is mild - Seeds are sown on raised beds each about 1 metre wide, near the edges in hills spaced about 30 to 45 cm away. The seed rate is about 5 kg per hectare (2-4 seeds per hill). When the plant start growing, supports are needed.

The basal dose of formyard manure at the rate of 40 to 50 tonnes per hectare is applied. Ammonium sulphate (100 kg/ha) should be applied as top dressing when the plants start bearing. Irrigation every 4 or 5 days is practised during the hot season.

The tender fruits at the half-grown stage are harvested (The fruits do not keep well for more than two days). The yield varies from 15,000 to 25,000 kg per hectare.

LEAFY VEGETABLES

Fenugreek (*Trigonella foenum - graecum* L. *methi*) is native to Eastern Europe and Ethiopia. The plant was very popular in ancient Rome as a food and as a medicine. The green leaves are used as a vegetable. The seeds are used as a condiment, in pickles and in indigenous medicine.

Kasturi methi (champa methi) and *Marwari methi* are scented varieties (the leaves are scented).

Cultural practices

Fenugreek is a cool season crop. Heavier types of soils are said to promote high yields. The seeds are sown in September to mid-November and February-March inthe plains and March-April and October in the hills. 20 to 25 kg seed is required to sow in one hectare. Seeds are sown broadcast and the bed surface is raked to cover the seed. The beds are irrigated immediately after sowing.

Good early growth is promoted by basal application of 15 tonnes of bulky manure and a side dressing of ammonium sulphate at 100 kg per hectare, when the first cutting is over. A frequency of 7 to 10 days will be sufficient to irrigate the fields.

The crop is cut after 45 days of sowing when the plants are 15 to 18 cm high. The plants are cut 2 cm above ground, leaving the stabs which produce new stalks, and are cut in turn. The crop is bundled, basketed and taken to the market. It is kept cool and moist. An early disposal is necessary.

7,000 to 9,000 kg of green leaves may be obtained from an hectare of land.

Purslane (*Portulaca oleracea* L. var. *sativa, kulfa*)

Purslane is a tiny creeper with small, succulent, fleshy leaves and tender stems. The tender *stems and leaves* are edible; used as salad and also cooked with meat, gram etc.

Ecology

It is a hot weather crop and is available early in summer in the market. It requires warm climate for good leaf growth. Moderately fertile soils are suitable. Land is prepared thoroughly with 2-4 ploughings before sowing seeds.

Cultural practises

March to June is the common season for sowing. 2.5 to 4 kg of seed per hectare is sown broadcast in beds after mixing with sand and raked to cover. The beds are watered. A basal dressing of 12 to 15 tonnes of bulky manure is given per hectare.

The crop is pulled out after about eight weeks of sowing. These are bundled, packed and sent to the market immediately.

The yield is about 7,500 kg per hectare.

SALAD CROPS

Celery (*Apium graveolens var - dulce*)

Celery (*Apium graveolens* var - *dulce*) is native to Mediterranean region. *Cereriac* is a turnip rooted celery which is especially suited for use in soup but is not eaten raw.

Ecology

The crop requires a long, cool season for its growth. It does not stand *severe freezing climate*. Plenty of sunshine during day and cool nights are beneficial. Muck or peat soils are used due to their high water holding capacity and friability. Alternatively any rich, friable, well drained and deep soil is good.

Cultural practises

The seeds (fruits) are sown from July to September in well prepared beds and need a partially shady and cool situation. The seeds take a long time for germination. The seeds may be kept moist for a few days before sowing to help in germination. 300 to 400 gm seeds are required for raising the crop in an area of one hectare. The seeds are lightly covered and seedlings come out in 20 to 60 days and are thinned to about 1-2 cm spacing. After 45 to 60 days, seedlings are transplanted in well manured and prepared fields to a spacing of 45 x 15 cm. Fertilisers up to 1 tonne per hectare are applied once after transplanting

and 1 or 2 dressings are given at intervals of 20 days. Irrigation is necessary.

The harvest is taken up 120 to 150 days after sowing. The plants are forked up and the roots are cut. The average yield of about 25,000 kg per hectare is obtained **Endive** (*Cichorium endivia* L. *Kuru salad*). The plant is native to East Indies. The tender succulent leaves are eaten boiled and as a salad. The plant grows well in a richy moist soil. Seeds are sown from August to November in seed beds or directly in the field. Nursery plants are set out in rows 45 cm apart and 24 cm in the row. For blanching the leaves of fully grown plants are drawn together and tied to exclude light. Water during irrigations is kept away from the heart of the plants.

Cress

Cress (halion; *Lepidium sativum* L.) is native to Iran. It is grown on any soil, throughout the year. The *leaves* are used as salad and garnish. Successive sowings are made beginning from July to February in plains and from March to September in the hills. Seed rate is 6 kg per hectare. The plants are ready for use in a few days after sowing.

Upland cress (*Barbarea* sp.)

This plant resembles the cress. The *leaves* are used as salad and garnish. Seeds are sown in the cool season at a spacing of 30 x 15 cm.

Water cress (*Rorippa nasturtium* var. *aquaticum* Syn. *Nasturtium officinale*)

Water cress is a perennial aquatic plant grown along margins of streams etc. It has a pleasant and pungent flavour. *Leaves* are edible.

Seeds are sown in the nursery which is kept wet and later on seedlings are transplanted. The beds are flooded when the plants have grown to about 5 cm but they should not be submerged. The plants are thinned to a spacing of 15 x 15 cm. They grow rapidly.

Propagation by cuttings is also practised for which period of sowing is during July-November.

DRUMSTICK

Drumstick (*Moringa oleifera*, Moringaceae) is a multi-purpose vegetable. The plant is indigenous to sub-Himalayan tract extending from the river Chenab to Sarda in the south. A small or middle-sized tree, bark corky, wood soft, root pungent and young parts slightly hairy; leaves 3-pinnate compound; sometimes 45 cm long, rachis slender; thickened and articulated at the base; and deciduous, drumstick bears white-coloured flower.

Leaves, flowers and pods are used as vegetables. Fresh drumstick leaves are nutritious, contain vitamin A 11,300 IV/100 g protein content is 6.7 g/100 g; contains high amount of calcium and phosphorus. Two alkelvids *moringine* and *moringinine* are present in the plant parts (Fig. 21.2).

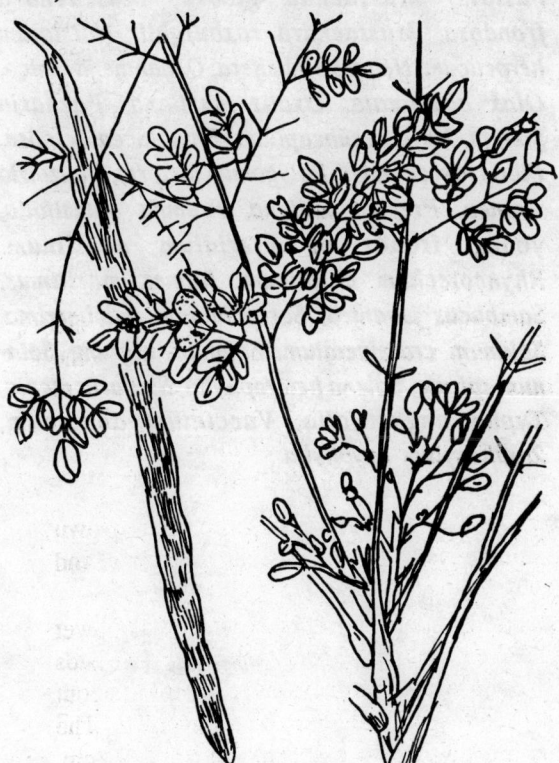

Fig. 21.2. Drumstick, *Moringa pterygosperma*.

Root bark

It is easily digestible and astringent to the bowels.

It is used to prevent enlargement of the spleen, tuberculous glands in the neck, to destroy tumours, ulcers, earaches shuttering of ear; and as a fementation to relieve spas.

Stem bark

Bark removes all kinds of pain. It is aphordisiac, authelmintic and useful to cure eye diseases. The stem bark is used in Ayurveda to cure hallucinations, dry tumours and also to prevent cough and asthma.

Flowers : Used to cure inflammations, muscle diseases, cough tumours, and also to prevent cough and asthma.

Flowers : Used to cure inflammations, muscle diseases, cough tumours, and enlargement of spleen.

The seeds yields the *"ben oil"* of commerce, which is greatly valued by watchmers and used generally as a lubricant for fine machinery, while perfumers hold it in high esteem, because of its property of absorbing and retaining fugntive odours.

The medicinal properties of this plant have long been appreciated in India. All parts of the plant have been used in the treatment of ascites, rheumatism, venomous bites, and as a cardiac and circulatory stimulant. The pungent root of the young tree has rubefacient and vesicant properties, and has the odour and flavour of the true horse-radish, for which it is used as a substitute. The root-bark has also rubefacient and vesicant properties.

Root

Root is bitter and tonic to the body and the lungs.

Cultivation

Drumstick grows well on all types of soils, humus rich forest soil being the most ideal. The plant grows well on sandy loam soils in Uttar Pradesh and sandy soils of western coasts of Kerala. A spacing of 5 m x 5 m between trees is quite adequate. The plant is propagated through seeds as well as stem cuttings. Being a cross-pollinated crop, propagation through cutting is preferred to maintain genetic purity and dwarfening of trees. Planting of cutting is done after rains around

August-September. It takes 2-2½ years for flowering and fruiting. The plant is in full bloom around February. The pods are matured by May.

ACID MELONS

The fruits of *Cucumis melo* L., Cucurbitaceae are low in acid, sweet or bland fruits, consumed avidly, cooked in a variety of dishes in South India. Chromosome number is $2n = 24$. The fruits are *beautifully golden yellow*, smooth and oval. The plant is a monoecious vine possess female flowers which are slightly larger than male flowers and have pubescent ovary. Fruits are borne singly on the secondary branches, which are 2-3 in number. Each vine yields 3 or 4 fruits totally weighing 1.5 kg in 70 to 75 days of age of the plant. *The fruits have good keeping quality.* They are found to be fresh and appealing even 3 to 4 months after harvest.

Fruit

The fruit skin very thin; fruit flesh has pleasant flavour, white, crisp, 2 cm thick, contains 0.384 per cent citric acid, pH is 5.5. The seeds are embedded in a compact moist jelly in the seed cavity.

The fruit is resistant to the attack of fruit fly.

Varieties

Desavali, Guntur, Kurnool, Macherla Green, Macherla White, Nakka Dosa, Ongole etc.

Green vegetables of tribals

Alocasia macrorrhiza, Ardisia crispa, Ardisia polycephala, Ardisia solanacea, Asystasia neesiana, Bambusa tulda, Calamus rotang, Casearia esculenta, Casearia glomerata, Cirsium lipskyi, Cissus adnata, Cissus discolor, Cissus repens, Clerodendrum colebrookianum, Clerodendrum indicum, Codonopsis sp., *Conocephalus suaveolens, Cyanotis tuberosa, Dendrocalamus hamiltonii, Embelia gamblei, Embelia naguchia, Embelia subcoriacea, Emhydra fluctuans, Garcinia lancaefolia, Hedyotis capitellata, Houttuynia cordata, Hygrophila salicifolia, Hygrophila spinosa, Ipomoea aquatica, Lasia spinosa, Leea indica, Lysimachia candida, Maesa chisia, Malva verticillata, Medinella rubicunda, Meliosma pinnata, Merremia umbellata, Monochoria hastata, Mussaenda glabra, Mussaendra frondosa, Mussaendra roxburghii, Natsiatum herpeticum, Nelumbo mucifera, Oenanthe javanica, Olax acuminata, Oxalis martiana, Paedaria foetida, Pavetta subcapitata, Phytolacca acinosa, Picris hieracioides, Polygonum bistorta, Pouzolzia vininea, Premna latifolia, Premna obtusifolia,* young fronds of *Pteridium aquilinum, Rhyncotechum ellipticum, Rumex maritimus, Sambucus javanica, Sarcochlamys pulcherrima, Solanum crassipetalum, Solanum nigrum, Solanum spirale, Solena heterophylla, Stellaria media, Typha angustifolia, Vaccinium donianum, Zanthoxylum oxyphylla.*

22

Plant Insecticides

Natural insecticides have been extracted from pyrethrum (*Chrysanthemum coccineum, C. cinerarifolium* and *C. marschallii* Asteraceae); from tobacco (*Nicotiana tabacum,* Solanaceae); insecticidal compounds from pyrethrum are sold as *dalmation* or *Persian insect powder. Pyrethrins,* the active compounds, are esters that break down into acids and alcohols. Pyrethrins stun insects, but not instantly kill them. Pyrethrins are usually mixed with other substances that ensure death of insects.

Derris elliptica (barbasco), Fabaceae, *Lonchocarpus nicou* (tuba), Fabaceae contain *rotenone,* a isoflavamoid compound which is fatal to insects but comparatively harmless to humans because it leaves no residue.

The seed oil of *Croton tiglium*, Euphorbiaceae; leaves of neem (Fig. 22.1); leaves of *Anabasis aphylla,* Chenopodiaceae; *Haplophyton cimicidum*, Apocynaceae; roots of *Heliopsis longipes,* Asteraceae; fly agaric, *Amanita muscaria,* Amanitaceae yield powerful insecticides.

Pyrethrum is a valuable anti-malarial insecticide; pyrethrin spray is used effectively against house-flies, mosquitoes, moths, cockroaches, bed bugs, ticks and lice. Pulverized flowers are used for manufacturing mosquito-repellent smoulder coils.

Rotenone is useful for the control of pests of crop plants. It is used effectively against Mexicin bean beetle, wooly apple aphid, European corn borer, pea aphid, house-fly, mosquito, cockroach etc.

Fig. 22.1. Neem, *Azadirachta indica.*

Microbial insecticides

In 1835 Bassi, demonstrated for the first time that a microorganism, the fungus *Beauveria bassiana,* could cause an infectious disease in an animal (the silkworm). Metchnikoff (1879) and Krassilstschik (1888) found that the fungus *Metarrhizium anisopliae* can be used to control the grain weevil and the sugar beet curcukio. White and Dutky

(1940) succeeded in demonstrating control of the Japanese beetle by distributing spores of the milky disease bacteria, *Bacillus papillae*. The use of viruses to control insect pests was stimulated by the efforts of Balch and Bird (1944) and Thompson (1949).

Entomopathogen	Target insect
Bacteria	
Bacillus popillae	Japanese beetle
Bacillus thuringiensis	Alfalfa caterpillar, many caterpillars
Coccobacterium acridiorum	Grasshoppers
Serratia marcescens	Termites
Viruses	
Nuclear polyhedrosis	European spruce sawfly Alfalfa caterpillar, Wattle bagworm, Cabbage looper, Boil worm - budworm complex
Cytoplasmic polyhedrosis	Pine processionate worm
Granulosis	Cabbage worm, red-banded leaf roller

(Contd.)

Entomopathogen	Target insect
Fungi	
Entomophthora spp.	Brown-tail moth, Spotted alfalfa aphid
Beauveria spp.	Chinch bug Colorado potato beetle White grubs
Metarrhizium anisopliae	Corn borer, Sugarbeet curculio, Froghopper
Aeschersonia spp.	White fly, scale insects
Protozoa	
Thelohania hyphantriae	Fall web worm
Mattesia grandis	Boll weevil
Malameba lacustiae	Grasshoppers

Commercial insecticides

International minerals and chemicals corporation (US) prepares "agritrol", "thuricide", "parasporin", biotrol BTB, entobacterin 3, "bankthane L69" etc., from the bacterium *Bacillus thuringiensii*. "Virex" is prepared from *Heliathis virus*.

23

Economic Uses of Microorganisms

Microorganisms are used in the manufacture of many industrial products. In various branches of the food industry (dairy, bakery, meat and vegetables) a variety of cultures produce specific flavours or ferment products during some stage in the processing of certain food products.

Cheese industry

Bacteria : *Streptococcus lactis, S. cremoris* combined with *Leuconostoc* species to produce starters for the manufacture of cultured or fermented milk products. *Lactobacillus acidophilus* is used in the preparation of acidophilus milk as *L. bulgaricus* is used to prepare Bulgarian milk. A mixture of *L. bulgaricus* and *Streptococcus theromophilus* is used to produce yogurt and Swiss and Romano cheeses. In Swiss cheese the latter two species are usually combined with *Propionibacterium shermanii*. Generally only lactic cultures which contain strains of *S. lactis* or *S. cremoris* in combination with *Leuconostoc* sp. are used in the manufacture of American cheese and its varieties.

Fungi : To enhance flavour, aroma, body texture, composition and appearance of cheese *Penicillium* sp. is used. The Societe de Roquefort manufactures mold-ripened cheese from ewe's milk on a large scale. Roquefort cheese is a major French export. *Penicillium roqueforti* is the mold utilized for flavour production in Roquefort cheese.

Penicillium camemberti is the mold which produces the characteristic flavour of Brie, Camembert and Neufchatel cheeses. *P. caseicolum* is also used to produce Camembert Cheese, particularly in the French province of Normandy. There are many cheese varieties in which typical flavour development is principally dependent upon the internal and or external growth of mold. Blue, Gorgonzola, Stilton, Wensleydale, Gammelost, Nu-World, Mycella, Niva, Camembert, Neufchatel, Brie etc., are produced in many nations including the United States, Denmark, Norway, Italy, England, Czeckoslovakia, Argentina, Canada, New Zealand and Australia.

Flavour production by molds in oriental foods

The use of molds for the production of flavourful foods is widely practiced in the Orient. Soybeans and/or rice serve as the raw material for most of the common products. *Shoyu,* or soy sauce, and *miso,* or soybean paste, are common articles of diet in Japan. *Sufu* or Chinese cheese, is fermented from *tofu,* or soy curd. *Tempeh,* ragi, and *angkak,* or Chinese red rice, are other important types of fermented foods.

Shoyu and *miso* have been produced in Japan for more than a thousand years. Over one million bushels of whole soybeans and seven and one-half million bushels of soymeal are utilized for shoyu production each year. The per capita consumption of *shoyu* in Japan is about 27 g per day or nearly

10 litres per year. *Shoyu* is also produced in China, Indonesia, Philipines and United States. *Aspergillus oryzae* is the principal mold used in the manufacture of *shoyu* and *miso*. *Aspergillus soyae* is also utilized in *shoyu*. *Saccharomyces rouxii* is the key organism in the second stage of *miso* fermentation; but species of *Lactobacillus* and *Hansenula* may also be involved.

Koji culture production : Molds used by the Japanese for many types of food and beverage fermentations are cultured or bulky, fibrous carried, usually rice, wheat orbran, occasionally barley or soybeans. These mold cultures are termed *Koji*. Sufu (Chinese cheese).

Sufu or Chinese cheese, is a popular food in China and Formosa. *Sufu,* the soybean cheese, predates Christian history. The process of producing *tofu*, the bland, unfermented soybean curd, is reported as the invention of the Liu Ap, A Chinese emperor, who ruled from 179 to 122 BC. The fermentation of *sufu* from *tofu* is reported in annals of the Ching Dynasty (1644 to 1912).

Actinomucor and *Mucor* species are utilized in *sufu* production. Factory production in Formosa utilizes *Actinomucor elegans* (termed by Wai (1964) as *Rhizopus chenesis* var. *chungyuen*). Home-produced *sufu* in Formosa utilizes *Mucor hiemalis* and *M. silvaticus*.

Tampeh : *Tempeh*, a popular food in Indonesia and adjoining areas, results from a food fermentation primarily practised in Indonesian homes. Soybeans are washed and soaked overnight in tap water. The seed coats are removed and the beans are boiled in excess water for about 30 minutes. After thorough draining and cooling, the beans are inoculated with spores the mold *Rhizopus* strain. Small patties of the inoculated beans are tightly packed in leaves, often banana leaves. After 20 to 24 hours of incubation at 31 to 32°C. The pressed bean patties are completely bound with pure white mycelial growth. *Tempeh* is usually thinly sliced and fried in oil. Freshly prepared *tempeh* is a perishable product.

Rhizopus oligosporus, R. stolonifer, R. oryzae, R. arrhizai, and *R. formosaensis* have been isolated from *tempeh* fermentations. *R. oligosporus* is the typical fermenter in *tempeh*.

Ragi : Ragi is a fermentation product of Indonesia and China. It is formed from rice flour which has been fermented by mold, yeast, and bacteria, resulting in white, ball-like shapes about 3 cm in diameter. Ragi is not eaten as such, but instead serves as a sparater in the manufacture of several other foods from rice, molasses, or cassava. The flavours produced are characterized as "sweet-sour" or "sweet-alcoholic". *Rhizopus oryzae, Chlamydomucor oryzae,* and *Hansenula anomala* are predominant in *ragi* fermentation.

Ang-Kak production

Ang-kak, or red rice, is primarily used in the Orient to colour food. It is an article of commerce in the Philippines, Indonesia, and China. Washed, soaked rice is thoroughly drained and autoclaved, cooled, and inoculated with the mold *Monascus purpureus* spores. After 3 weeks incubation at 25 to 32°C, the rice grains, not deep purplish red, are oven-dried at 40°C. The dried red rice may be ground into a powder. Corn may also be used as a substrate for the ang-kak fermentation.

Meat aging and curing

Williams described processes which utilize the growth of the mold *Thamnidium elegans* to tenderize and impart flavour to meat, usually beef. Leistner et al. (1965) reported that molds and yeasts play an important role in the flavour development of certain cured meats. Surface growth of *Asperigillus* and *Penicillium* molds aids in typical flavour of "country cured" ham. *Penicillium* and *Scopulariopsis* molds function in flavour development of "fermented" sausage of Hungerian type. The growth of yeast species, both on the surface and internally, contributes to the typical flavour of "fermented" Italian-type sausage. *Debaryomyces* species predominate.

24

Forage Crops

Forage crops (fodder crops) are a diverse group of plants used directly or in a preserved form (such as hay or silage) for feeding ruminant livestock. **Tropical and subtropical fodder crops** are *Andropogon* species (bluestems); *Dichanthium annulatum* (diaz bluestem); *Dichanthium aristatum* (brahman bluestem); *Hyparrhenia hirta, H. rufa, Chloris gayana* (rhodes grass); *Digitaria sanguinalis* (hairy crab grass); *Digitaria decumbens, Digitaria longifolia* (pangola grass); *Panicum maximum* (guinea grass); *Panicum miliaceum* (broomcorn millet); *Panicum purpurescens (para grass)*; *Paspalum dilatatum* (dallis grass); *Paspalum notatum* (bahta grass); *Paspalum urvellei* (vasey grass); *Pennisetum americanum* (pearl millet); *Pennisetum ciliare* (buffel grass); *Pennisetum purpureum (napier grass*, elephant grass); *Tripsacum laxum* (Guatemala grass); *Themeda triandra.*

Fodder legumes : Butterfly pea (*Centrosema plumieri, C. pubescens, C. virginianum*, sesban (*Sesbania sesban, S. cinerascens*) buggarweed (*Desmodium tortuosum*); tick clover (*Desmodium discolour, D. heterophyllum, D. heterocarpon*, (*D. triflorum*), wild popinac (*Lucaena glauca, L. leucocephala*); Velvet beans, *Bengal bean* (*Mucuna deeringiana*); *Macroptilium atropurpureum, M. geophilum; Kudzu vine* (*Puraria lobata*); kudzu (*Pueraria phaseoloides*); wild lucerne, stylo (*Stylosanthes guianensis, S. capitata, S. humilis, S. hanata, S. scabra.*

Temperate fodder crops are rye grass (*Lolium perenne, L. multiflorum*); timothy grass (*Phleum pratense*), orchard grass (*Dactylis glomerata*);

maize (*Zea mays*); rye (*Secale cereale*); wheat grass (*Agropyron fragile* sub sp. *sibiricum*; crested wheat-grass (*Agropyron cristatum*); slender wheat-grass (*Elymus trachycaulus*); bent-grass (*Agrostis stolonifera*); *A. capillaris* bluestem (*Andropogon* species); oatgrass (*Arrhenatherum elatius*); carpet grass (*Axonopus affinis*); blue grama (*Bouteloua gracilis*); sidioats (*Bouteloua curtipendula*); smooth brone (*Bromus inermis*; resote grass (*Bromus willedenowii*); buffalo grass (*Buchloe dactyloides*); rhodes grass (*Chloris gayama*); Bermuda grass (*Cynodon dactylon*); Canada wild rye (*Elymus canadensis*); Siberian wild rye (*Elymus sibiricus*); blue wild rye (*Elymus glaucus*); love grass (*Eragrostis curvula*); taff (*Eragrostis teff*); tall reed fescoe (*Festuca arundinacea*); sheep's fescoe (*Festuca ovina*); meadow fescoe (*Festuca pratensis*); Canary reed grass (*Phalaris arundinacea, P. aquatica* var. *stenoptera*; Texas bluegrass (*Poa arachnifera*); Canada bluegrass (*Poa compressa*); Kentucky blue grass (*Poa pratensis*).

Sedges : *Cortex lyngbyeis, Carex nigra, Carex norvegica, Carex rariflora.*

Legumes : Kidney vetch (*Anthyllis vulneraria;* crown vatch (*Coronilla varia*); lucerne, alfalfa (*Medicago sativa*); yellow trefoil (*Medicago lupulina*); white sweet *clover* (*Melilotus alba*); yellow sweet *clover* (*Melilotus officinalis*); esparcet (*Onobrychis vicifolia*); sirradella (*Ornithopus sativus*); red clover (*Trifolium pratense*); white clover (*Trifolium repens*); strawberry clover (*Trifolium fragiferum*); alsike clover (*Trifolium subterraneum*); broad bean (*Vicia faba*), spring vetch (*Vicia sativa*); hairy vetch (*Vicia villosa*), Crucifers; Kales (*Brassica oleracea, Brassica campestris*) rape (*Brassica napus*).

A new *Brassica* cultivar named *gobhia sarson* (*Brassica campestris* var. gobhia sason 2n = 20) has been tried as fodder crop. This cultivar gives high yield of green leafy fodder of good quality.

In winter, *berseem* (*Trifolium alexandrium*) is the main fodder crop of Haryana. *Daenanath grass* (*Pennisetum pedicellatum*) popular in Bihar, gives an yield of 700-800 q/ha of green fodder. Deenanath grass grows on waste lands. It is a perennial grass. 10 kg seed is required to sow one hectare. *Brachiaria brizantha* gives an yield of 50 q/ha green fodder in winter months (January-February). *Mucuna* a legume gives an yield of 20 q/ha of green fodder.

Fodder oats (var. Kent) has been found to be the most suitable winter fodder, having a yield potential of 40 to 50 tonnes per ha in Sikkim Himalayas. The crop attains bloom stage in about 100 days. The crop can be harvested at 50 per cent bloom.

The oldest fodder legume is probably lucerne which is native to western Asia. Typical grass genera are *Andropogon, Hyparrhenia* and *Themeda* in Africa; *Cynodon* and *Dichanthium* in India and *Paspalum* in Brazil.

Guar orcluster bean Cyamopsis tetragonoloba, Fabaceae is a hardy leguminous fodder crop grown during summer season. Guar green forage is a useful raw material for paper, food, textiles, waste water treatment and mineral processing. Pods are used as vegetable. Seeds have high protein content. Guar meal has 40 per cent protein for feeding the livestock. Guar gum is exported and earns considerable foreign exchange. Guar plant conserves the soil from wind erosion and improves the soil fertility.

Shaftal or Persian clover (*Trifolium resupinatum*, Fabaceae) is a promising fodder for alkali soils. Moth being the most drought resistant kharif legume, is a good source of forage for arid and semi-arid regions. The plant is grown on light sandy soils under barani conditions.

Rice bean (*Vigna umbellata*) proved to be an important forage crop of eastern India. It can have a prolonged vegetative phase and is sensitive to photoperiod. To meet scarcity of green forage

during two lean periods, i.e. April-June and November-December rice-bean variety K-1 can be grown successfully.

Perennial forage grasses **Rhodes grass** (*Chloris gayana*), **Blue Panic** (*Panicum antidotale*) *Panicum laevifolium*, **Karnal grass** (*Diplachne fusca*) are good fodders for milch cattle. Karnal and Rhodes grass are suitable for growing in alkali soils also, after the addition of gypsum to the soil.

Bhimal (*Grewia oppositifolia*) is a tree for fodder, fuel and fibre. The feeding of the fodder (leaves and soft twigs) from the tree is almost a universal practice throughout the year in the hills.

Golden timothy *Mazungula grass. Setaria sphacelata* is a grass resistant to drought and cold; an ideal forage grass for Himalayas. *Bermuda and para grasses* are suitable to grow on alkali soils. *Hybrid napier N.B.21* (*Pennisetum purpureum*), *para grass* (*Brachiaria mutica*), *guinea grass* (*Panicum maximum*) and *bermuda grass* (*Cynodon dactylon*) are good forage yielders.

Lucerne (*Medicago sativa*) originally a native of south western Asia, is now grown practically all over the world as a forage crop.

In rainfed areas, during kharif, fodder crops such as fodder sorghum, pearl millet, cow pea, cluster bean (guar), horse gram (moth) are sown in June end.

KOOBABUL

Koobabul is a promising fodder tree. Kobabul is a multipurpose leguminous plant of medium height. It is being grown in tropical and sub-tropical regions of the world. It is a drought resistant perennial tree. The green fodder is rich in protein but it may cause toxicity in cattle etc. Symptoms of mimosine (leucinol), a free aminoacid) in sheep and cattle are excessive salivation loss of weight, loss of hair and wool, skin lesions, enlarged thyroid glands (goitre). Due to higher consumption of *mimosine* present up to an extent of 3 to 5 per cent of the protein on dry basis. The mimosine content can be reduced by treating the meal (i) *By heat treatment* : Moist heating at 70-100°C for 72 hr. may reduce the mimosine content up to 50 per cent while steaming for 2 hr. can also serve the purpose. (ii) *Supplementation of ferrous sulphate* : Addition of ferrous sulphate at 0.02 per cent on dry basis also reduces the mimosine toxicity, as ferrous salt hinders the absorption of mimosine and help in its excretion in faeces.

Koobabul foliage can be used as green manure; in Central America and Indonesia, green pods are used as vegetable. The plant acts as shade tree, strong wind break and also checks soil erosion. Provides fuel wood, timber, paper pulp, charcoal, gum and tannins, can act as a natural wool shearer.

Atriplex : The climatic conditions of Western Rajasthan are very hard and dry. With combination of poor edaphic factors and biotic pressure on the land, a major portion of it is unproductive and remains fallow due to sand dumes or saline conditions. *Atriplex* (saltbushes) are perennial, evergreen, shrubs which have ability to remove high concentration of salts from soil and thus making useless waste lands into productive soils. *Atriplex nummularia* lindle is a bluish grey shrub grows well in deep soils as well as in sandy-loam and more drought resistant than other species. *Atriplex canescens (Pursh)* Nutt known as fourwing salt-bush is a shrub with 4-winged fruit. All these species are palatable and liked by cattle, sheep and goats.

Atriplex is cultivated by seed sowing or by cutting. Seedlings are first raised in nursery for 2 to 5 months and then transplanted in the field, preferably after rains. Transplanting is done in a well prepared levelled field. Plant to plant distance should be 1.5 m. Row to row distance should be 3 m.

A. canescens and *A. halinus* start flowering after 6 months. *A. nummularia* starts flowering in the second year within 2-3 years, the plant attains a height of 150 cm, the time when plants should be used for browsing and grazing by the animals.

SUBABUL

Subabul is a fast growing perennial, ever green and thornless leguminous shrub. It can be grown in the humid and sub-humid tropics and within pH range of 6.5 to 8.5. Under irrigated conditions on fertile loams, it can yield 100 tonnes/ha/year nutritious fodder.

As a tree, subabul grows to a height of 5 m and 3.5 cm diameter at breast height in the first year and 15 m height and 25 cm DBH within 7 years. Apart from fuel, the wood can be used for building huts and for making furniture, charcoal, paper, pulp, cellophone and light agricultural implements.

Mimosine : There is a belief that mimosine, a non-protein amino acid is subabul, causes hair fall, enlarged thyroids, excessive salivation, conjunctivitis and general deterioration in the health of cattle as reported from Australia. But feeding of shoots 1.0 to 1.25 m to cattle in India caused no adverse effect. There is evidence that in the rumen of Indian, Indonesian and Philippine cattle, mimosine is broken down into non-toxic compounds.

ANDROPOGON PUMILUS

Grasses have an important role in the utilization of land resources. *Andropogon pumilus* is a tall, handsome grass growing up to 2.5-3.0 m high, remains green even in the month of December, when other local grasses drug away. The grass is easily cultivated. It is propagated by seeds root clumps, during the rainy season.

The land is prepared by ploughing and cross ploughing. If possible compost or FYM can be applied at 5-10 tonnes/ha at the time of land preparation to obtain higher herbage yield. Application of urea at the rate of 250 kg/ha may also be given at the time of planting. June-July is the best time for planting seed or root clumps. Planting should be done at a distance of 25-30 cm apart. The grass establishes in 40-45 days, first cutting is available in August-September and subsequent cuttings can be taken in October-November and January. From second year it gives profuse growth and fodder yield. The grass gives a yield of 400-600 quintals of green herbage annually. The inflorescence is also very soft and palatable to the animals at all stages of its growth.

Miscellaneous

VEGETABLE IVORY

It is a common knowledge that ivory is obtained from elephant tusks. However, in plant world two palms provide a cheap substitute for ivory which is called 'vegetable ivory'. Like the elephant ivory vegetable ivory is also put to many prolific uses including making of biliard balls, buttons, chessmen inlays, toilet articles, toys and fancy articles and even ornaments.

The plants furnishing vegetable ivory are *Phytelephus macrocarpa* (Ivorynut palm) and *Metroxylon amicarum* (Carolina ivory palm) which are native to South America and Carolina Island respectively. Endosperm of their seed is the source of ivory.

Phylelephus macrocarpa a member of Plamaceae is a dioecious (i.e. male and female flowers are found on separate plants) plant with creeping stem of about 20 feet. Female trees bear fruit as very large containing a cluster of 6-7 berry like parts each containing a hard white seed. Seeds filled with milky endosperm which is edible as early stages.

Metroxylon amicarum a member of Palmaceae is a monoecious (male and female flowers separate but on the same plant) plants about 60 feet high. Fruits are ovoid, 3-4 inches long, covered with glossy spherical scales. Seeds are filled with milky endosperm which is edible as early stages but as the fruits gain maturity endosperm becomes hard and resembles ivory obtained from the elephants. Endospermic ivory is made up of cellulose. One seed of *Phytelephus* and *Metroxylon* weighs about 2.5 and 1.26 respectively.

Flowering of food stuffs: In India food items - sweet meat, flattened rice, cereal and pulse products, confectionery, fish, vegetables, cut-fruits, milk and milk products, *velpuri*, turmeric powder and shells of eggs are variously coloured with the percentage of added colouring matter in each is as follows :

Content of colouring matter (per cent)	
Sweet meats	80-90
Flattened rice	45
Cereal and pulse products	20
Confectionery	15-20
Fish	15-17
Vegetables	15-17
Cut-fruits	15
Milk, milk products	15
Velpuri	14
Turmeric powder	10
Eggs (on shells)	5-6

Metanil yellow, Orange II, Rhodamine B, Auramine, Congo red, Malachite green, Copper sulphate are some of the non-permitted colours used.

PETROCROPS

Development of crops which will produce something resembling petroleum is the subject of major research by world scientists. Some species of *Euphorbia* are reported to be important source of petro-chemicals.

Pobreng kahoy (*Euphorbia tirucalli*) is reported to have produced 15 barrels of oil per acre annually. In Philippines this species has **been** discovered to grow on any type of soil with good drainage. It is a small xerophytic tree but can grow to a height of 20 feet.

Gopher plant (*Euphorbia lathyris*) is believed to be an important source of petro-chemicals and seems to offer good prospects as energy source to be exploited as *gasoline agriculture* in view of the increasing prices and decreasing availability of the petroleum products.

Euphorbia abyssinica and *Euphorbia resinifera* are other species which can successfully be utilized as energy source in India.

Glossary

Abortifacient : An agent that produces abortion.

Alterative : A drug which corrects disordered process of nutrition and restores the normal function of an organ or of the system.

Amenorrhoea : Abnormal suppression of menses.

Anaemia : A deficiency of blood or of red blood cells.

Angina pectoris : A disease of the heart marked by severe constricting pains in the chest.

Anodyne : A drug that relieves pain.

Anthelmintic : A drug that kills intestinal worms.

Antihydrotic : A drug which checks sweating.

Antilithic : A drug which counteracts the development of stone.

Antiperiodic : A drug that cures periodic attacks.

Antipyretic : A drug which reduces fever.

Antiscorbutic : A drug which cures scurvey.

Antispasmodic : A drug which counteracts spasmodic disorders.

Aphrodisiac : A drug which promotes sexual desire.

Aromatic : A drug which is fragrant, spicy and mildly stimulant.

Asthma : A chronic disorder of the bronchial tubes.

Astringent : A drug which checks secretion or bleeding.

Beriberi : A deficiency disease caused by lack of vitamins especially B_1.

Bronchitis : An inflammation of the air passage.

Calculus : A hard and solid concretion formed in the body, especially in the urinary organs; it may be sand, gravel or stone.

Cancer : Any malignant growth.

Caries : Decay of teeth.

Carminative : A drug which relieves flatulence.

Cathartic : A drug which induces active movement of the bowels.

Cholagogue : A drug which promotes flow of bile.

Colic : Pain due to spasmodic contraction of the abdomen.

Congestion : An abnormal collection of blood in the blood vessels of any organ or part of the body.

Conjunctivitis : Inflammation of the conjunctiva, the mucous membrane covering the eyeball and lining the eyelids.

Dandruff : An inflamed condition of the scalp characterized by the presence of white scales in the hair due to the exfoliation of the horny cells of the scalp.

Demulcent : An agent having a soothing effect on the skin and mucous membranes.

Deobstruent : A drug that removes an obstruction to secretion or excretion by opening the natural passages or pores of the body.

Diabetes : A wasting disease of metabolism; abundant sugar is present continuously in the urine.

Diaphoretic : A drug that induces copious perspiration.

Diuretic : A drug which increases the secretion and discharge of urine.

Dropsy : A disease marked by an excessive collection of a watery fluid in the tissues or the cavities of the body.

Dysentery : An infectious disease of which the chief symptoms are acute diarrhoea and discharge of mucus and blood.

Dysmenorrhoea : Usually painful and difficult menstruation.

Dyspepsia : Indigestion.

Eczema : A skin disease accompanied by swelling, redness and exudation of lymph.

Elephantiasis : A disease of the skin caused by a tiny worm and attended with hypertrophy of the affected parts.

Emetic : A drug which induces vomiting.

Emmenagogue : A drug which promotes menstruation or regulates the menstrual periods.

Emollient : A drug which allays irritation of the skin and alleviates swelling and pain.

Enteritis : Inflammation of the intestines.

Epilepsy : A chronic nervous disorder marked by attacks of unconsciousness or convulsions.

Expectorant : A drug that promotes the removal of catarrhal matter and phlegm from the bronchial tubes.

Febrifuge : An agent used for reducing fever.

Flatulence : A disorder in which there is an excessive collection of the gas in the stomach.

Galactagogue : An agent that promotes secretion and flow of milk.

Gleet : A chronic discharge from the urethra.

Goitre : A chronic enlargement of the thyroid gland.

Gonorrhoea : An infectious venereal disease marked by an inflammatory discharge from the genital organs.

Haemoptysis : Spitting of blood from the lungs or bronchial tubes.

Haemorrhage : Bleeding, especially prefuse, from any part of the body.

Heartburn : A burning feeling in the regions of the chest and stomach, generally due to indigestion.

Hepatitis : Inflammation of the liver.

Hysteria : A disease in which a physically healthy patient has lost control over acts and feelings and suffers from imaginary ailments.

Intermittent fever : Fever which is marked by intervals of normal temperature between periods of rise of temperature.

Jaundice : A diseased condition in which there is a yellowish staining of the tissues and excretions with bile.

Lactagogue : Galactagogue.

Laryngitis : Inflammation of the larynx.

Leprosy : A chronic wasting disease caused by germ; the disease generally results in mutilations and deformities.

Leucoderma : A condition of the skin in which there is loss of pigment wholly or partially.

Lithontriptic : A drug used for removing calculi of stones formed in the urinary system.

Malaria : A recurrent disease marked by bouts of shivering, sudden rise of temperature and general aching of the body.

Narcotic : A drug which induces deep sleep.

Nausea : A feeling that vomiting is about to take place.

Nephritis : Inflammation of the kidney.

Neuralgia : Pain felt above a nerve.

Ophthalmia : Conjunctivitis.

Orchitis : Inflammation of the testicles.

Paralysis : A disease in which there is loss of power of voluntary movement in any part of the body.

Pectoral : A drug to cure disorders of the chest.

Pharyngitis : Inflammation of the pharynx.

Phythisis : Consumption; tuberculosis of the lungs.

Piles : An inflammed condition of the veins in the rectal region.

Pneumonia : Inflammation of the lungs.

Prophylactic : An agent that prevents disease.

Refrigerant : A drug which relieves feverishness or produces a feeling of coolness.

Rheumatism : An indefinite term used for pains in the muscles, joints and certain tissues.

Rubefacient : A mild counter-irritant.

Scabies : An itching skin disease caused by a mite.

Scorbutic : Suffering from scurvy.

Scurvy : A deficiency disease due to lack of vitamin C.

Sedative : A drug which reduces excitement, irritation and pain.

Sialagogue : A drug which promotes salivation.

Soporific : A drug that induces sleep.

Stomachic : A drug that strengthens the stomach and promotes its action.

Styptic : An agent which checks bleeding.

Syphilis : A chronic venereal disease.

Tetanus : An infectious disease, marked by painful contraction in the muscles.

Tonsillitis : Inflammation of the tonsils.

Ulcer : An open sore of the skin.

Vermifuge : A drug which expels intestinal worms.

Whooping Cough : An acute infectious disease of coughing.

Some Other Important Medical Terms

Acne : A pimple-like eruption of the sebaceous glands of the skin, with accumulation of yellow secretion and black overgrowth of the horny layer of the skin.

After-Pains : Painful contraction of the womb after child birth.

Alopecia : A disease of the scalp resulting in complete or partial baldness.

Anasarca : Dropsy.

Antacid : A drug which neutralizes the acidity of the gastric juice.

Antiphlogistic : A drug which counteracts inflammation.

Aperient : A mild purgative.

Aphthae : Minute white ulcers on the tongue and in the mouth.

Apoplexy : Sudden loss of consciousness with some paralysis.

Ardor urine : A burning sensation on urinating.

Ascaris : Intestinal parasitic round worms.

Atony : Lack of muscular power.

Bechic : A remedy for cough.

Bedsores : Ulceration of any part of the body exposed to pressure of bed-ridden patient.

Bronchorrhoea : Excessive discharge from the bronchial mucous membrane.

Colic : Pain due to spasmodic contraction of the abdomen.

Contusion : An injury to the soft parts without breaking the skin.

Cystitis : Inflammation of the bladder.

Depilatory : An agent that removes or destroys hair.

Diphtheria : An infectious disease of the throat and the air passage.

Discutient : A drug which disperses or absorbs a tumour or any coagulated fluid in the body.

Dysuria : Painful and difficult urination.

Fistula : An abnormal channel which connects one cavity of the body with another, or which opens out from a cavity to the surface of the body.

Freckles : Coloured spots on the exposed parts of the skin.

Glycosuria : A diseased condition of the urine in which sugar is excreted.

Hemiplegia : Paralysis of one side of the body.

Hepatic : Pertaining of the liver.

Hernia : Rupture; Protrusion through its covering of any organ of the body.

Hypnotic : A drug which induces sleep.

Hypochondriasis : A mental disorder in which the patient is tormented by meloncholy views, particularly about its health.

Itch : An infectious skin disease, caused by a mite, without specific lesions and marked by excessive itching.

Lithontriptic : A drug used for removing stones formed in the urinary systems.

Monorrhagia : Abnormally excessive menstruation.

Metrorrhagia : Bleeding from the womb.

Migraine : Periodic attack of headache affecting one side of the head.

Night-blindness : A disease in which the patient is incapable of seeing in the dark.

Otitis : Inflammation of the ear.

Psoriasis : A common chronic inflammation of the skin, marked by rounded reddened patches which are covered with dry silvery scales.

Pyrrhoea : A disease marked by purulent discharge from the gums.

Ringworm : A parasitic skin disease usually marked by red, scaly, circular patches.

Sciatica : An inflammation of the sciatic nerve at the back of the thigh.

Stomatitis : Inflammation of the month.

Typhoid fever : An acute infectious disease characterized by ulceration of the intestines, eruption of rose-coloured spots, and a typical course of temperature.

Urethritis : Inflammation of the urethra.

Vulnerary : A drug which promotes healing of wounds.

Research Institutes and Societies in India Related to Economic Botany

The Indian Agricultural Research Institute (IARI), New Delhi.

The Central Arid Zone Research Institute (CAZRI), Jodhpur (Rajasthan).

The Cotton Technological Research Laboratory (CRTL), Matunga, Bombay (Maharashtra).

The Indian Grassland and Fodder Research Institute (IGFRI), Jhansi (Uttar Pradesh).

The Institute of Horticultural Research (IHR), Bangalore (Mysore).

The Jute Agricultural Research Institute (JTRL), Calcutta (West Bengal).

The Indian Lac Research Institute (ILRI), Namkum, Ranchi (Bihar).

The Central Plantation Crops Research Institute (CPCRI), Kudlu, Kasaragod (Kerala).

The Central Potato Research Institute (CPRI), Simla (Himachal Pradesh).

The Central Rice Research Institute (CRRI), Cuttack (Orissa).

The Central Soil Salinity Research Institute (CSSRI), Karnal (Haryana).

The Indian Institute of Sugarcane Research (IISR), Rai-Bareli Road, Lucknow (Uttar Pradesh).

The Sugarcane Breeding Institute (SBI), Coimbatore (Tamil Nadu).

The Central Tobacco Research Institute (CTRI), Rajahmundry (Andhra Pradesh).

The Central Tuber Crops Research Institute (CTCRI), Trivandrum (Kerala).

The Indian Veterinary Research Institute (IVRI), Izatnagar (Uttar Pradesh).

The National Dairy Research Institute (NDRI), Karnal (Haryana).

The Indian Council of Agricultural Research (ICAR), Krishi Bhavan (New Delhi).

The Indian Institute of Science (IIS) (Bangalore).

The Birbal Sathi Institute for Paleobotany (BSIPB) (Lucknow).

The Central Inland Fisheries Research Institute (CIFRI), Barrackpore (West Bengal).

The Central Marine Fisheries Research Institute (CMFRI), Cochin (Kerala).

The Central Institute of Fisheries Technoigy (CIFT), Ernakulam, Cochin (Kerala).

The Central Sheep and Wool Research Institute (CSWRI), Auikangar (Rajasthan).

The Institute of Agricultural Research Statistics (IARS), New Delhi.

All India Coordinated Rice Improvement Project (AICRIP), Rajendra Nagar, Hyderabad (Andhra Pradesh).

All India Coordinated Sorghum Improvement Project (AICSIP), Rajendra Nagar, Hyderabad (Andhra Pradesh).

International Crops Research Institute for Semi-Arid Tropics (ICRISAT), Patancheru, Hyderabad (Andhra Pradesh).

Bhabha Atomic Research Centre (BARC), Trombay, Bombay.

Birla Institute of Science and Technology, Nagda (Madya Pradesh).

Bose Institute, Calcutta (West Bengal).

Vallabhai Patel Chest Institute (VPCI), Delhi.

Botanical Survey of India, Madan Street, Calcutta (West Bengal).

Central Botanical Laboratories, Allahabad.

Central Coir Research Institute, Kalavoor, Kerala.

Central Food Technological Research Institute, Mysore.

Central Indian Medicinal Plants Organization, C/o National Botanical Gardens, Lucknow (Uttar Pradesh).

Central Public Health Engineering Research Institute, Nagpur (MP).

Council of Scientific and Industrial Research (CSIR), Rafi Marg, New Delhi.

High Altitude Research Laboratory, Gulmarg, Jammu & Kashmir.

Indian Association for the Cultivation of Science, Jadavpur, Calcutta.

Indian Jute Mills Association Research Institute, Paratola Road, Calcutta.

Indian National Scientific Documentation Centre, Hillside Road (INSDOC), New Delhi.

Institute of Jute Technology, Ballygunge Circular Road, Calcutta.

Rubber Board and Rubber Research Institute of India, Kottayam (Kerala).

Sri Ram Institute of Industrial Research, University Road, Delhi.

Tata Institute of Fundamental Research, Kulaba, Bombay.

Technological Institute of Textiles, Bhiwani (Haryana).

Technological Research Laboratory (Cotton), Matunga, Bombay.

Agricultural Society of India, Calcutta.

Ahmedabad Textile Industry's Research Association Navrangpura, Ahmedabad.

Botanical Society, University of Saugar, Sagar.

Botanical Society of Bengal, Calcutta.

Horticultural Society of India, Krishi Bhawan, New Delhi.

Indian Academy of Sciences, Hebbal, Bangalore.

Indian Food Preserver's Association, Chakraberia Lane, Calcutta.

Indian Dairy Science Association, Hosur Road, Bangalore.

Indian Jute Industries Research Association, Taratola Road, Calcutta.

Indian Leather Technologist's Association, Calcutta.

Indian Phytopathological Society, IARI, New Delhi.

Indian Plywood Industries Research Association, Tumkur Road, Bangalore.

Indian Pulp & Paper Technical Association, Sirpur Kagaznagar, Andhra Pradesh.

Indian Society of Agronomy, IARI, New Delhi.

Indian Society of Genetics and Plant Breeding, IARI, New Delhi.

Institute of Advancement of Science & Culture, Hauzkhas Enclave, New Delhi.

Inventions Promotion Board, 39 Ring Road, Mulchand Hosital Corner, New Delhi.

National Academy of Sciences, India, 5 Lajpat Rai Road, allahabad.

National Institute of Science of India, Bahadur Shah Zafar Marg, New Delhi.

South India Textile Research Association, Coimbatore (Tamil Nadu).

Sugar Technologist's Association of India, Kanpur.

Society of Environmental Engineers, LRDE,Bangalore.

Bombay Textile Research Association, Bombay-Agra Road, Bombay.

Indian Association of Biological Sciences, University Road, Delhi.

Hindustan Antibiotics, Pimpri, Poona.

Indian Dairy Science Association, Greater Kailash, New Delhi.

IndianPharmaceutical Association, Kalyan, Bombay.

Indian Research Society, 78 Serpentine Lane, Calcutta.

Indian Rubber Manufacturer's Research Association, 7 Homji St., Bombay.

Indian Science Congress Association, Calcutta.

Indian Standards Institute, Bahadur Shah Zafar Marg, New Delhi.

International Society of Plant Morphologists, University of Delhi, Delhi.

National Council of Education Research & Training, Sri Aurobindo Marg, New Delhi.

Palynological Society of India, University Road, Calcutta.

Pharmacy Council of India, Temple Lane, New Delhi.

Society of Biological Chemists, India, Bangalore.

National Research Development Corporation of India, 61 Ring Road,Lajpat Nagar, New Delhi.

National Botanical Research Institute, Ranapratap Marg, Lucknow.

National Chemical Laboratory, Poona.

Central Drug Research Institute, Lucknow.

National Environmental Engineering Research Institute (NEERI), Nehru Marg, Nagpur.

Indian Institute of Experimental Medicine, Raja Subodh Mullic Road, Calcutta.

Central Fuel Research Institute, Dhanbad Distt., Bihar.

Central Leather Research Institute, Adayar, Madras.

Central Institute of Medicinal & Aromatic Plants, Sitapur Road, Lucknow.

National Institute of Oceanography, Dona Paula, Goa.

Indian Institute of Petroleum, Dehradun.

Regional Research Laboratory, Bhubaneshwar (Orissa).

Regional Research Laboratory, Hyderabad.

Regional Research Laboratory, Jammu Tawi (J & K).

Regional Research Laboratory, Jorhat (Assam).

Central Salt & Marine Chemicals Research Institute, Bhavnagar (Gujarat).

Central Scientific Instruments Organisation, Chandigarh.

Industrial Toxicology Research Centre, Lucknow.

National Institute of Nutrition, Jamia Osmania, Hyderabad.

Oil Technological Research Institute, Anantapur (Andhra Pradesh).

Indian Oilseeds Development Council, Himayatnagar, Hyderabad.

Forest Research Institute, Dehradun.

The Research Association, Tocklai Experimental Stn., Jorhat (Assam).

Soil Conservation Society of India, Patna.

Soya Production & Research Association, Bareilly (Uttar Pradesh).

Grain Storage Research & Training Centre, Hapur (Uttar Pradesh).

Tea Research Association, Calcutta.

Central Inland Fisheries Research Institute, Madras (Tamil Nadu).

Industrial & Scientific Research Association, Madras.

Pulp & Paper Research Institute, Jayakaypur, Koraput Distt., Orissa.

Dalmia Institute of Scientific & Industrial Research, Rajgangpur, Sundargarh Distt., Orissa.

Wool Research Association, Matunga, Bombay.

Indian Rubber Manufacturers Research Association, Thane (Maharashtra).

Appendix II

1. MICROCHEMICAL TESTS

1. Cellulose

Plant materials	(A) filter paper, (B) pith
Chemicals	Iodine, Sulphuric acid (50%)
Procedure	Soak for a few minutes in Iodine, mount in few drops of 50% of sulphuric acid. Blue colouration.

2. Lignin

Plant materials	Match shavings; match sticks, wood shavings
Chemicals	(A) phloroglucin, hydrochloric acid
	(B) 1% neutral aq. potassium permjanganate, hydrochloric acid, ammonium hydroxide (sodium bicarbonate)
Procedure	(A) sections placed in 1% alcoholic solution of phloroglucin, cover with coverslip, allow 25% hydrochloric acid to diffuse. Red-violet colouration.
	(B) treat with 1% neutral aq. solution of potassium permanganate for 12 to 20 minutes, wash thoroughly with 2% hydrochloric acid, wash with water many times, add a few drops of either ammonium hydroxide or sodium bi-carbonate, deep red colour develops in lignified elements of the deciduous plants.

3. Suberin

Plant materials	Bottle cork, natural cork.
Chemicals	Sudan IV, alcohol 50%
Procedure	Leave the fresh sections to stain for 20 minutes, wash excess of stain with 50% alcohol, transfer the section to glycerin, imparts red colour.

4. Mucilage

Plant material	Linseed testa
Chemicals	Copper sulphate (10%), potassium hydroxide (10%)
Procedure	Soak in 10% copper sulphate, wash in water and transfer to 10% potassium hydroxide, stained bright blue

5. Protein

Plant materials	Gram flour, legumes, soyabeans.
Chemicals	(A) nitric acid, ammonium hydroxide
	(B) sodium hydroxide (20%), copper sulphate (1%)
Procedure	(A) Xanthoproetic test : Treat a section or a suspension of the tissue in water, add concentrated nitric acid, yellow colour, add a few drops of cone. ammonium hydroxide, colour changes to orane.
	(B) Biruet test : Add to the test solution 1 cc of 20% sodium hydroxide and one drop of 1% copper sulphate so-

lution, violet colour indicates the presence of proteins.

6. Fatty oils

Plant materials	Seeds of almonds or soyabean
Chemicals	Sudan III (70%), glycerine
Procedure	Section the material, stain for a few minutes in Sudan III in 70% alcohol, wash thoroughly in water, mount in glycerin. Red colour indicates the presence of fatty oil.

7. Inulin

Plant material	Roots of *Dahlia*.
Chemicals	Orcin (conc. alcoholic solution), sulphuric acid.
Procedure	Section material, mount in a drop of a stain orcin, diffuse a drop of sulphuric acid through the coverslip, orange colour appears, if not, heat gently.

8. Sugar

Plant materials	Potato tuber, wheat flour, carrot, sugar.
Chemicals	(A) Fehlings solution 'A' (copper sulphate 34.6 gms dissolved in 500 ml water), Fehling 'B' (potassium sodium tartarate 173 gms, sodium hydroxide 50 gms dissolved in 500 ml water).
	(B) Benedict's solution : A. sodium citrate 173 gms, 100 gms of sodium carbonate, 800 ml of water by warming, make 850 ml solution. B. dissolve 17.3 gms of copper sulphate in water, make upto 100 ml.

Add B to A slowly with constant stirring.

Procedure	(A) Fehling's test : warm about 2-3 ml of Fehling's solution in a test tube and a few drops of sugar solution, boil, brown-red precipitate is formed.
	(B) Benedict's solution : To 5 ml of reagent add 1-2 ml of sugar solution, boil, red, yellow or green precipitate formed.

9. Carbohydrates (Non-reducing)

Plant materials	Starch, Sucrose, beet root.
Chemicals	Conc. HCl, sodium carbonate
Procedure	Add equal volume of hydrochloric acid to the sugar, bring to boil, neutralise with sodium carbonate or sodium bi-carbonate, repeat the stages for the reducing sugars (as in item 8).

10. Barfoed's test for sugars

Plant material	Sugars
Chemical	Crystalline copper acetate 13.3 gm in 200 ml water, add 1.9 gm glacial acetic acid.
Procedure	Add the soltion to sugar, formationof precipitate.

11. Latex

Plant materials	Latex from *Calotropsis*, members of Euphorbiaceae, Apocynacae.
Chemicals	Sucrose, conc. sulphuric acid.
Procedure	Prepare alcoholic extract of the latex, mix sucrose and the acid in equal amounts freshly, pinkish-purple colour with the latex.

Appendix III

Detection of food adulteration

Food stuffs	Adulterant	Method of detection
1. Cereals & pulses	(a) Foreign matter	Take a known quantity (50 gms) and pick up all foreign matter by hand or forceps. Weigh the amount and calculate the percentage (it should not exceed 4%)
	(b) Insect infection	(a) Take a known quantity (50 gms). Pick up all damaged grains. Calculate percentage (it should not exceed 1%)
	(b) Put the grains in water.	Infested grains float on the surface). Calculate the percentage.
2. Pulses (arhar, massor and chana)	Khesari dal	Pick up khesari dal (triangular and gray coloured seeds) and calculate percentage as before (it should not exceed 1%)
3. Turmeric (Haldi)	Lead chromate	Weigh 2 gms of Haldi powder, reduce it to white ash in crucible (600°C for 4 hours). Cool. Add 5 ml of 1 : 7 dil. H_2SO_4 and filter. Add a few drops of 0.2% diphenyl carbazide (alcoholic). Pink colour indicates the presence of lead chromate.
4. Milk	(a) Water added or fat removed	Specific gravity determination of milk by lactometer
	(b) Starch	Iodine test
5. Tea leaves	Artificial colour	(a) Place the leaves on white paper. The appearance of yellow or reddish colour over the paper will show the presence of artificially added colour
		(b) Spread a little slaked lime on glass plate. Sprinkle a little tea dust on the lime. Any colour (e.g. red, orange, etc.) other than greenish yellow (due

(Contd.)

		to presence of naturally occurring chlorophyll) indicates the presence of coal tar dye.
6. Chillies	Coloured saw dust, brick powder, talcum powder	Ash a spoonful of chillies. Abundant amount of ash indicates adulteration
7. Oil	(a) Mineral oil (Hodès test)	Take 2 ml of sample. Add an equal amount of N/2 alcoholic potash. Heat for 15 minutes in boiling water bath. Cool and add about 10 ml of water. Presence of turbidity indicates presence of mineral oil
	(b) Argemone oil Add nitric acid.	If red colour appears, it indicates the presence of argemone oil
8. Pure ghee	Vanaspati ghee	Bodoudouin test (for the presence of sesame oil). To 5 ml of melted ghee add 0.1 gm of sucrose dissolved in 5 ml of dilute HCl. Shake well and keep for 15 minutes. A permanent pink colour indicates the presence of sesame oil
9. Sweets	Metanil yellow	Dissolve a little sample in water, shake and transfer the water extract to another tube. Add dilute HCl. A violet red colouration indicates the presence of metanil yellow

Appendix IV
List of Economically Useful Plants :
Ready Reckoner

1. CEREALS & MILLETS (FARN : POACEAE)

A. Tropical grain crops

Brachiaria ramosa
Echinochloa frumentacea
Eleusine coracana (Ragi)
Oryza sativa (Rice)
Panicum miliaceum
Panicum miliare
Paspalum scrobiculatum (Kodo millet)
Pennisetum americanum (Pearl millet)
Setaria italica (Foxtail millet)
Sorghum vulgare (Great millet)
Zea mays (Maize)

B. Winter or temperate or sub-tropical grain crops (Fam. Poaceae)

Avena sativa (Oats)
Hordeum vulgare (Barley)
Secale cereale (Rye)
Triticum aestivum (Wheat)

2. PULSES (Fam. FABACEAE)

A. Pulses under cultivation in plains

Cajanus cajan (Redgram)
Cicer arietinum (Bengal gram)
Dolichos biflorus (Horse gram)
Lablab purpureus (Hyacinth bean or field lab-lab)
Phaseolus aconitifolius (Dew gram)
Phaseolus aureus (Green gram)
Phaseolus mungo (Black gram)
Phaseolus sublobatus (Climbing black gram)
Phaseolus trillobus (Pillipersara)
Vigna unguiculata (Cowpea)

B. Pulses under cultivation on the hills or in winter in the plains

Glycine max (Soybeans)
Lathyrus sativus (Kesari)
Lens esculenta (Masuri)
Phaseolus lunatus (Lima bean)
Pisum sativum (Peas, field pea)
Vicia faba (Broad bean)

VEGETABLES

I. FOR LOW AND MEDIUM ELEVATIONS

A. Leguminous vegetables (Fam : Fabaceae)

Canavalia ensiformis (Sword bean or Jack bean)
Cyamopsis tetragonoloba (Cluster bean)
Lablab purpureus var. lignosus (Field lab-lab)
Dolichos lab-lab var. typicus (Garden lab lab)
Pachyrhizus erosus (Yarn bean)
Phaseolus lunatus (Limabean)
Phaseolus vulgaris (French bean)
Psophocarpus tetragonolobus (Goa bean)
Stizolobium indicum (velvet bean)
Vicia faba (broad bean)
Vigna unguiculata (cowpea)

B. Cucurbitaceous vegetables (Fam : Cucurbitaceae)

Benincasa hispida (Ashgourd)
Cucumis sativus (Cucumber)
Coccinia indica (Donda)

Cucumis pubescens
Cucurbita maxima (Pumpkin)
Cucurbita moschata (Musk melon)
Cucurbita pepo (Vegetable marrow)
Lagenaria siceraria (Bottle gourd)
Luffa cylindrica (Sponge gourd)
Luffa acutangula (Ribbed gourd)
Momordica charantia (Bitter gourd)
Momordica tuberosa
Momordica dioica (Bur cucumber)
Sechium cesula (Chow-chow)
Trichosanthes anguina Linn. (Snake gourd)

C. Tuberous vegetables

Amorphophalus campanulatus (Araceae, Elephant foot yam)
Alocasia indica (Giant taro)
Coleus paruiflorus (Lamiaceae, Country potato)
Colocasia antiquorum (Taro or edible yam)
Decalepsia hamiltonii (Asclepiadaceae, Sarasaparilla)
Dioscorea alata (Dioscoreaceae, Greater yam)
D. esculenta (Dioscoreaceae, Leeser yam)
D. oppositifolia (Betal yam)
Helianthus tuberosus (Jerusalem artichoke)
Ipomoea batatas (Sweet potato)
Manihot esculenta (Tapioca)
Solanum tuberosum (Potato)

D. Green leaf vegetables

Allium cepa (Onion)
Alocasia indica (Giant taro)
Alternanthera triandra (Amaranthaceae)
Amaranthus tricolor (Amaranthaceae)
Amaranthus causatus (Amaranthaceae)
Amaranthus polygamus (Amaranthaceae)
Basella alba (Indian spinach)
Basella rubra (Indian spinach)
Beta vulgaris (Beet root)
Boussingaultia baselloides
Chenopodium album (Igweed)
Hibiscus cannabinus (Deccan hemp)
Hibiscus subdariffa (Rosella)
Lactuca sativa (Lettuce)

Moringa oleifera (Drumstick)
Pisonia alba (Tree lettuce)
Portulaca oleraceae (Indian purselane)
Rumex vasicarius (Sarrel, Duck sorrel)
Sauropus androgynus (Euphorbiaceae)
Sesbania grandiflora (Sesban)
Solanum nigrum (Black nightshade)
Spinach oleraceae (Garden spinach)
Talinum triangulare (Ceylon spinach)
Trigonella foenum-graecum (Fenugreek)

E. Other miscellaneous vegetables

Allium cepa (Liliaceae)
Artocarpus communis (Bread fruit)
Brassica oleracea var. *rapa* (Turnip)
Brassica deracea var. *gongyloides* (Knol-knol)
Calonyction bono-nox (Moon flower)
Hibiscus esculentus (Lady's finger)
Lycopersicon esculentum (Tomato)
Mangifera indica (Mango)
Moringa oleifera
Musa paradisiaca (Plantain)
Raphanus sativus (Radish)
Solanum melongena (Brinjal)

II. TEMPERATE OR SUBTROPICAL VEGETABLES

Allium ascalonicum (Shallotis)
Allium cepa
Allium ampeloprasum (Leek)
Beta vulgaris
Brassica oleracea var. *botrytis* (Cauliflower)
B. oleracea var. *bullata* (Cabbage)
B. oleracea var. *italica* (Sprouting broccde)
B. oleracea var. *gemmifera* (Brussels sprouts)
B. oleracea var. *caulorapa* (Knol-knol)
B. oleracea var. *rapa* (Turnip)
Capsicum annum (Vegetable type of chillies)
Cichorium endimia (Endive)
Cucumis sativus (Cucumber)
Cynara scolymus (Globe artichoke)
Daucus carota (Carrot)
Helianthus tuberosus
Lycopersicon esculentum (Tomato)
Phaseolus multiflorus (Carlet runner)

Phaseolus vulgaris (French or kidney bean)
Pisum sativum Linn.
Raphanus sativus (Radish)
Rheum rhapporticum (Rhubarb)
Sechium edule (Chow chow)
Solanum tuberosum (Potato)
Tragopogon porrifolius Linn. (Oyster plant)
Vicia faba (Broad bean)

Salad crops

Asparagus officinalis (Asparagus)
Apium graveolens (Celery)
Lactuca sativa (Lettuce)
Lepidium sativum (Garden cress)
Nasturtium fontanum (Water cress)
Pastinaca sativa (Parsnip)
Rumex vesicarius (Sarrel)
Spinacea oleracea
Tropaeolum majus (Indian cress)

4. FRUITS

A. Fruits mainly of the plains

Aegle marmelos (Bilva, Shriphala)
Anacardium occidentale (Cashew nut)
Borassus flabellifer (Palmyra)
Carissa carandus
Citrullus vulgaris (Water melon)
Cucumis melo (Musk melon)
Feronia limonia (Wood apple)
Grewia asiatica (Phalsa)
Phoenix dactylifera (Date palm)
Pithecellobium dulce (Madras thorn)
Psidium guava (Guava)
Zizyphus jujuba (Zizyphus)

B. Fruits common to the plains and humid zones

Achras zapota (Sapota)
Ananas comosus (Pine apple)
Annona cherimolia (Cherimoyer)
Annona muricata (Sour sop)
Annona reticulata (Bullis heart)
Annona squamosa (Custard apple)
Artocarpus communis (Bread fruit)
Averrhoa bilimbi

Averrhoa carambola (Carambola)
Carica papaya (Papaya)
Chrysophyllum cainito (Star apple)
Citrus aurantifolia (Lime)
Citrus aurantium (Sour orange)
Citrus grandus (Pummalo)
Citrus maderaspatana
Citrus medica (Citron)
Citrus medica (Fingered citron)
Citrus paradisi (Grape fruit)
Citrus pennisesiculata
Citrus sinensis (Sweet orange)
Emblica officinalis (Amla)
Eriobotrya japonica (Loquat or Japan plum)
Syzygium jambos (Rose apple)
Ficus carica (Fig)
Mangifera indica (Anacardiaceae)
Musa paradisiaca (Musaceae)
Cicca acida (Star gooseberry)
Punica granatum (Pomegranate)
Vitis vinifera (Grape wine)

C. Fruits mainly of the humid zones

Durio zibethinus (Durian)
Garcinia mangostana (Mango steen)
Lansium domesticum (Langsat)
Litchi chinensis (Litchi)
Monstera deliciosa (Monstera)
Nephelium lappacum (Rambutan)
Persea americana (Avocado pear)

D. Fruit trees of the hills

Annona cherimolia
Citrus grandis
Citrus limon
Citrus medica
Citrus medica var. *sarcodactylis*
Citrus paradisi
Citrus reticulata (Mandarin orange)
Cyphomandra betacea (Tree tomato)
Diospyros kaki (Persimmon)
Eugenia uniflora (Surinam cherry)
Feijoa sellowiana (Feijoa)
Fragaria vesca (Strawberry)
Mangifera indica

Musa paradisiaca
Morus alba (Mulberry)
Passiflora edulis (Passion fruit)
Physalis peruviana (Cape gooseberry)
Prunus armeniaca (Apricot)
Prunus domestica (European plum)
Prunus persica (Peach)
Prunus salicina (Japanese plum)
Psidium cattleianum (Strawberry guava)
Pyrus communis (Pear)
Malus sylvestris (Apple)
Rubus cerasus (Rasp berry)
Aleurites moluccana (Belgaum Walnut, Candlenut)
Anacardium occidentale
Arachis hypogaea (Ground nut)
Buchanania angustifolia (Buchnan)
Castanea sativa (Sweet or spanish chest nut)
Cola acuminata (Cola nut)
Corylus avellana Linn. (Indian hazel nut)
Buryale ferox (Gorgon nut)
Juglans regia (Wal-nut)
Pistacea vera (Pistachio-nut)
Prunus communis (Almond)
Terminalia catappa (Country almond)
Trapa bispinosa (Singara nut; filbertnut)

6. Beverages

Camellia sinensis (Tea) (Theaceae)
Coffee arabica (Arabian coffee, coffee) (Rubiaceae)
Coffea robusta (Robusta coffee)
Theobroma cacao Linn. (Cacao) (Sterendiaceae)

II. OIL YIELDING PLANTS

1. Plants yielding edible oils

Aleurites moluccana (Kakoona oil)
Anacardium occidentale (Kernal oil)
Arachis hypogaea (Ground nut oil)
Brassica campestris (Sarson or colza oil)
Brassica juncea (Mustard oil)
Buchanania lanzan (Chironji oil)
Carthamus tinctorius (Safflower)
Cocos nucifera (Coconut oil)
Elaeis guineensis (Palm oil)

Eruca sativa (Taramira)
Garcinia morella (Gamboge)
Gossypium arboreum (Cotton)
Guizotia abyssinica (Niger)
Helianthus annuus (Sunflower oil)
Juglans regia (Walnut oil)
Linum usitatissimum (Linseed oil)
Madhuca indica (Doli oil or Mahua butter)
Madhuca longifolia
Olea europaea (Olive oil)
Papaver somniferum (Poppy oil)
Pistacia vera Linn. (Pista oil)
Prunus armeniaca (Apricot oil)
Ricinus communis (Castor oil)
Schleichera oleosa (Macassar oil)
Sesamum indicum (Gingelly oil or Till oil)
Theobroma cacao (Cacao butter)

2. Plant yielding oil for illumination

Aleurites moluccana (Kakoona oil)
Aphanamis polystachya (Meliaceae)
Arachis hypogaea (Groundnut oil)
Argemone mexicana (Mexicin poppy seed oil)
Brassica campestris (Sarson or Colza oil)
Calophyllum inophyllum (Pinnay or Dumba oil)
Calophyllym elatum (Guttiferae)
Cannabis sativa (Hemp seed oil)
Carthamus tinctorius (Safflower)
Celastrus paniculatus (Black oil)
Cerbera manghas (Dabur)
Cocos nucifera Linn. (Coconut oil)
Eruca sativa Mill. (Brassicaeae)
Garcinia morella (Gambogei)
Gossypium arboreum Linn. (Cotton seed oil)
Guizotia abyssinica Cass. (Niger oil)
Helianthus annums (Sunflower oil)
Hibiscus cannabinus (Malvaceae)
Jatropha curcas Linn. (Purgin nut oil)
Linum usitatissimum (Linseed oil)
Madhuca butgracea (Doli oil)
Madhuca indica (Doli oil)
Madhuca longifolia (Doli oil)
Mesua ferrea (Guttiferae)
Pongamia pimata (Pongam oil)
Putranjiva roxburghii (Putranjiva)

Ricinus communis (Castor oil)
Schleichera oleosa (Macassar oil)
Sesamum indicum (Gingelly or Till oil)
Theobroma cacao (Cacao butter)
Vateria indica (Indian copal)
Xanthium strumarium (Asteraceae)

3. Plant yielding oil for soaps and candles

Arachis hypogaea (Groundnut oil)
Butyrospermum parkii (Shea butter)
Gossypium arboreum (Cotton seed oil)
Guizotia abyssinica (Niger oil)
Madhuca indica (Doli oil)
Olea europaea (Olive oil)

4. Plants yielding oil for varnishes and paints

Aleurites fordii (Euphorbiaceae) (Tung oil)
Aleurites montane (Euphorbiaceae) (China wood oil)
Cannabis sativa (Cannabinaceae) (Hempseed oil)
Carthamus tinctorius Linn. (Asteraceae) (Safflower oil)
Gossypium arboreum (Malvaceae) (Cotton seed oil)
Lium usitatissimum (Linaceae) (Linseed oil)
Nicotiana tabacum (Solanaceae) (Tobacco seed oil)
Papaver somniferum (Papaveraceae) (Poppy oil)
Sapium sebiferum (Euphorbiaceae) (Chinese tallow)

5. Plant yielding oil formedicinal purposes

Anacardium occidentale Linn. (Anacardiaceae) (Kernal oil)
Azadirachta indica (Meliaceae) (Neem oil)
Brassica juncea (Brassicaceae) (Mustard oil)
Calophyllum inophyllum (Guttiferae) (Keena tel)
Cocos nucifera (Arecaceae) (Coconut oil)
Croton tiglium (Euphorbiaceae) (Croton oil)
Diospyros peregrina (Ebenaceae) (Diospyros oil)
Guizotia abyssinica (Asteraceae) (Niger oil)
Gynocardia odorata (Flacourtiaceae) (Chaulmoogra oil)

Hydnocarpus laurifolia (Flacourtiaceae) (Soorty oil)
Jatropha curcas Linn. (Euphorbiaceae) (Purgingnut)
Kokoona zeylamica (Celastraceae) (Kokoon oil)
Madhuca longifolia (Sapotaceae) (Doli oil)
Olea europaea (Oleaceae) (Olive oil)
Ricinus communis (Euphorbiaceae) (Castor oil)
Sesamum indicum (Pedaliaceae) (Gingelly oil)

6. Essential oil yielding plants

Acorus calamus Linn. (Araceae) (Calamus oil)
Aquilaria agallocha (Thymelaceaceae) (Agar)
Bursera delpechiana (Bursereaceae) (Lignaloe or linaloe oil)
Cananga odorata (Annonaceae) (Cananga soil)
Carum carvi (Aplaceae) (Caraway oil)
Trachyspermum ammi (Apiaceae) (Ajowan oil)
Chenopodium ambrosioides (Chenopodiaceae) (wormseed oil or chinapodi oil)
Cinnamomum camphora (Lauraceae) (Camphor oil)
Citrus aurantium (Rutaceae) (Neroli oil)
Citrus aurantium burgtamia (Rutaceae) (Burgomet oil)
Citrus lemon (Rutaceae) (Lemon oil)
Coriandrum sativum (Apiaceae) (Coriander oil)
Cuminum cyninum (Apiaceae) (Cumin oil)
Cymbopogon citratus (Poaceae) (Lemon grass oil)
Cymbopogon flexuosus (Poaceae) (Malabar lemon grass or ginger grass oil)
Cymbopogon martini var. *sofia* (Poaceae) (Sofia oil)
Cymbopogon martini var. *motia* (Poaceae) (Matya oil)
Elletaria cardamomum (Zingiberaceae) (Cardamom oil)
Eucalyptus globulus (Myrtaceae) (Eucalyptus oil)
Gaultheria fragrantissima (Ericaceae) (Winter green oil)
Jasminum officinale (Oleaceae) (Jasminum oil)
Lavandula vera (Lamiaceae) (Lavender oil)
Melaleuca leucadendron (Myrtaceae) (Cajeput oil)

Mentha piperita (Lamiaceae) (Peppermint)

Mentha spicata Linn. (Spear mint)

Myristica fragrans (Mytisticaceae) (Nutmeg oil)

Majorana hortensis (Lamiaceae) (Origanum oil)

Pelargonium odoratissimum (Geraniaceae) (Geranium oil)

Pimenta racemosa (Myrtaceae) (Bay oil)

Pimpinella anisum (Apiaceae) (Aniseed oil)

Podostemon heyneanus (Lamiaceae) (Pacholi)

Rosa damascena (Rosaceae) (Rose oil or Attar)

Santalum album (Santalaceae) (Sandal wood oil)

Thymus vulgaris (Lamiaceae) (Thymus oil)

Trachyspermum averi (Apiaceae) (Ajoran oil)

Vetiveria zizanioides (Poaceae) (Vetiver or khurkhus oil)

Zingiber officinale (Zingiberaceae) (Ginger oil)

Plants yielding sugars and starches

Acer negundo (Aceraceae) (Sugar maple of nebraska)

Acer rubrum (Aceraceae) (Swamp maple of Pennsylvania)

Acer saccharinum (Aceraceae) (The sugar made of Northern States and California)

Beta vulgaris (Chenopodiaceae) (Sugar beet)

Borassus flabellifer (Arecaceae) (Palmyrah)

Caryota arens (Arecaceae) (Sagopalm of India; Palm sugar of Southern Ceylon)

Cocos nucifera (Arecaceae) (Coconut palm)

Helianthus tuberosus (Asteraceae)

Madhuca indica (Sapotaceae)

Phoenix sylvestria (Arecaceae) (Date palm)

Saccharum officinarum (Poaceae) (Cane sugar)

Sorghum vulgare var. saccharatum (Poaceae) (Sargo or Chinese sorghum)

II. PLANTS YIELDING STARCHES

Aesculus indica (Hippocastanaceae) (Indian Horse chestnut)

Alocasia indica (Araceae) (Giant taro)

Amorphophallus campanulatus (Araceae) (Elephant-foot yam)

Arenga pinnata (Aracaceae) (Sago palm of Malaya)

Arracacia esculenta (Apiaceae) (Peruvian carrot)

Avena sativa (Poaceae)

Canna orientalis (Cannaceae) (Indian shot)

Caryota urens (Arecaceae)

Colocasia antiquorum (Araceae)

Commelina benghalensis (Commelinaceae)

Curcuma angustifolia (Zingiberaceae) (Wild or East Indian arrow root)

Dioscorea alata (Dioscoreaceae)

Eulophia campestris (Orchidaceae) (Salep)

Fagopyrum esculentum (Polygonaceae) (Buck wheat)

Helianthus tuberosus (Asteraceae) (Jerusalem artichoke)

Hordeum vulgare Linn. (Poaceae)

Ipomoea batas (Convolvulaceae)

Manihot esculenta (Euphorbiaceae)

Maranta arundinacea (Marantaceae) (Arrowroot)

Metroxylon rumphii (Arecaceae) (Sago starch)

Metroxylon sagus (Arecaceae) (Sago starch)

Musa paradisiaca (Banana or plantain)

Nelumbo nucifera (Nymphaeaceae) (Sacred lotus)

Oryza sativa (Poaceae)

Pueraria tuberosa (Fabaceae)

Solanum tuberosum (Solanaceae) (Potato)

Tacca pinnatifeda (Taccaceae) (South Sea arrow root)

Triticum aestivum (Poaceae)

Zea mays (Poaceae)

IV. FIBRES

A. Plant yielding fibres-grouped on morphological basis of the fibres

1. Fibre from vegetable hair

Salmalia malabarica (Bombacaceae) (Red silk cotton)

Ceiba pentandra (Bombacaceae) (Kapok)

Gossypium arboreum (Malvaceae)

2. Bast fibres

Abutilon indicum (Malvaceae) (Indian mallow)
Boehmeria nivea (Urticaceae) (Ramie)
Cannabis sativa (Cannabinaceae)
Corchorus capsularis (Tiliaceae) (Jute)
Corchorus olitorius (Tiliaceae) (Jute)
Crotalaria juncea (Fabaceae) (Sunnhemp)
Helicteres isora (Sterculiaceae) (Screw tree)
Hibiscus cannabinus (Malvaceae)
Hibiscus sabdariffa (Malvaceae)
Hibiscus tiliaceus (Malvaceae)
Linum usitatissimum (Linaceae) (Flax)
Sesbania speciosa (Fabaceae)
Sida rhombifolia (Malvaceae)
Urena lobata (Malvaceae) (Cadillo)

3. Leaf fibres

Agave sisalana (Amaryllidaceae) (Sisal fibre)
Agave vera (Amaryllidaceae) (Railway aloe)
Agave wightiana (Amaryllidaceae)
Boehmeria nivea (Urticaceae) (Ramie)
Cannabis sativa (Cannabinaceae) (Hemp)
Musa textilis (Musaceae) (Manilla hemp)
Sansevieria spp. (Agavaceae) (Bow string hemp)
Yucca gloriosa (Liliaceae)

3. Fibre for brush making

Agave sisalana (Amaryllidaceae)
Borassus flabellifer (Arecaceae)
Caryota urens Linn. (Arecaceae)
Cocos nucifera Linn. (Arecaceae)
Luffa cylindirica (Cucurbitaceae) (Vegetable sponge)
Pandanus tectorius (Pandanaceae) (Screw pine)
Phoenix sylvestris (Arecaceae)

4. Stuffing fibres

Salmalia malabarica (Bombacaceae)
Calotropis gigantea (Asclepidadaceae) (Madar)
Ceiba pentandra (Bombacaceae)
Cochlospermum religiosum (Cochlospermaceae) (Silk cotton)
Cryptostegia grandiflora (Asclepiadaceae) (Rubber vine)

Aerva tomentosa (Amaranthaceae) (Javanese wool plant)
Cocos nucifera (Arecaceae)
Gossypium arboreum (Mlvaceae)
Typha angustata (Typhaceae) (Cat's tails)

5. Matting fibres

Cyperus corymbosus (Cyperaceae) (Madras mat or Chinese mat grass)
Cyperus madaccensis (Cyperaceae)
Cyperus pangorei (Cyperaceae) (Madras mat)
Pandanus tectorius (Pandanaceae)
Typha angustata (Typhaceae)
Calamus rotang (Arecaceae) (Cane palm)

V. MEDICINAL PLANTS, NARCOTICS AND MASTICATORIES

A. Medicinal plants

Abrus precatorius (Fabaceae) (Indian liquorice)
Abutilon indicum (Malvaceae) (Country mallow)
Acacia arabica Willd. (Mimosaceae) (Gum arabic kikar)
Acalypha indica (Euphorbiaceae) (Indian acalypha)
Achyranthes aspera (Amaranthaceae) (Prickly chaff)
Aconitum ferox (Ranunculaceae) (Indian aconita)
Adansonia digitata (Bombacaceae) (Monkey bread tree)
Adhatoda vasica (Acanthaceae) (Vasaka)
Aegle marmelos (Rutaceae) (Bilva)
Alangium salvifolium (Alangiaceae) (Sage leaved alangium)
Allium sativum (Liliaceae) (Garlic)
Aloe barbadensis (Agavaceae) (Indian aloe)
Alstonia scholaris (Apocyanaceae) (Ditabark)
Alternanthera toriandra (Amaranthaceae)
Anacardium occidentale (Anacardiaceae)
Anamirta cocculus (Menispermaceae) (Fish berry)
Andrographis paniculata (Acanthaceae) (Great chirata)
Argemone mexicana (Papaveraceae) (Mexican poppy)

Aristolochia bracteata (Aristolochiaceae) (Brachtiated birthwort)

Aristolochia indica (Aristolochiaceae) (Indian birthwort)

Asparagus racemosus (Liliaceae)

Asteracantha longifolia (Acanthaceae)

Atalantia monophylla (Rutaceae) (Wild lime)

Atropa belladona (Solanaceae) (Deadly nightshad; Belladonna)

Madhuca butyracea (Sapotaceae) (Indian butter tree)

Berberis aristata (Berberidaceae) (Barberry)

Borhaavia diffusa (Nyctaginaceae) (Spreading hogweed)

Brassica nigra (Brassicaceae) (Black mustard)

Bursera delpechiana (Bursaraceae) (Linaloe)

Caesalpinia crista (Caesalpiniaceae) (Physic nut)

Calonyction bononox (Convoluvlaceae) (Moon flower)

Calophyllum inophyllum (Guttiferae) (Alexandrian lamel)

Calotropis procera (Asclepiadaceae) (Giant swallow-wort)

Cannabis sativa (Cannabinaceae)

Trachyspermum ammi (Amaranthaceae)

Cassia alata (Leguminosae) (Ringworm shrub)

Cassia angustifolia (Caesalpiniaceae) (Tinnevelly sema)

Cassia auriculata (Caesalpiniaceae) (Tanner's cassia)

Cassia nigricans (Caesalpiniaceae)

Cassia tora (Caesalpiniaceae)

Cedrus deodara (Pinaceae) (Cedrus pine)

Centratherum anthelminticum (Asteraceae) (Purple fleabane)

Cephaelis ipecacuanha (Rubiaceae) (Ipecacuanha)

Chenopodium anthelmintica (Chenopodiaceae) (Worm seed)

Cinnamonum zeylamicum Blume (Lauraceae) (Cinnamomum)

Cissus quadrangularis (Vitaceae) (Adamant creeper)

Citrullus colocynthis (Cucurbitaceae) (Colocynth)

Cleome icosandra (Capparidaceae)

Clerodendron inerme (Verbenaceae) (Smooth volkamenia)

Clerodendron serratum (Verbenaceae) (Beetle killer)

Clitoria ternatea (Fabaceae) (Butterfly pea)

Coccinium fenestratum (Menispermaceae) (Tree turmeric)

Crocus sativus (Iridaceae) (Saffron)

Croton tiglium (Euphorbiaceae)

Curcuma aromatica (Zingiberaceae) (Wild turmeric)

Delonix elata (Caesalpimiaceae) (Tiger bean)

Dodonaea viscosa (Sapindaceae) (False bog myrtle)

Eclipta alba (Asteraceae)

Emblica fischerii (Euphorbiaceae) (Myrobalamemblic)

Emblica officinalis (Euphorbiaceae)

Enicostemma littorale (Gentianaceae) (Indian gentian)

Eucalyptus globulus (Myrtaceae)

Euphorbia tirucalli (Euphorbiaceae)

Evolvulus alsinoides (Convolvulaceae)

Feronia limonica (Rutaceae) (Wood apple)

Ficus racemosa (Moraceae) (Insect fig)

Ficus hispida (Moraceae) (Devil fig)

Foeniculum vulgare (Apiaceae) (Fennel)

Garcinia indica (Guttiferae) (Cocum)

Gardinia gemmifera (Rubiaceae) (Gummy cape jasmine)

Gentiana lutea (Gentianaceae) (Common gentian)

Gloriosa superba (Liliaceae)

Glycosmis pentaphylla (Rutaceae)

Glycyrrhiza glabra (Fabaceae) (Liquorice)

Gmelina arborea (Verbenaceae)

Gmelina asiatica (Verbenaceae) (Small cashmeri tree)

Gnetum scandens (Gnetaceae, Gnetales) (Joint fir)

Gracilaria lichenoides (Rhodophyceae, algae) (Agar agar)

Grangea maderaspatana (Asteraceae)

Guizotia abyssinica (Asteraceae) (Niger)

Gymnema sylvestre (Asclepiadaceae) (Small Indian ipecacuanha)

Cleome gynadra (Capparidaceae)

Gynocardia odorate (Flacourtiaceae)

Hackelochloa granularis (Poaceae)

Helicteres isora (Sterculiaceae) (Indian screw tree)

Heliotropium indicum (Boraginaceae) (Scorpion tail)

Helleborus niger (Ranunculaceae) (Black hellebore)

Hemidesmus indicus (Asclepiadaceae) (Indian sarasaparilla)

Hibiscus rosa-sinensis (Malvaceae) (Shoe flower)

Holarrhena antidysenterica (Apocyanaceae)

Hydnocarpus kuzzii (Flacourtiaceae)

Hyoscyamus niger (Solanaceae) (Henbane)

Ichnocarpus frutescens (Apocyanaceae) (Slander fruited wip[er dogbane)

Exogonium purga (Convolvulaceae) (Jalap)

Kaempferia galanga (Zingiberaceae) (Galanga)

Kedrostis rostata (Cucurbitaceae)

Kokoona zeylanica (Celastraceae) (Kokoon oil plant of Singhalese)

Leucas aspera (Lamiaceae)

Lippia nodiflora (Verbenaceae)

Martynia annua (Pedaliaceae) (Tiger's claw)

Mimosa pudica (Mimosaceae) (Sensitive plant)

Mimusops elengi (Sapotaceae) (West Indian meddlar)

Mimusops hexandra (Sapotaceae) (Indian ape flower)

Mirabilis jalapa (Nyctaginaceae) (Four O'clock plant)

Mitragyna parvifolia (Naneleaceae)

Mollugo peruviana (Ficoidaceae)

Myristica fragrans (Myristicaceae) (Nutmeg)

Myrtus communis (Myrtaceae) (Common myrtle)

Nardostachys jatamansi (Valerianaceae) (Indian spikenard)

Naregamia alata (Meliaceae) (Goanese ipecacuanha)

Nasturtium fontaneana (Brassicaceae) (Water cress)

Nephelium lappaceum (Sapindaceae) (Rambuttan)

Neptunia oleracea (Fabaceae)

Nigella sativa (Ranunculaceae) (Small fennel)

Notonia grandiflora (Asteraceae) (Common fleshy regweed)

Nymphaea pubescens (Nymphaeaceae) (Water lily)

Ochna squarrosa (Ochnaceae) (Golden champak)

Ocimum basilicum (Lamiaceae) (Basil)

Ocimum gratissimum (Lamiaceae) (Shrubby basil)

Ocimum sanctum (Lamiaceae) (Sacred basil)

Oenothera biennis (Onagraceae) (Evening primrose)

Ophiorrhiza mungos (Rubiaceae) (Mangoose plant)

Oroxylum indicum (Bingnoniaceae) (Indian trumpet flower)

Oxalis corniculata (Oxalidaceae) (Indian sorrel)

Oxystelma esculentum (Asclepiadaceae) (Cotton milk plant)

Pedalium murex (Pedaliaceae)

Pergularia daemia (Asclepiadaceae) (Hedge cotton)

Peristrophe bicalyculata (Acanthaceae)

Anethum sowa (Apiaceae)

Phyllanthus niruri (Euphorbiaceae)

Pimpinella anisum (Apiaceae)

Piper cubeba (Piperaceae) (Cubeb)

Piper longum (Piperaceae) (Long pepper, Water soldier)

Plantago ovata (Plantaginaceae) (Isphagul)

Plumbago zeylanica (Plumbaginaceae) (White flowered leadwort)

Plumeria rubra Linn. var. *acutifolia* (Apocyanaceae) (Pagoda tree)

Premna serratifolia (Verbenaceae)

Psoralia corylifolia (Fabaceae)

Pterocarpus marsupium (Fabaceae) (Bijasar)

Punica granatum (Punicaceae)

Quercus infectoris (Fagaceae) (Gali)

Rauwolfia serpentina (Apocyanaceae) (Common snake dogbane) (Sarpagandha)

Rejoua dichotoma (Apocynaceae)

Rhinacanthus nasutrus (Acanthaceae) (Snake jasmin)

Rhizophora mucronata (Rhizophoraceae) (Mangroove)

Rubia cordifolia (Rubiaceae) (Indian madder)

Ruta graveolans (Rutaceae) (Garden rue)

Salicornia brachiata (Chenopodiaceae) (Crab or frog grass)

Sansevieria roxburghiana (Agavaceae) (Bow string hemp)

Saussurea lappa (Asteraceae) (Costus)

Scilla indica (Liliaceae) (Indian squill)

Semecarpus anacardium (Anacardiaceae) (Marking nut free)

Sida cordifolia (Malvaceae) (Kungyi)

Smilax ferox (Liliaceae) (China root)

Smilax officinalis (Liliaceae) (Sarasaparilla)

Smilax zeylanica (Liliaceae)

Solanum nigrum (Solanaceae) (Black night shade)

Solanum xanthocarpum (Solanaceae)

Sphaeranthus indicus (Asteraceae)

Spilanthes acmillar (Asteraceae)

Sterculia foetida (Sterculiaceae) (Foetid tree)

Streblus asper (Moraceae) (Demon tree)

Swertia chirata (Gentianaceae) (Chirata)

Symplocos racemosa (Symplocaceae) (Losh tree)

Tamarindus indica (Caesalpiniaceae) (Tamarind)

Tamarix aphylla (Tamaricaceae)

Taraxacum officinale (Asteraceae) (Dandelion)

Terminalia arjuna (Combretaceae) (White winged myrobalan)

Terminalia belerica (Combretacae) (Bellric myrobalan)

Terminalia chebula Retz. (Combretaceae) (Chebulic myrobalan)

Tinospora corsifolia (Menispermaceae) (Moon creeper)

Toddalai asiatica (Rutaceae) (Forest pepper)

Tragia involucrata (Euphorbiaceae) (Climbing nettle)

Trianthema decandra (Ficoidaceae)

Tribulus terrestris (Zygophyllaceae) (Small caltrops)

Trichodesma indicum (Boraginaceae)

Trichosanthes cucumerina (Cucurbitaceae)

Trichosanthes bracteata (Cucurbitaceae)

Trigonella foenumgraecum (Fabaceae) **(Fenugreek)**

Triumfetta rhomboidea (Tiliaceae) (Parroquetiburr)

Tylophora indica (Asclepiadaceae) **(Emetic swallor wort)**

Uvaria lagopoides (Fabaceae) (Sweet smelling harefoot)

Urginea indica (Liliaceae) (Indian squill)

Vanda roxburghii (Orchidaceae) (Vanda)

Vangueria spinosa (Rubiaceae) (Epinous honey thorn)

Vateria indica (Dipteroocarpaceae) (White dammar)

Ventilago madraspatana (Rhamnaceae) (Buck thorn climber)

Vernonia cinerea (Asteraceae) (Ash coloured fleabane)

Vetiveria zizamoides (Poaceae)

Vitex negundo (Verbenaceae) (Negundo)

Walsura piscidia (Meliaceae) (Ochre flowered fish poison cedar)

Withania somnifera (Solanaceae) (Aswagandha)

Xanthium strumarium (Asteraceae) (Burr weed)

Zingiber cassumnar (Zingiberaceae)

Zingiber officinale (Zingiberaceae) **(Ginger)**

Zinziber zerumbet (Zingiberaceae)

B. Narcotics

Cannabis sativa (Cannabinaceae)

Datura stramonium (Solanaceae)

Erythroxylum coca (Erythroxylaceae)

Hyoscyamus niger (Solanaceae)

Amanita muscaria (Poisonous mushroom) **(Fungi)** (Fly agaric of Europe, Asia and America)

Banisteriopsis caapi (Malpighiaceae) **(Cappi** of Brazil)

Lophophora williamsii (Caclaceae) **(Reyote or** mescal button of Mexico)

Rivea corymbosa (Convolulaceae) **(Olaluigui** of Mexico)

C. Masticatories

Areca catechu (Arecaceae) **(Arecanut)**

Nicotiana tabacum (Solanaceae) **(Tobacco)**

Piper betle (Piperaceae) **(Betel leaf)**

VI. SPICES, CONDIMENTS AND SEASON-ING HERBS

Allium cepa (Liliaceae)

Allium sativum (Liliaceae)

Allium schoenoprasum (Liliaceae) (Chivas)

Alpinia galanga (Zingiberaceae)

Apium graveolens (Apiaceae)

Brassica juncea (Brassicaceae)

Capsicum annum (Solanaceae)

Carum carvi (Apiaceae) (Caraway)

Trachyspermum ammi (Apiaceae)

Carum nigrum (Apiaceae) (Black caraway)

Cinnamonum zeylanicum (Lauraceae)

Coriandrum sativum (Apiaceae)

Crocus sativus (Iridaceae) (Saffron)

Cuminum cyminum (Apiaceae)

Curcuma longa (Zingiberaceae) (Turmeric)

Cucurma zedoaria (Zingiberaceae) (Zerumbet)

Elettaria cardamomum (Zingiberaceae)

Syzygium aromaticum (Myrtaceae)

Foeniculum vulgare Mill. (Apiaceae)

Hemidesmus indicus (Asclepiadaceae) (Sarasaparilla)

Mentha arvensis (Lamiaceae)

Mentha piperata Linn. (Lamiaceae) (Peppermint)

Mentha spicata (Spearmint)

Murraya koenigii (Rutaceae) (Curry leaves)

Myristica fragrans (Myristicaceae)

Nigella sativa (Ranunculaceae)

Papaver somniferum (Papaveraceae)

Petroselinum crispum (Apiaceae) (Parsley)

Anethum sowa (Apiaceae) (Dill)

Pimenta officinalis (Myrtaceae) (Allspice)

Pimpinella anisum (Apiaceae) (Aniseed)

Piper nigrum (Piperaceae) (Pepper)

Quercus infectoria (Fagaceae)

Rosmarinus officinalis (Lamiaceae) (Rosemary)

Salvia officinalis (Lamiaceae) (Sage4)

Trigonella foenum-graecum (Fabaceae)

Vanilla planifolia (Orchidaceae) (Vanilla)

Zingiber officinale (Zingiberaceae)

VII. PLANTS YIELDING RUBBER, TANNINS, DYES, ETC.

1. Rubber yielding plants

Cryptostegia grandiflora (Asclepiadaceae)

Castilloa elastica (Moraceae) (Panama or Ula rubber)

Ficus elastica (Moraceae) (Indian rubber)

Funtumia elastica (Apocynaceae) (Lagos rubber)

Heavea brasiliensis (Euphorbiaceae) (Pararubber)

Manihot glazionii (Euphorbiaceae) (Ceararubber)

Manihot dichotoma (Euphorbiaceae) (Jequie Manicoba rubber)

Manihot piauhyensis (Euphorbiaceae) (Remano Manicoba rubber)

Manihot heptaphylla (Euphorbiaceae) (Manicoba rubber)

Parthenium argentatum (Asteraceae) (Guayule)

Raphionaeme utilis (Periplocaceae) (Bitinga rubber)

Taraxacum officinale (Asteraceae) (Dandilion)

2. Plants useful for tanning purposes

A. Bark tans

Acacia arabica (Minosaceae) (Black babul)

Acacia decurrens (Mimosaceae) (Green wattle)

Acacia decurrens var *dealbata* (Mimosaceae) (Silver wattle)

Acacia decurrens var. *mollis* (Leguminosae) (imosaceae) (Black wattle)

Acacia pycnantha (Mimosaceae) (Goldaowattle)

Acacia leucophloea (Mimosaceae) (White babul)

Cassia auriculata (Caesalpiniaceae) (Tanner's cassia)

Cassia fistula (Caesalpiniaceae) (Indian lamburnum)

Ceriops tagal (Rhizophoraceae) (Mangrove-Kandal)

B. Fruit tans

Caesalpinia coriaria (Caesalpiniaceae)

Caesalpinia digyna (Caesalpiniaceae)

Terminalia chebula Retz. (Combretaceae)

Zizyphus xylopyrus (Rhamnaceae) (Woody fruited jijibe)

C. Leaf tans

Anogeissus latifolia (Combretaceae) (Button tree)

Emblica officinalis (Euphortriaceae)

Lawsonia inermis (Lythraceae)

*Uncaria gambir*Roxb. (Rubiaceae) (Gambier)

3. Plant yielding dyes

A. Wood dyes

Acacia catechu (Mimosaceae) (Catechu)

Artocarpus heterophyllus (Moraceae) (Jack fruit)

Caesalpinia sappan Linn. (Caesalpiniaceae) (Sappan)

Mahonia leschenaultii (Berberidaceae) (Holly-leaved berbery)

Morinda tinctoria (Rubiaceae) (All dye, Marinda)

Pterocarpus santalinus (Fabaceae) (Santalin)

Rubia cordifolia (Rubiaceae)

B. Bark dyes

Acacia arabica (Minaosaceae)

Acacia leucophloea (Mimosaceae)

Cassia auriculata (Caesalpimiaceae)

Erythrina variegata (Fabaceae)

Garcinia tinctoria (Guttiferae) (Egg tree)

Morinda tinctoria (Rubiaceae)

C. Root dyes

Mallotus philippinensis (Euphorbiaceae) (Kamala tree)

Morinda citrifolia (Rubiaceae)

Morinda tinctoria (Rubiaceae)

Oldenlandia umbellata (Rubiaceae) (Chayroot)

D. Flower dyes

Butea monosperma (Fabaceae) (Flame of the forest)

Cedrella tonna (Meliaceae) (Singapore cedar)

Chloroxylon swietenia (Rutaceae) (Satin wood)

Erythrina variegata (Fabaceae)

Questions

Give an account of drugs obtained from plants.

Assume that you and a small group of your friends are transported to an uninhabited island in the Indian Ocean to live and that you are permitted to take the seeds or propagative organs of 12 species of plants with you. List them in order of preference (mention the botanical name and family) and give reasons for your selection.

By means of outline sketches only, show the morphology of the economically important of any five of the following plants : saffron, opium, poppy, cocoa, black pepper, citrus, coffee, and clove; give the botanical name of each source and comment on the uses.

State the botanical names, families and morpholoy of the parts from which the following are obtained : (a) cotton, (b) jute, (c) coir, (d) hemp. How would you differentiate one from the other?

Name five important drugs plants of India and write detailed account of any two of them.

List the chief fatty oil-yielding crops in India. Classify the oils on the basis of their drying properties and mention their uses. What do you mean by the term hydrogenation.

By simple sketches only, show the morphology of the economically important parts of the following : clove, cashew, coriander and black pepper. Mention their latin names and uses.

What are essential oils? Give a brief account of the botany, cultivation, extraction and uses of important essential oils produced in India.

Give the method of extraction, properties and economic uses of the following : (a) groundnut oil, (b) linseed oil and (c) castor oil.

Name any four fibre yielding plants, giving their families. Describe the methods of fibre extraction from them and their commercial uses.

Give the methods of cultivation of any one of the following : (a) cotton, (b) rubber.

Explain how the latex is processed to get rubber?

The whole coconut tree is economically useful. Justify the above statement.

Write what do you know of the source and economic importance of the cork.

Enumerate various conventional and non-conventional methods employed for the improvement of cereals. List the high yielding varieties produced in India.

Give an account of the origin, botany, regions of cultivation, nutritional value and improvement of potato.

Write notes on :

(i) Rubber cultivation and its extraction

(ii) Improvement of sugarcane

(iii) Drug-yielding plants

(iv) Morphology and uses of pepper or clove

(v) Name any three plants which are the sources of economically important products. Name these products and their uses.

(vi) Which legume do you regard as nutritionally superior? List points in support of your answer.

Write what do you know of the source and economic importance of coconut.

List the botanical names of six plants from Cruciferae (Brassicaceae) used as vegetables and mention the morphology of the useful part in each.

Name the species of cottons exploited for commercial purposes. How would you distinguish between the Old World and New World cottons?

How is opium obtained and how is it used?

Write note on the following :

(a) *Oryza*

(b) *Rauwolfia*

(c) *Cannabis*

Give botanical names of three important timber-yielding plants of India.

Name the sources of any one of the following products; mention the plant part from which they are obtained and indicate the main areas of cultivation of these plants in India.

(a) Ephedrine

(b) Tobacco

(c) Clove

(d) Saffron

Describe the medicinal importance of *Adhatoda*, *Zingiber* and *Penicillium*.

What is the morphology of the edible part in (i) orange, (ii) potato, (iii) papaya?

Distinguish between any one pair of the following :

(a) Ring porous and diffuse porous.

(b) Sap wood and heart wood.

(c) Beedi and cigarette

(d) Green tea and black tea

(e) Lint and fuzz

Name a plant each yielding a worldwide commodity which owes its origin to (i) epidermis, (ii) tuberous stem, (iii) periderm, and (iv) whole leaf.

What plant materials (mention Latin names) are used in the manufacture of essential oils.

Write short note on the following :

(a) Indigenous raw materials of paper production

(b) Wood seasoning

(c) Molasses

(d) Papain

How would you distinguish of the following pairs :

(a) *Curcuma* from *Zingiber*

(b) Starch of wheat from that of rice and maize

(c) Essential oil from fatty oil .

(d) Fibres of cotton from those of jute

Give the botanical names, morphological nature of plant part/s yielding economic products, and uses of the following :

(a) Sugarcane

(b) Jute

(c) Potato

(d) Wheat

Write botanical names of any six plants in which either the swtem or the leaf is of economic importance.

Explain giving suitable reasons for the following.

Farmers generally use leguminous plants, with modulated roots, for rotation of crops.

Write briefly : Extraction of essential oils from ose and Sandal wood.

In India cotton textile industry is largely continued to Bombay and other parts of Western India, whereas jute industry is mostly restricted to Calcutta and other parts of West Bengal. Comment.

Write short note on the following :

(a) Bye-products of sugar industry

(b) Extraction of essential oils

(c) Important Indian spices

Cannabis sativa hits the front lines in many of the leading magazines of the world. Comment.

Write a short note on Soybean.

Distinguish between the following pairs :

(a) Pine wood and teak wood

(b) Fennel and Coriander

(c) Essential oil and fatty oil

(d) Flue curing and sun curing

Name the principal states in India in which the following are cultivated :

(i) Jute

(ii) Coconut

(iii) Tea

(iv) Mango, and

(v) black peppers

How does milling affect the quality of rice.

Name drugs of plant origin which could be used in the following : Malaria, hypertension, rheumatism, heart disorders and dysentery.

Write what you know of the source and economic importance of Ginger.

Write briefly : Tobacco and health hazards.

Write briefly : Harvesting, grading and processing of tea.

Write briefly : Origin of wheat.

Write briefly : Principal uses of wood.

Write briefly : Extraction of sugar and by-products of sugar industry.

What are the principal sources of vegetable oils in India. List the four major uses of oils.

Give the botanical name of the plant yielding any one of the following product and explain its morphology and economic importance :

(a) Raserpine

(b) Coffee

(c) Rose-wood

Name a useful plant product that is derived from (i) epidermis, (ii) whole leaf, (iii) mesocarp, (iv) periderm, (v) tuberous root, and (vi) heart wood.

Mention some algae as source of food to human beings.

Mention the importance of Gramineae (Poaceae) as a source of food.

Write notes on : Economic uses of tannins.

Write any three economic uses of Euphorbiaceae members.

Outline the processing of coffee by wet method.

What kind of soil will be preferred for growing rubber.

How will you start to process the Paddy grains soon after harvesting them (Mention only the first two steps).

Mention 5 rubber yielding plants and their families, cultivated in India.

The natural habit of tea is tree but under cultivation it is a shrub. How is it so?

Mention the other uses of Paddy, other than the grains being used as a staple food.

Give the technical name of the fruit and mention the morphology of the useful parts and their actual use in the fruits of the following : (a) Mango, (b) Cashew, and (c) Jack fruit.

State what you know about the morphological nature of the following : (i) Apple, (ii) Cauliflower, (iii) Radish.

Give the technical name of the fruit and mention the morphology of the edible portion of the following : (i) Tomato, (ii) Rice, (iii) Coco, (iv) Apple and (v) Groundnut.

Which is the active principle in coffee?

How is the mesocarp of coconut fruit used in industry?

Name any two economic uses of cotton other than using it in textile industry.

Arrange the following in the order of their food value : tapioca, groundnut, soyabean, potato, rice and *Citrus*.

Name five plant products which earn foreign exchange for our country.

Mention the titles and names of authors of four books on economic botany.

What is the nature of the chief food reserve in (i) tapioca, (ii) mango, (iii) soyabean, (iv) maize.

Which of the Indian states are the leading producers for Cane sugar, Sandal wood, Potato, Jute and Teak.

What are the 'active' chemical principles in the following : Coffee, Ergot, Opium, Rauwolfia and Tobacco?

What plant materials (mention Latin names) are used in the manufacture of the following :

(a) Sugar

(b) Spice powder for flavouring foods

(c) Paper

(d) Soap

Name the Research Institute along with its location where improvement work in any one of the following crop plants is being carried out in India :

(a) Jute

(b) Wheat

(c) Sugarcane

(d) Rice

(e) Potato

Give the botanical name from which the following is obtained :

(a) Caffeine

(b) Papain

(c) Morphine

(d) Hashish

Which of the Indian states are leading producers of the following :

(a) Tea

(b) Tobacco

(c) Rubber

(d) Banana

(e) Belladonna

(f) Deodar

What is the centre of the origin of the following:

(a) Coffee

(b) Cocoa

(c) Clove

(d) Cinchona

(e) Groundnut

(f) Black pepper

(g) Eucalyptus

(h) Mango

(I) Red pepper

(j) Soybean

Mention one important source (botanical name) for the following :

(a) Atropine

(b) Cigar

(c) Match sticks

(d) Tapioca

How is hydrogenation of oil done?

In what ways does paddy differ from rice?

Where did maize originate?

What makes a wood heavy?

Give the botanical name and the morphology of the product used in :

(a) Cinchona

(b) Groundnut

(c) Rubber

(d) Sorghum

(e) Cardamum

(f) Caster

Describe how *Cichorium intybus* is used as an economically important product.

Name any two fibre-yielding plants.

Give the botanical name which yields the following : (a) Rose wood, (b) Satin wood, (c) Sandal wood, (d) Laurel wood.

Name two essential oil-yielding plants.

Describe the morphology of the useful part in: (a) Ganja, (b) Taro, (c) Cardamom, (d) Groundnut, (e) Pepper, (f) Coriander, (g) Cloves.

Name the plant which yields quinine. How is it extracted?

Give the botanical name for the following : (a) Bengal gram, (b) Green gram, (c) Black gram, (d) Red gram.

Give the botanical name of the following : (a) The plant produces Aconitum, (b) The plant produces cabbage, (c) The plant produces Melia oil (d) The plant produces Bengal gram.

Name a plant for which desuckering is done.

Mention one cultivated species and one wild species of sugarcane.

Give the use of bagasse.

What is meant by retting?

Which plant produces chicory?

Give the name of the plant which produces coffee.

Describe the morphology of the useful parts of cotton.

Give the names of any two major rice research stations, their location in India.

Describe the morphology of the useful part in groundnut and cotton.

Arrange in order of highest protein value : (a) Soybean, (b) Potato, (c) Rice, (d) Wheat, (e) Sugarcane.

Give one example each of plant/plant-products used in the manufacture of following : (a) Macroni, (b) rum, (c) morphine, (d) haematoxylin.

MULTIPLE CHOICE

1. Botanical name of jute is :
   (a) *Cannabis sativa*
   (b) *Linum usitatissimum*
   (c) *Corchorus capsularis*
   (d) None of the above

2. Teak wood which is obtained from *Tectna grandis* belongs to the family of :
   (a) Labiatae (Lamiaceae)
   (b) Verbenaceae
   (c) Rubiaceae
   (d) Apocynaceae

3. digitalis purpurea belongs to the family of :
   (a) Papilionaceae
   (b) Acanthaceae
   (c) Scrophulariaceae
   (d) Solanaceae

4. Sunhemp is obtained from :
   (a) *Crotalaria juncea*
   (b) *Cannabis sativa*
   (c) *Musa textilis*
   (d) *Cocos nucifera*

5. Cloves are obtained from *Eugenia aromatica* and the morphology of the product being :
   (a) product petioles

    (b) dried pedicels

    (c) dried flower buds

    (d) dried seeds

6. Botanical name of the plant that yields pararubber :

    (a) *Ficus elastica*

    (b) *Hevea brasiliensis*

    (c) *Manihot glaziovii*

    (d) *Castilla elastica*

7. Reserpine, a drug is extracted from :

    (a) *Rauwolfa serpentina*

    (b) *Ferula asafoetida*

    (c) *Atropa belladona*

    (d) *Digitalis purpurea*

8. Plant parts useful for extraction of opium from *Papaver somniferum* are :

    (a) young seedlings

    (b) old leaves

    (c) unripe fruits

    (d) riened seeds

9. Botanical name of cauliflower is :

    (a) *Brassica oleracea* var. *botrytis*

    (b) *Brassica oleracea* var. *gongylodes*

    (c) *Brassica oleracea* var. *capitata*

    (d) *Brassica oleracea* var. *gemmifera*

10. The common name of *Eleusine coracana* is:

    (a) Ragi

    (b) Barley

    (c) Wheat

    (d) Oats

11. The cotton fibre from cotton plant is obtained from :

    (a) roots

    (b) stems

    (c) seeds

    (d) leaves

12. The commercial jute fibres are :

    (a) Phloem fibre

    (b) Xylem fibres

    (c) Interxylary fibres

13. Cotton fibre is derived from :

    (a) Phloem fibres

    (b) Epidermal hairs on seed

    (c) Outgrowth on the stem

    (d) Sclerenchymatous cells

14. Paddy is suitable for cultivation in :

    (a) Red soils

    (b) Dry soils

    (c) Black soils

    (d) Irrigated soils

15. The economic product of tobacco plant is :

    (a) Flowers

    (b) Leaves

    (c) Roots

    (d) Stems

16. Fibre yielding plant is :

    (a) *Triticum*

    (b) *Pennisetum*

    (c) *Gossypium*

    (d) *Rauwolfia*

17. The quality of tobacco depends mainly on:

    (a) Curing process

    (b) Variety and curing process

    (c) Climatic conditions in which plant grows

    (d) Nutrition of the plant in the field

18. Opium is obtained from :

    (a) dried leaves

    (b) roots

    (c) latex from unripe capsules

    (d) seeds that are fried

19. Coir is obtained from :

    (a) roots of date palm tree

    (b) leaf bases of areca nut palm

    (c) mesocarp of coconut

    (d) leaves of *Cocos nucifera*

20. The centre of origin of rice plant is :

    (a) India

    (b) Indo-Malayan region

    (c) India and Africa

    (d) Africa

21. Rubber is collected from :

    (a) Crushing the stem of *Euphorbias*

    (b) Tapping the stem of *Carica papaya*

    (c) Tapping the stem of *Hevea brasiliensis*

    (d) Crushing the fruits and collecting the latex of *Achras sapota*

22. Essential oils are those :

    (a) oils which are essential for human beings

    (b) oils which are essential to the plants which produce them

(c) oils which are used as lubricants

(d) oils which yield perfumes

23. Asafoetida is obtained by :

(a) exudation from the stem

(b) extraction from the fruits

(c) extraction from the root

(d) extracted from the leaves

24. Short fibres are known as :

(a) lint

(b) fluff

(c) fuzz

(d) flint

25. Long fibres are known as :

(a) flint

(b) lint

(c) fluff

(d) fuzz

26. The bark of *Cinnamomum zeylanicum* is used as spice because of :

(a) A sweet flavour

(b) Aromatic oils secreted

(c) Ethereal oils secreted

27. Corms and root tubers of Araceae have a pungent aerial taste because they contain :

(a) Aromatic oil

(b) Oxalate crystals

(c) Alkaloid

(d) Latex

28. The composition of the cotton fibre is :

(a) cellulose

(b) callose

(c) chitin

(d) pectin

29. Tea can be grown :

(a) at sea level

(b) sea shore

(c) in elevated areas

(d) in dry places

30. If tea leaves are kept in hot water for longer period, the liquid becomes bitter because :

(a) of the defect in the tea leaves

(b) the volatile oil in the leaves dissolves out

(c) the thein dissolves out

(d) the tannin dissolves out

31. Tea and coffee can be classified as :

(a) distilled beverages

(b) non-alcoholic beverages

(c) fermented beverages

(d) alcoholic beverages

32. The products of commercial importance, yielded by coconut palm are :

(a) latex and oil

(b) rubber and wood

(c) oil and fibre

(d) fibre and rubber

(e) medicines

33. Potatoes are usually propagated by regetative means because :

(a) they do not produce seeds

(b) by this method it is possible to maintain genetic quality

(c) by this method medicine of diseases may be reduced

(d) potato seeds have long dormancy period

34. The banana plant :

(a) tree

(b) shrub

(c) herb

35. The true stem of banana is :

(a) Bulb

(b) Rhizome

(c) Corm

36. The major composition of banana fruit is :

(a) starch

(b) protein

(c) glucose

(d) fat

37. Commercial tea is prepared from :

(a) stem

(b) leaves

(c) flowers

(d) root

(e) bark

38. Rubber is obtained from :

(a) cell sap

(b) gum

(c) resin

(d) latex

39. The commercial coffee is prepared from :

(a) leaves

(b) flowers

(c) fruits

(d) root

(e) seed

40. Most of the rubber plants belong to the family :

(a) euphorbiaceae

(b) Cannaceae

(c) Rubiaceae

(d) Rutaceae

41. Latex cells occr in :

(a) xylem

(b) cambium

(c) bark

(d) cortex

42. The artificial ripening of banana is done by :

(a) Keeping them in a room having temperature below 0°C

(b) Keeping them at room temperature

(c) Keeping them in a room where the temperature is high

43. Coconut oil derived from the copra can be classified as :

(a) essential oil

(b) drying oil

(c) semi-drying oil

(d) vegetable fat

44. What is the source of chewing gum :

(a) sugar from sugar-cane

(b) Gum arabic form *Acacia* spp.

(c) Latex from *Achras sapota*

(d) Fluid from *Musa* spp.

ANSWERS

1. c	2. b	3. c	4. a	5. c	6. b	7. a	8. c	9. c	10. a
11. c	12. b	13. b	14. d	15. b	16. c	17. b	18. c	19. c	20. c
21. c	22. d	23. a	24. c	25. b	26. b	27. b	28. a	29. c	30. b
31. b	32. c	33. d	34. a	35. b	36. a	37. b	38. d	39. c	40. a
41. d	42. a	43. d	44. c						

Index